I0056336

ABOUT THE AUTHOR

Dr. I.M. Mujtaba is a Senior Lecturer in Chemical Engineering in the School of Engineering, Design & Technology at the University of Bradford and is a member of the University Senate. He is a Fellow of the IChemE, a Chartered Chemical Engineer and is currently the Secretary of the IChemE's Computer Aided Process Engineering Subject Group.

Dr. Mujtaba is actively involved in many research areas like: dynamic modelling, simulation, optimisation and control of batch and continuous chemical processes with specific interests in distillation, industrial reactors, refinery processes and desalination. He has published more than 50 technical papers in major Engineering Journals, International Conference Proceedings and Books. He is a co-editor of the book *"Application of Neural Networks and Other Learning Technologies in Process Engineering"* published by the Imperial College Press, London in 2001. He has several ongoing research collaborations and consultations with industries and academic institutions in the UK, Italy, Hungary, Malaysia and Thailand.

Dr. Mujtaba obtained his BSc in 1983 and MSc in 1984 all in Chemical Engineering from the Bangladesh University of Engineering & Technology (BUET). He studied at Imperial College, London with the Commonwealth Scholarship and received his PhD and DIC in 1989.

Dr. Mujtaba was a Lecturer and Assistant Professor at BUET during 1984-1990. From 1990-1994 he worked as a Research Fellow at the Centre for Process Systems Engineering, Imperial College, London.

SERIES ON CHEMICAL ENGINEERING

Series Editor: Jerry Y. S. Lin *(Arizona State Univ., USA)*

Series on Chemical Engineering   Vol. 3

Batch Distillation
Design and Operation

I.M. Mujtaba
University of Bradford, UK

Imperial College Press

Published by

Imperial College Press
57 Shelton Street
Covent Garden
London WC2H 9HE

Distributed by

World Scientific Publishing Co. Pte. Ltd.
5 Toh Tuck Link, Singapore 596224
USA office: 27 Warren Street, Suite 401-402, Hackensack, NJ 07601
UK office: 57 Shelton Street, Covent Garden, London WC2H 9HE

British Library Cataloguing-in-Publication Data
A catalogue record for this book is available from the British Library.

Series on Chemical Engineering — Vol. 3
BATCH DISTILLATION
Design and Operation

Copyright © 2004 by Imperial College Press

All rights reserved. This book, or parts thereof, may not be reproduced in any form or by any means, electronic or mechanical, including photocopying, recording or any information storage and retrieval system now known or to be invented, without written permission from the Publisher.

For photocopying of material in this volume, please pay a copying fee through the Copyright Clearance Center, Inc., 222 Rosewood Drive, Danvers, MA 01923, USA. In this case permission to photocopy is not required from the publisher.

ISBN-13 978-1-86094-437-6
ISBN-10 1-86094-437-X

To my parents: Professor M. Ishaque and Mrs. R. Ishaque
My wife: Nasreen
And my children: Sumayya, Maria, Hamza and Usama

Foreword

Batch distillation process is around us for many centuries. It is perhaps the oldest technology for separating/purifying liquid mixtures and is the most frequently used separation method in batch processes. In batch distillation, the main concerns (issues) for the researchers and process engineers in the last four decades were: (i) the design of alternative and suitable column configurations, (ii) the development of mathematical models in line with the development of numerical methods, (iii) the formulation and solution of dynamic optimisation problems for optimal design, operation and control, (iv) the development of off-cut recycling strategies, (v) the use of batch distillation in reactive and extractive mode and most recently (vi) the use of artificial neural networks in dynamic modelling, optimisation and control.

Although there are several books on distillation in general where batch distillation is only briefly introduced, there is only one book currently available in the market that is solely dedicated to Batch Distillation. It addresses some of the issues mentioned above using short-cut methods, simplified models and Maximum Principle based optimisation techniques and therefore is a good book to start with for the undergraduate students and for the practitioners in process engineering whose interests lie in the basics of batch distillation and in the preliminary design of batch distillation columns and operations.

In the last 25 years, with continuous development of faster computers and sophisticated numerical methods, there have been many published work that have used detailed mathematical models with rigorous physical property calculations and advanced optimisation techniques to address all the issues mentioned above. These have been the motivating factors to write this book in which excellent and important contributions of many researchers around the globe and those by the author and co-workers are accommodated.

This book is structured in 12 chapters highlighting the major developments in the last 25 years. Moreover, in comparison to the materials in the existing book on batch distillation and the materials available in other distillation books, the new materials included in this book are:

- State Task Network (STN) representation of operating sequence for binary and multicomponent batch distillation
- Simple to detailed mathematical models for conventional and unconventional batch distillation processes
- Maximum Principle to sophisticated SQP based nonlinear optimisation techniques
- Short-cut to rigorous methods for optimal design and operation
- Binary to multi-component off-cut recovery and optimal recycling strategies.

vii

- Modelling and optimisation of batch reactive and extractive distillation processes
- Inverted, middle vessel and multivessel batch distillation column operations
- Use of continuous distillation columns for batch distillation
- Neural Network based hybrid dynamic modelling and optimisation methods for conventional and unconventional column configurations

I certainly believe that this book will be beneficial and will be a good reference book for the undergraduate and postgraduate students, academic researchers, batch processing industries, industrial operators, chemists and engineers for many years to come.

Dr. Iqbal M. Mujtaba
Senior Lecturer in Chemical Engineering
School of Engineering, Design and Technology
University of Bradford, Bradford BD7 1DP, UK
Email: I.M.Mujtaba@bradford.ac.uk
September 2003.

Contents

Acknowledgements

Alhamdulillah- all praises be to almighty Allah who made it possible for me to write this book.

About sixty percent of our own work included in this book was carried out during my stay at the Imperial College London in the Department of Chemical Engineering and in the Centre for Process Systems Engineering (CPSE) between 1985 and 1994. I am greatly indebted to Professor Sandro Macchietto who was very kind to accept me as his PhD student in 1985 and allowed me to work with him until 1994. He greatly influenced and motivated me to explore batch distillation in depth. I owe to Professor Pantelides for his generous help and support in mathematical modelling and advanced numerical methods. I am grateful to Professor Sargent and Professor Perkins (ex directors) for allowing me to work at the CPSE from 1990 to 1994. I would like to express my gratitude to Dr. Chen who helped me in great deal in understanding and learning SQP based optimisation techniques. Finally, I would like to acknowledge sincerely the financial support provided by the ACU during 1985-1988 in the form of Commonwealth Scholarship to carry out the research in batch distillation.

The remaining forty percent of our own work included in this book was carried out at the University of Bradford during 1995-2002. I am grateful to Professor Bailes, Professor Benkreira and Dr. Slater for their continuous support and motivation in my work on batch distillation. I owe to Professor Coates for his encouragement to get on with this book and complete it as soon as possible. The continuous supports to my batch distillation research by Professor Macchietto, Professor Pantelides and the Process Systems Enterprise (PSE) Ltd. during 1995-2002 are gratefully acknowledged.

Some of the materials included in this book are due to international collaborations with the University of Malaya (Malaysia) and the University of Padova (Italy). Dr. Hussain (University of Malaya) influenced and motivated me in applying Neural Network Techniques in dynamic modelling and optimisation, I am grateful for his support. I would like to thank Professor Barolo (University of Padova) for his involvement in the work on MVC columns and for inviting me to Padova to give lectures on recent developments in batch distillation in 1997. I would like to acknowledge the UK Royal Society for the financial supports to carry out the collaborative work with Dr. Hussain (Malaysia) since 1999 and Professor Perkins for his kind support in the process. In this regard, I am also grateful to Professor Day (Dean) for supporting me to continue my research collaboration with Malaysia.

Special thanks are due to Professor Urmila Diwakar (University of Illinois at Chicago, USA), Dr. Eva Sorensen (UCL, London), Dr. Peter Lang (Technical University Budapest, Hungary) and Dr. Ben Betlem (University of Twente, Netherlands) for many useful technical discussions in the past. I would also like to thank my past and current postgraduate students and post-doc researchers Dr. Greaves, Mr. Tran, Mr. Avignon, Mr. Miladi, Mr. Magbary and Dr. Milani for their contributions in some of the work presented in this book.

Most of our own work included in this book is computational and I owe to a large extent to: (a) late Dr. John Flower of the University of Leeds who encouraged me in computational research during his visit to Bangladesh University of Engineering & Technology (BUET) in 1984 and (b) Professor J. Zaman who taught me the basics of batch distillation, computer programming and supervised my MSc research project on process modelling and simulation. I am also grateful to Professor N. Ahmed, Professor I. Mahmud, Professor K. Rahman, Professor A.K.M.A. Quader and Dr. T. Mahmud for teaching me, with great care, Separation Processes, Transport Phenomena and Process Design at both undergraduate and postgraduate levels at BUET during 1979-1984. All of them largely contributed in shaping up of my research in batch distillation.

I am indebted to my father Professor M. Ishaque (ex graduate of Imperial College) and mother Mrs. R. Ishaque for encouraging me to study at Imperial College, London and for their countless love and blessings throughout my life and career.

I owe to my wife Nasreen for her great support and continuous encouragement in writing this book. Despite her fulltime work schedule as a social and community worker, she freed me from my household commitments during the writing up of this book. The wonderful time and fun I had with my children Sumayya, Maria, Hamza and Usama during the research and writing up of this book deserve great appreciations. It would be impossible to write this book without them around me.

Finally, special thanks are due to Imperial College Press, London for giving me over 3 years to complete this book and for publishing it. I sincerely acknowledge their support and help. Also I would like to thank all other publishing companies (including organisations and societies) who have kindly granted me permissions to reproduce some of the Tables and Figures in this book without which the book would be incomplete.

Batch Distillation:
Design and Operation

CHAPTER 1

1. INTRODUCTION

1.1. Batch Processes

In the 1950s, chemical engineers might have the impression that the ultimate mission of the engineers was to transform old-fashioned batch processes into modern continuous ones (Rippin, 1983). With such a perspective it is surprising to find that, today, fifty years later a significant proportion of the world's chemical production by volume and a much larger proportion by value is still made in batch plants and it is unlikely that this proportion will decline in the near future. Parakrama (1985) reported that 99 batch processes were in operation in 74 UK manufacturing companies. Among these, 80% plants were producing chemicals in steady or growing markets. Moreover, many more products, which could be manufactured continuously, are in fact made in batch plants on economic grounds (Rippin, 1991).

Batch production is usually carried out in relatively standardised types of equipment, which can easily be adapted and if necessary reconfigured to produce many other different products. It is particularly suitable for low volume, high value products such as pharmaceuticals, polymers, biotechnologicals or other fine chemicals for which annual requirement can be manufactured in few days or few batches in an existing plant. The flexibility of the production arrangements can also cope with the fluctuations or rapid changes in demand, which is often characteristic of products of this type. Other factors (Shah, 1992) which favours batch processing are:

- increased global competition in the bulk chemicals sector
- need to produce customer specific products
- seasonal demands of certain products

Where small amounts of different products must be produced, it is usually more economically efficient to manufacture them in a common facility such as multipurpose batch plant, rather than operating one plant per product.

3

1.2. Distillation

Distillation separates two or more liquid components in a mixture using the principle of relative volatility or boiling points. The greater the difference in relative volatility the greater the nonlinearity and the easier it is to separate the mixture using distillation. The process involves production of vapour by boiling the liquid mixture in a still and removal of the vapour from the still by condensation. Due to differences in relative volatility or boiling points, the vapour is rich in light components and the liquid is rich in heavy components.

Often a part of the condensate is returned (*reflux*) back to the still and is mixed with the outgoing vapour. This allows further transfer of lighter components to the vapour phase from the liquid phase and transfer of heavier components to the liquid phase from the vapour phase. Consequently, the vapour stream becomes richer in light components and the liquid stream becomes richer in heavy components. Different types of devices called plates, trays or packing are used to bring the vapour and liquid phases into intimate contact to enhance the mass transfer. Depending on the relative volatility and the separation task (i.e. purity of the separated components) more *trays* (or more packing materials) are stacked one above the other in a cylindrical shell to form a *column*.

The distillation process can be carried out in *continuous*, *batch* or in *semi-batch (or semi-continuous)* mode.

1.3. Continuous Distillation

Figure 1.1 shows a typical continuous distillation column. The liquid mixture (*feed*), which is to be separated into its components, is fed to the column at one or more points along the column. Liquid runs down the column due to gravity while the vapour runs up the column. The vapour is produced by partial vaporisation of the liquid reaching the bottom of the column. The remaining liquid is withdrawn from the column as *bottom product* rich in heavy components. The vapour reaching the top of the column is partially or fully condensed. Part of the condensed liquid is refluxed to the column while the remainder is withdrawn as the *distillate product*. The column section above the feed tray rectifies the vapour stream with light components and therefore is termed as *rectifying* section. The column section below the feed tray strips heavy components from the vapour stream to the liquid stream and is termed as *stripping* section.

The readers are directed to Smith (1963), Seader and Henley (1998), Perry et al. (1997), McCabe et al. (2001), Gani and co-workers (1986a, 1986b, 1988, 2000),

Perkins and co-workers (1996, 1999, 2000, 2001) for detailed account of modelling, design, operation, control and synthesis of continuous distillation processes.

1.4. Batch Distillation

Batch distillation is, perhaps the oldest operation used for separation of liquid mixtures. For centuries and also today, batch distillation is widely used for the production of fine chemicals and specialised products such as alcoholic beverages, essential oils, perfume, pharmaceutical and petroleum products. It is the most frequent separation method in batch processes (Lucet et al., 1992).

The essential features of a conventional batch distillation (CBD) column (Figure 1.2) are:

- A bottom receiver/reboiler which is charged with the feed to be processed and which provides the heat transfer surface.
- A rectifying column (either a tray or packed column) superimposed on the reboiler, coupled with either a total condenser or a partial condenser system.

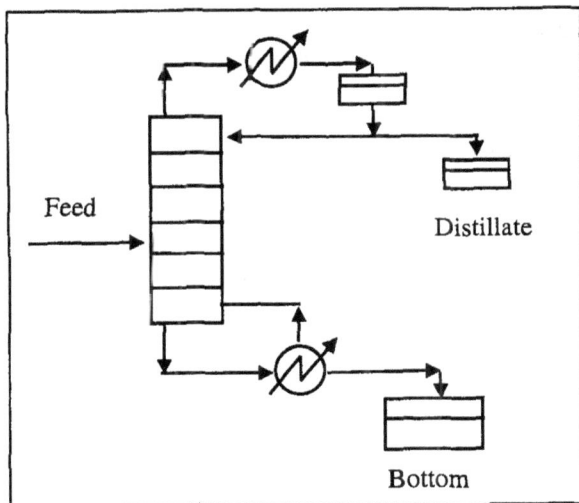

Figure 1.1. Continuous Distillation Column

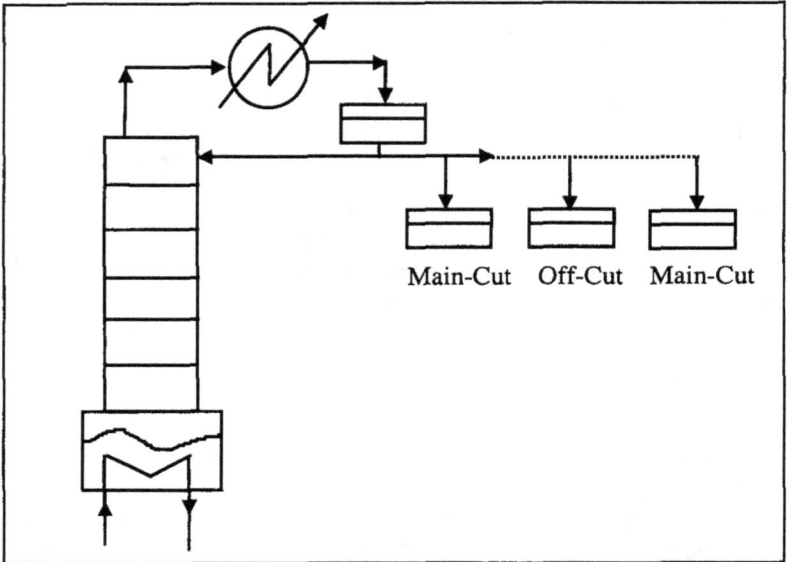

Figure 1.2. Conventional Batch Distillation (CBD)

- A series of product accumulator tanks connected to the product streams to collect the main and or the intermediate distillate fractions.

Operation of such a column involves carrying out the fractionation until a desired amount has been distilled off. The overhead composition varies during the operation and usually a number of cuts are made. Some of the cuts are desired products (main-cuts) while others are intermediate fractions (off-cuts) that can be recycled to subsequent batches to obtain further separation. A residual bottom fraction may or may not be recovered as product (Mujtaba, 1989).

Further details on batch distillation are provided throughout this book.

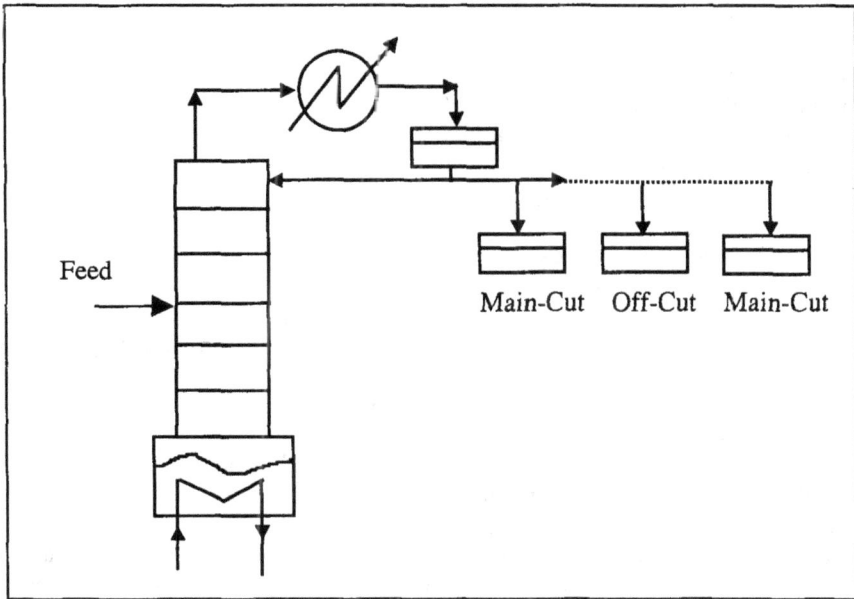

Figure 1.3. Semi-batch (Semi-continuous) Distillation Column

1.5. Semi-batch (semi-continuous) Distillation

Figure 1.3 shows a typical semi-batch (semi-continuous) distillation column. The operation of such columns is very similar to CBD columns except that a feed is introduced to the column in a continuous or semi-continuous mode. This type of column is suitable for extractive distillation, reactive distillation, etc. (Lang and co-workers, 1994, 1995; Mujtaba, 1999). Further details of semi-batch distillation in extractive mode of operation are provided in Chapter 10.

1.6. Advantages of Batch Distillation

The main advantages of batch distillation over a continuous distillation lie in the use of a single column as opposed to multiple columns and its flexible operation.

For a multicomponent liquid mixture with n_c number of components, usually $(n_c\text{-}1)$ number of continuous columns will be necessary to separate all the components from the mixture. For a mixture with only 4 components and 3 distillation columns there can be 5 alternative sequences of operations to separate all the components (Figure 1.4). For a mixture with only 5 components, 4 distillation columns can be sequenced in 14 different ways. The number of alternative operations grows exponentially with the number of components in the mixture. These alternative operations do not take into account the production of off-specification materials or provision for side streams (this would further increase dramatically the number of columns and or operational sequences).

On the other hand with CBD, only one column is necessary and there is only one sequence of operation (with or without the production of off-specification materials) to separate all the components in a mixture (Figure 1.2). The only requirements here are to divert the distillate products to different product tanks at specified times.

The continuous distillation columns are designed to operate for longer hours (typically 8000 hrs a year) and therefore each column (or a series of columns in case of multicomponent mixture) is dedicated to the separation of a specific mixture.

However, a single mixture (binary or multicomponent) can be separated into several products (*single separation duty*) and multiple mixtures (binary or multicomponent) can be processed, each producing a number of products (*multiple separation duties*) using only one CBD column (Logsdon et al., 1990; Mujtaba and Macchietto, 1996; Sharif et al., 1998).

Finally, in pharmaceutical and food industries product tracking is very important in the face of strict quality control and batch wise production provides the *batch identity* (Low, 2003).

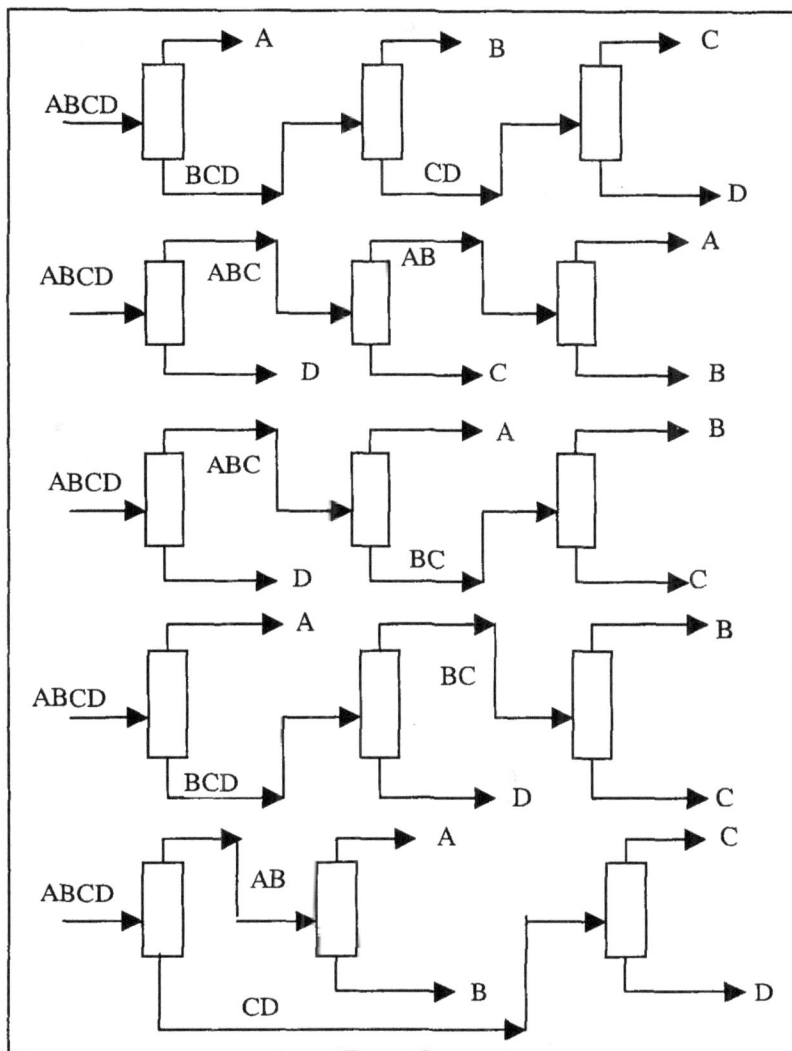

Figure 1.4. Alternative Separation Sequences for Quaternary Mixture Using Continuous Distillation Columns

References

Bansal, V., Ross, R., Perkins, J.D., Pistokopoulos, E.N., *J. Proc. Control.* **10** (2000) 219.

Gani, R. and Bek-Pedersen, E., *AIChE J.* **46** (2000), 1271

Gani, R., Ruiz, C.A. and Cameron, I.T., *Comput. chem. Engng.* **10** (1986a), 181.

Gani, R., Ruiz, C.A. and Cameron, I.T., *Comput. chem. Engng.* **10** (1986b), 199.

Ruiz, C.A., Cameron, I.T. and Gani, R., *Comput. chem. Engng.* **12** (1988), 1.

Lang, P., H, Yatim, P. Moszkowicz and M. Otterbein, *Comput. chem. Engng.* **18**, 11/12, 1057 (1994).

Lang, P., Lelkes, Z., Moszkowicz, P., Otterbein, M. and Yatim, H., *Comput. chem. Engng.* **19** (1995), s645.

Logsdon, J.S., Diwekar, U.M. and Biegler, L.T., *Trans. IChemE*, **68A** (1990), 434.

Low, K.H., *Optimal Configuration, Design and Operation of Batch Distillation Processes*, PhD Thesis, (University of London, 2003).

Lucet, M., Charamel, A., Champuis, A., Guido, D. and Loreau, J., Role of batch processing in the chemical process industry. *In Proceedings of NATO ASI on Batch Processing Systems Engineering*, Antalya, Turkey, May 29-June 7, 1992.

McCabe, W.L., Smith, J.C. and Harriot, P., *Unit Operations of Chemical Engineering* (6ᵗʰ edition, McGraw-Hill, 2001).

Mujtaba, I.M., *Optimal Operational Policies in Batch Distillation*. PhD Thesis, (Imperial College, University of London, 1989).

Mujtaba, I.M., *Trans. IChemE.* **77A** (1999), 588.

Mujtaba, I.M. and Macchietto, S., *J. Proc. Control.* **6** (1996) 27.

Perkins, J.D., Keynote Speech, *IChemE Research Event*, 2-3 April, Leeds, 1996.

Parakrama, R., *The Chemical Engineer*. September (1985), 24.

Perry, R.H., Green, D.W. and Maloney, J.O. eds., *Perry's Chemical Engineers' Handbook* (7ᵗʰ edition, McGraw-Hill, 1997).

Pilavachi, P.A., Schenk, M., and Gani, R., *Trans. IChemE*, **78** (2000), 217.

Rippin, D.W., *Comput. chem. Engng.* **7** (1983), 137.

Rippin, D.W., *Chem. Eng.* May (1991), 101.

Ross, R., Bansal, V., Perkins, J.D., Pistokopoulos, E.N., Koot, G.L.M. and van Schijndel, J., *Comput. chem. Engng.* **23** (1999), S875.

Ross, R., Perkins, J.D., Pistokopoulos, E.N., Koot, G.L.M. and van Schijndel, J., *Comput. chem. Engng.* **25** (2001), 141.

Seader, J.D. and Henley, E.J., *Separation Process Principles* (John Wiley & Sons, Inc., 1998).

Shah, N., *Efficient Scheduling Planning and Design of Multipurpose Batch Plants*, PhD thesis, ((Imperial College, University of London, 1992).

Sharif, M., N. Shah and C.C. Pantelides, *Comput. chem. Engng.* **22** (1998), S69.

Smith, B.D., *Design of Equilibrium Stage Processes* (McGraw-Hill, 1963).

CHAPTER 2

2. COLUMN CONFIGURATIONS

2.1. Conventional Column Configuration

In this configuration, the available separation section (tray or packed) is utilised in rectifying mode, with product cuts (recovered) and intermediate off-cut fractions (disposed of or recycled) collected as condensed distillate. A final residue bottom fraction may also be a desired product. Conventional column configuration has been discussed in section 1.4 of Chapter 1. Further details are given in later chapters.

2.2. Unconventional Column Configurations

Alternative configurations, collectively identified as unconventional columns, have been found in certain cases to be more advantageous. These are described below.

2.2.1. Inverted Batch Distillation Column

This type of batch distillation column (Figure 2.1) originally proposed by Robinson and Gilliland (1950) combines the feed charge and the condenser reflux drum and operates in an all-stripping mode with a small holdup reboiler. This type of column operates exactly as the conventional batch column except that products are withdrawn from the bottom. High boiling (heavy components) products are withdrawn first followed by the more volatile products. This type of operation is supposed to eliminate the thermal decomposition problems of the high boiling products.

Abrams *et al.* (1987), Mujtaba and Macchietto (1994) and Sorensen and Skogestad (1996) used such columns for batch distillation and compared their performances with conventional columns.

11

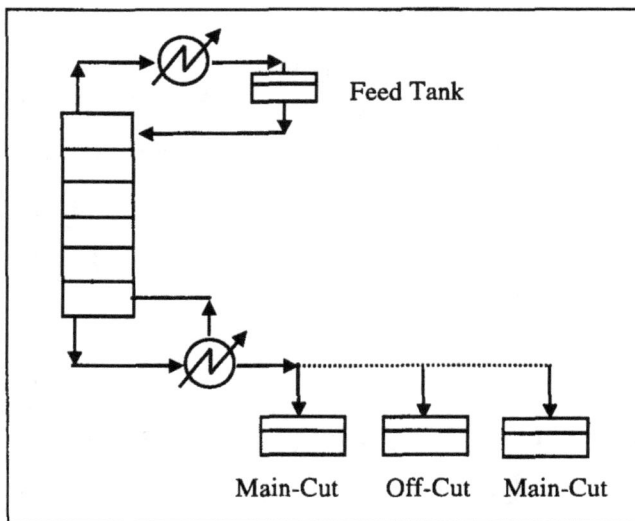

Figure 2.1. Inverted Batch Distillation (IBD)

2.2.2. *Middle Vessel Batch Distillation Column*

In a Middle Vessel Batch Distillation Column (MVC) (Figure 2.2) the separation section is divided, as in the usual continuous distillation column, into rectifying and stripping sections, with a feed tray in the middle. The essential features of this type of column are:

i)   The feed is supplied to a suitable location in the middle of the column. The reboiler holdup is kept to a minimum.

ii)   The liquid on the feed tray is recycled to the feed tank so that the composition of the liquid in the feed tank is close to that of the liquid on the feed tray.

iii)   Products or intermediate fractions can be withdrawn simultaneously from the top and bottom of the column.

Bortolini and Guarise (1970) presented such columns and methods for evaluating their performance with binary mixtures. Recent use of such column can be found in Hasebe et al. (1992), Mujtaba and Macchietto (1992), Mujtaba and Macchietto (1994), Barolo et al. (1996, 1998), Zamprogna et al. (2001), Greaves et al. (2003), Hilman et al. (1997), Safrit and Westerberg (1997) for nonideal, azeotropic, extractive and reactive separation of binary and multicomponent mixtures.

This type of column is inherently very flexible in the sense that it can be easily converted to a conventional or inverted batch distillation column by changing the location of the feed and by closing or opening appropriate valves in the product lines.

2.2.3. Multivessel Batch Distillation Column

A Multivessel Batch Distillation Column (MultiBD) has very similar configuration to that of a conventional batch distillation but with one or more intermediate charge/product vessels as shown in Figure 2.3. If the column operates at total reflux, the charges in each vessel will be purified as the distillation proceeds. However, the purity in each vessel will depend on the number of plates in each section of the column, vapour boil-up, the amount of initial charge in each vessel and the duration of operation. The top vessel will be richer in low boiling components while the bottom vessel will be richer in high boiling components.

Wittgens et al. (1996), Skogestad et al. (1997) and Furlong et al. (1999) used MultiBD columns for simulation, control and optimisation studies.

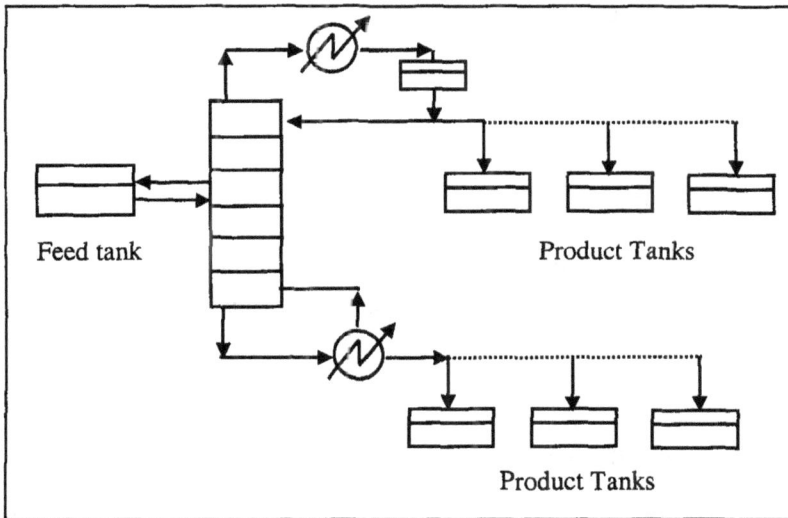

Figure 2.2. Middle Vessel Batch Distillation Column (MVC)

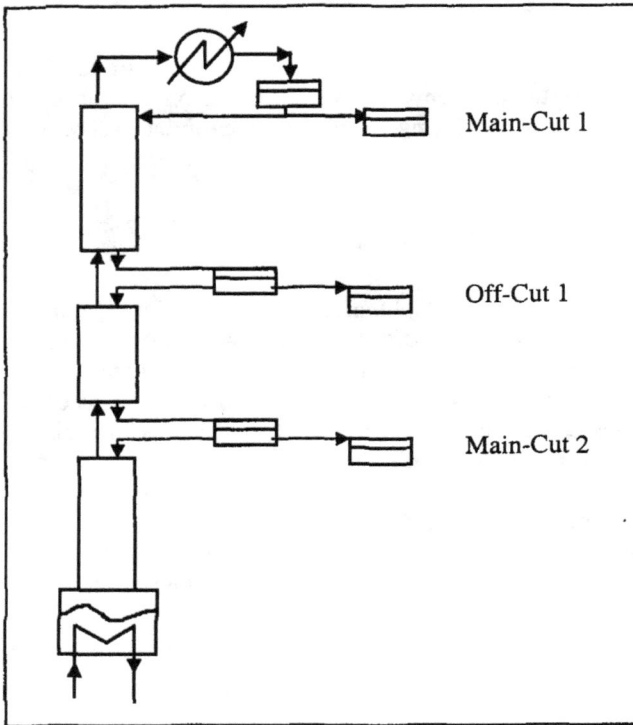

Figure 2.3. Multivessel Batch Distillation Column (MultiBD)

2.2.4. *Continuous Column for Batch Distillation*

Attarwala and Abrams (1974) considered the batch separation task operating in a single continuous column sequentially. The essential features of this type of column (Figure 2.4) are:

i)   The feed is supplied to a suitable location in the middle of the column from a feed tank continuously (as in continuous distillation). The reboiler and condenser holdups are kept to a minimum.

ii)   The column operates in a continuous mode and separate one component each pass as distillate and collect the residual in a storage tank.

iii)   The storage tank from the previous pass is used as a feed tank for the next pass and the process from step (ii) is continued. This process is continued until the final binary mixture is separated.

This type of operation is known as Single Pass Sequential Steady State (SPSSS) operation. Each pass is operated using one reflux ratio and the overhead or bottom composition remains constant (may have different values for different passes) throughout the operation (until the feed tank is empty). Also this operation mode allows use of an already built continuous distillation column (with little extra arrangements for feeding the column by alternate changing of the feed tank) as batch distillation and the single column can be used for separating all the mixtures sequentially as in the case of conventional batch distillation. However, in each pass the column will operate in continuous mode for a finite time depending on the total feed in the feed tank and on the feed flow rate (Mujtaba, 1997).

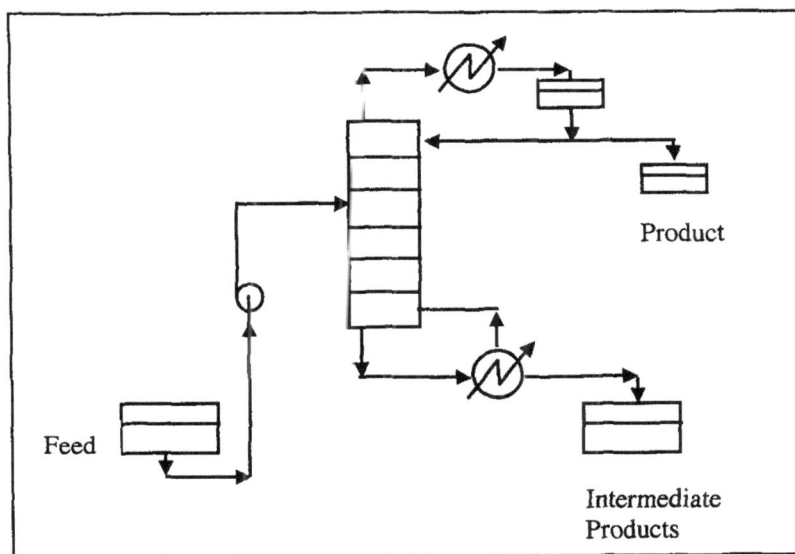

Figure 2.4. Batch Distillation Using Continuous Column

References

Abrams, H.J., Miladi, M. and Attarwala, F.T., Preferable alternatives to conventional batch distillation, Presented at the IChemE Symposium Series no. 104, Brighton, 7-9 September, 1987.

Attarwala, F.T., and Abrams, H.J., Optimisation techniques in binary batch distillation. *IChemE Annual Research Meeting*, London, 1974.

Barolo, M., Guarise, G. B., Ribon, N., Rienzi S. A., Trotta, A. and Macchietto, S., *Comput. chem. Engng.* **20** (1996), S37.

Barolo, M., Guarise, G. B., Rienzi S. A. and Trotta, A., *Comput. chem. Engng.* **22** (1998), S44.

Bortolini, P and Guarise, G.B. , *Quad. Ing. Chim. Ital.* **6** (1970), 150.

Furlonge, H.I., Pantelides, C.C. and Sorensen, E., *AIChE J.* **45** (1999), 781.

Greaves, M.A., Mujtaba, I. M., Barolo, M., Trotta, A. and Hussain, M. A., *Trans. IChemE.* **81A** (2003), 393.

Hasebe, S., Abdul Aziz, B.B., Hashimoto, I. and Watanabe, T., In *Proceedings IFAC Workshop*, London, 7-8 September (1992), 177.

Hilmen, E.K., Skogestad, S., Doherty, M.F. and Malone, M.F., *AIChE Annual Meeting*, 16-21 November, LA, USA, paper no. 201h (1997).

Mujtaba, I.M. and Macchietto, S., Optimal operation of reactive batch distillation. *Presented at the AIChE Annual Meeting*, Miami Beach, USA, Nov. 1-6, 1992.

Mujtaba, I.M., *Trans. IChemE.* **75A** (1997), 609.

Mujtaba, I.M. and Macchietto, S., 1994, *In Proceedings of IFAC Symposium ADCHEM'94*, Kyoto, Japan, 25 - 27 May, (1994), 415.

Robinson, C.S. and Gilliland, E.R., *Elements of Fractional Distillation*, (4th ed., McGraw Hill, New York, 1950).

Safrit, B.T. and A.W. Westerberg, *IEC Res.* **36** (1997), 436.

Sorensen, E. and Skogestad, S., *Chem. Eng. Sci.* **51** (1996), 4949.

Skogestad, S., Wittgens, B., Litto, R. and Sorensen, E., *AIChE J.* **43** (1997), 971.

Wittgens, B., Litto, R., Sorensen, E. and Skogestad, S., *Comput. chem. Engng.* **20** (1996), S1041.

Zamprogna, E., Barolo, M., Seborg, D.E., *Trans. IChemE*, **79A** (2001), 689.

CHAPTER 3

3. OPERATION

3.1. Representation of Operational Alternatives Using State Task Network

The batch distillation operation can be schematically represented as a State Task Network (STN). A state (denoted by a circle) represents a specified material, and a task (rectangular box) represents the operational task (distillation) which transforms the input state(s) into the output state(s) (Kondili et al., 1988; Mujtaba and Macchietto, 1993). For example, Figure 3.1 shows a single distillation task producing a main-cut 1 (D_1) and a bottom residue product (B_1) from an initial charge (B_0). States are characterized by the amount and composition of the mixture residing in them. Tasks are characterized by operational attributes such as their duration, the reflux ratio profile used during the task, etc.

Additional attributes of a distillation task are the set of values of all parameters (mainly operational) at the beginning and at the end of the task. The intermediate residue amount (B_1) and composition (x_{B_1}) respectively, are not the holdup and composition of the reboiler at the end of task 1, but the amount and composition which would be obtained if all holdups in the column at the end of task 1 were collected as the intermediate residue. Similarly, the initial holdups and composition in the column (task attributes) must be consistent (i.e. mass balance) with the amount and composition of the initial charge state in the STN. In the following some of the alternative STNs corresponding to often encountered binary and multicomponent batch distillation operations are discussed. Generalization to larger number of components is trivially simple. Only two types of tasks are considered, namely the production of a distillate product (*main-cut* module, Figure 3.2) and the production of a distillate by-product not meeting the distillate specifications (*off-cut* module, Figure 3.3).

17

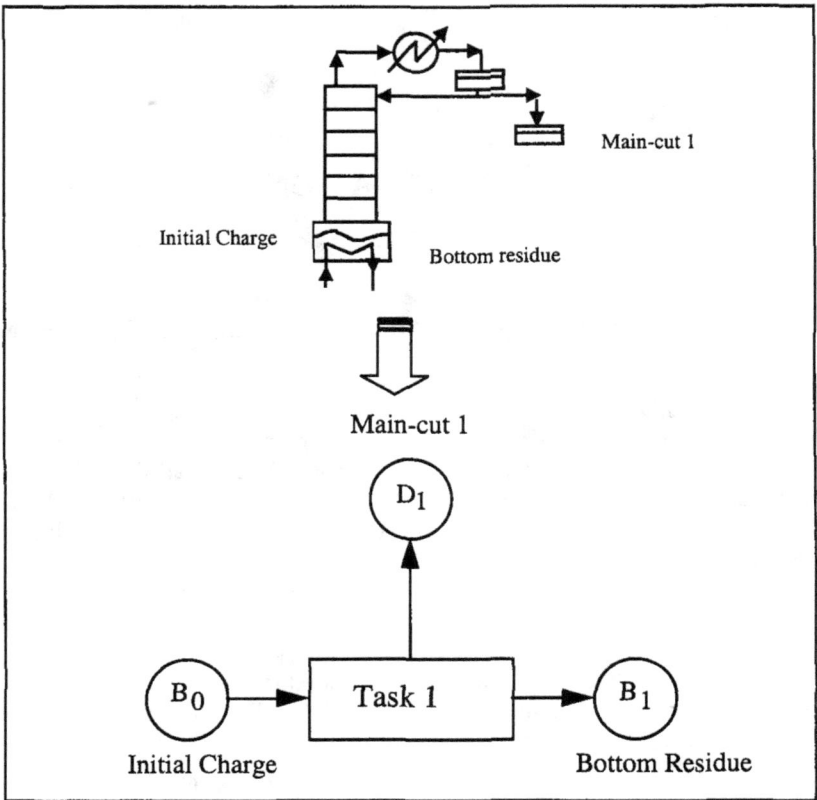

Figure 3.1. STN Representation of Batch Distillation Operation

3.1.1. Binary Mixtures

For binary mixtures there are only two basic operational alternatives:

i) The production of a single main-cut with certain purity constraints. The bottom residue may also be a valuable product and/or can be obtained with specified purity. The STN for this operation is shown in Figure 3.1. Coward (1967); Kerkhof and Vissers (1978); Mujtaba (1989); Mujtaba and Macchietto (1993) considered such operation.

ii) The production of a main-cut with specified purity followed by an off-cut. Figure 3.4 shows the STN for this operation that is generated by joining sequentially a main-cut module (Figure 3.2) and an off-cut module (Figure 3.3). The off-cut may be a valuable material and is usually stored for further separation or is recycled with the next batch. Its amount and composition are usually subject to optimisation. The bottom residual may or may not be a valuable material/product but may have to satisfy certain purity constraints due to environmental restrictions.

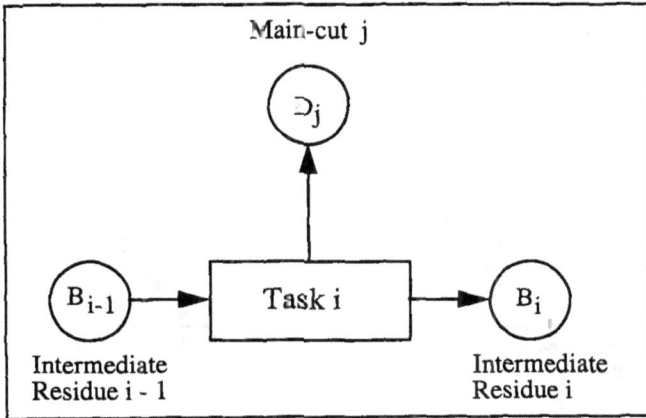

Figure 3.2. Main-cut Module in Batch Distillation

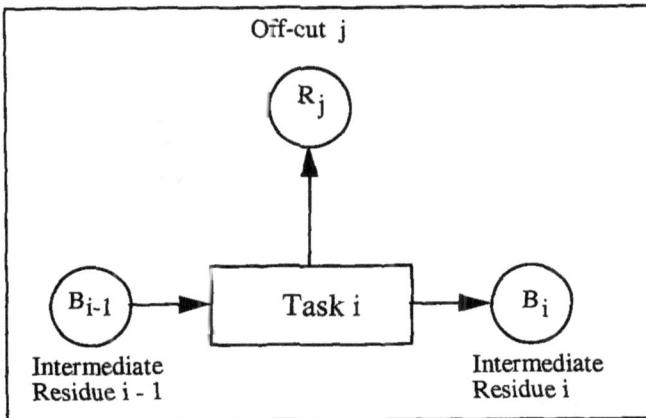

Figure 3.3. Off-cut Module in Batch Distillation

3.1.2. Multicomponent Mixtures

There can be several alternative STNs for multicomponent mixtures depending on the number of main-cuts and off-cuts to be produced. The two basic modules of Figures 3.2 and 3.3 can be combined as many times as required to describe the entire operation. A few alternative STNs for ternary batch distillation are given in Figures 3.5-3.7 as the combination of these two basic modules.

For mixtures with more than 3 components similar STNs may be generated to represent various operations, with additional states and distillation tasks defined as required. Other STNs can be generated when recycles of off-cuts or solvents are considered. Figure 3.8 shows the STN for a ternary batch extractive distillation with solvent recovery and recycling (Mujtaba, 1999).

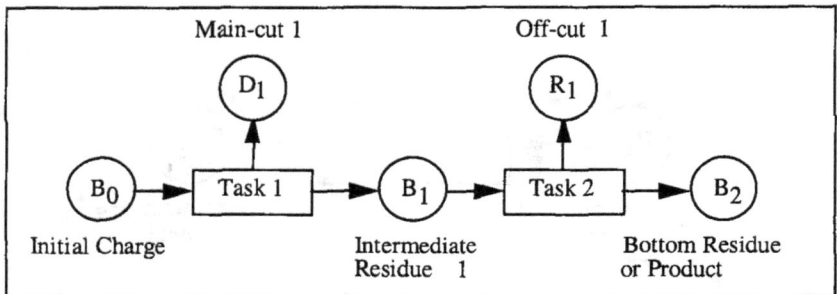

Figure 3.4. STN for Binary Batch Distillation

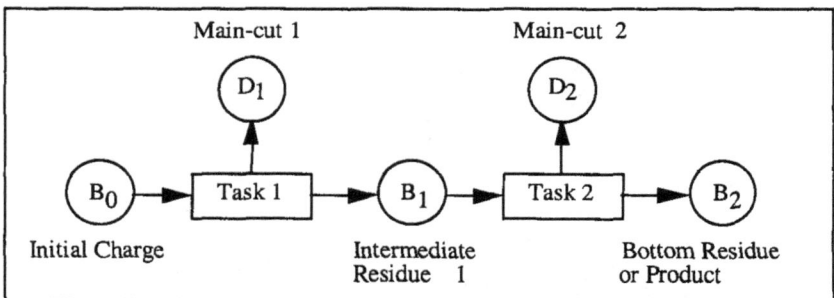

Figure 3.5. STN for Ternary Batch Distillation (with main-cuts only)

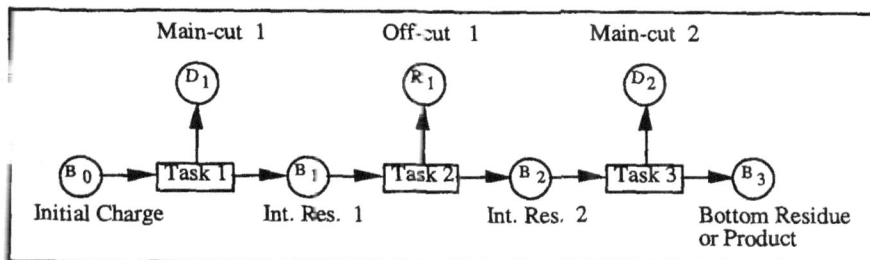

Figure 3.6. STN with Two Main-cuts and One Off-cut (Ternary Mixture)

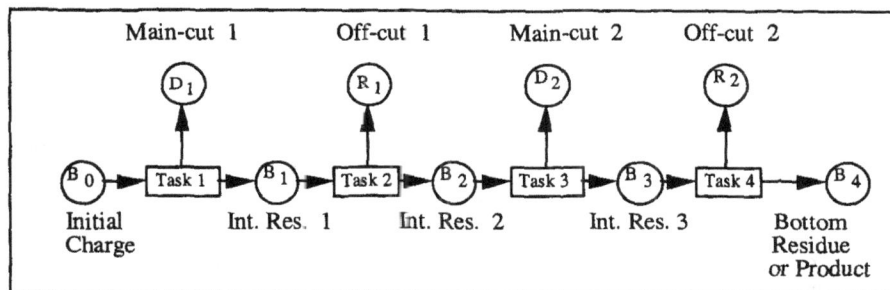

Figure 3.7. STN with Each Main-cut Followed by an Off-cut (Ternary Mixture)

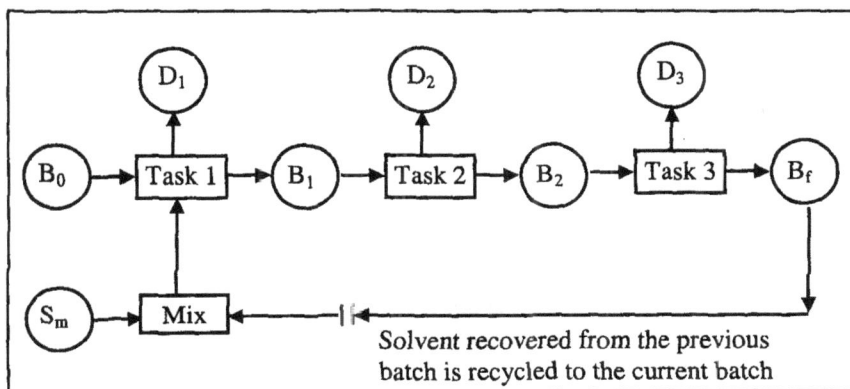

Figure 3.8. STN for Ternary Batch Extractive Distillation with Solvent Recovery and Recycle [Mujtaba, 1999][a]

---

[a] Reprinted with permission from IChemE. UK. Full reference is at the end of the chapter.

The main advantage of this representation is that all operations and intermediate conditions are defined in a formal and highly modular way. Thus it is possible to construct a large number of alternative operations by picking and joining from a small number of primitives. Similarly, it is possible to insert, extract and examine in detail, modify or eliminate an individual processing tasks without affecting the rest of the operation. Furthermore, the modular nature of the representation makes it easy to keep track of the degrees of freedom associated with the operation and to perform mass balances and other consistency checks involving the material states at the various stages of distillation (Mujtaba and Macchietto, 1993). Finally, it is possible to envisage the use of this representation to address "synthesis" problems where the problem is selecting the structure of the operation, i.e. the form of the distillation operation, from an initial superstructure. The use of such a formal and modular representation becomes clearly more advantageous as the number of components in the initial mixture increases and or if multiple separations are to be carried out using a single batch distillation column.

For example, Figure 3.9 shows a single batch distillation column processing M number of feed mixtures, each constituted of NC_m components (*Multiple Separation Duties*). For each mixture m, the procedure for processing a batch charge (*operation m*) is viewed as a sequence of NT_m distillation *tasks* to produce one or more main-cuts, possibly some intermediate off-cuts and a bottom residue, from a given feed (Mujtaba and Macchietto, 1996). For two separation duties (one binary and one ternary distillation) this is represented in the form of a STN in Figure 3.10. Further details are given in Chapter 6, 7 and 8.

3.2. Column Operation

A batch distillation column may be run using one of the specific operation strategies discussed in the following:

3.2.1. Constant Vapour Boilup Rate

In this mode of operation the vapour rate out of the reboiler is held constant throughout the operation by continuously increasing (conventional batch distillation column) or decreasing (inverted and middle vessel column) the reboiler heat duty as the reboiler composition changes (Robinson and Gilliland, 1950; Coward, 1967; Robinson, 1970; Mayur et al., 1970; Mayur and Jackson, 1971; Kerkhof and Vissers, 1978; Domenech and Enjalbert, 1981; Diwekar et al., 1987; Quintero-Marmol and Luyben, 1990; Farhat et al., 1990; Logsdon and Biegler, 1993; Mujtaba and Macchietto, 1996; Mujtaba, 1997; Barolo and Botteon, 1997).

3.2.2. Constant Condenser Vapour Load

Vapour load to the condenser is kept constant throughout the operation in this mode. Nad and Spiegel (1987) maintained this constancy of vapour load in their experimental column by making an enthalpy balance around the condenser system and running the column at piecewise constant reflux ratio. Mujtaba (1989, 1997, 1999), Mujtaba and Macchietto (1988a, 1988b, 1992a, 1992b, 1992c, 1993, 1996, 1997, 1998), Mujtaba et al. (1995), Macchietto and Mujtaba (1996), Walsh et al. (1995) and Wajge and Reklaitis (1999) also used this type of operation.

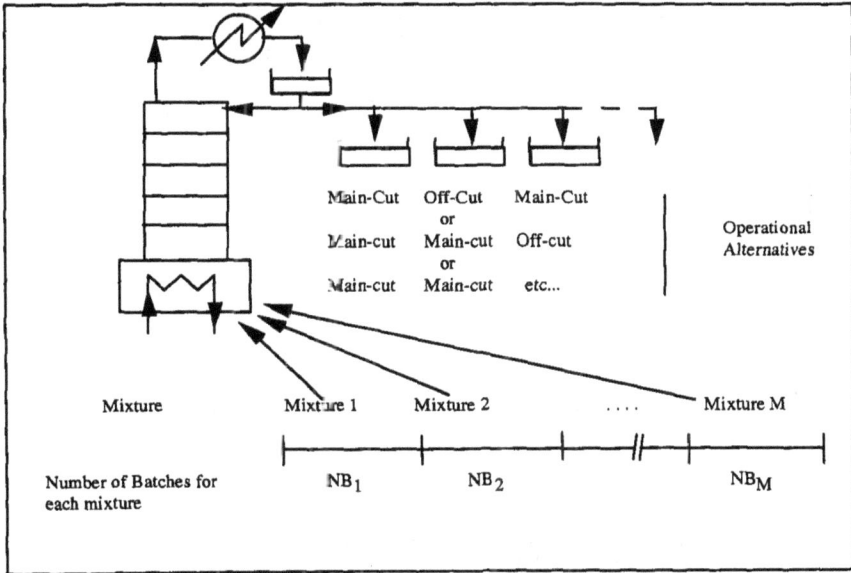

Figure 3.9. Multiple Separation Duties and Multiple Operational Alternatives

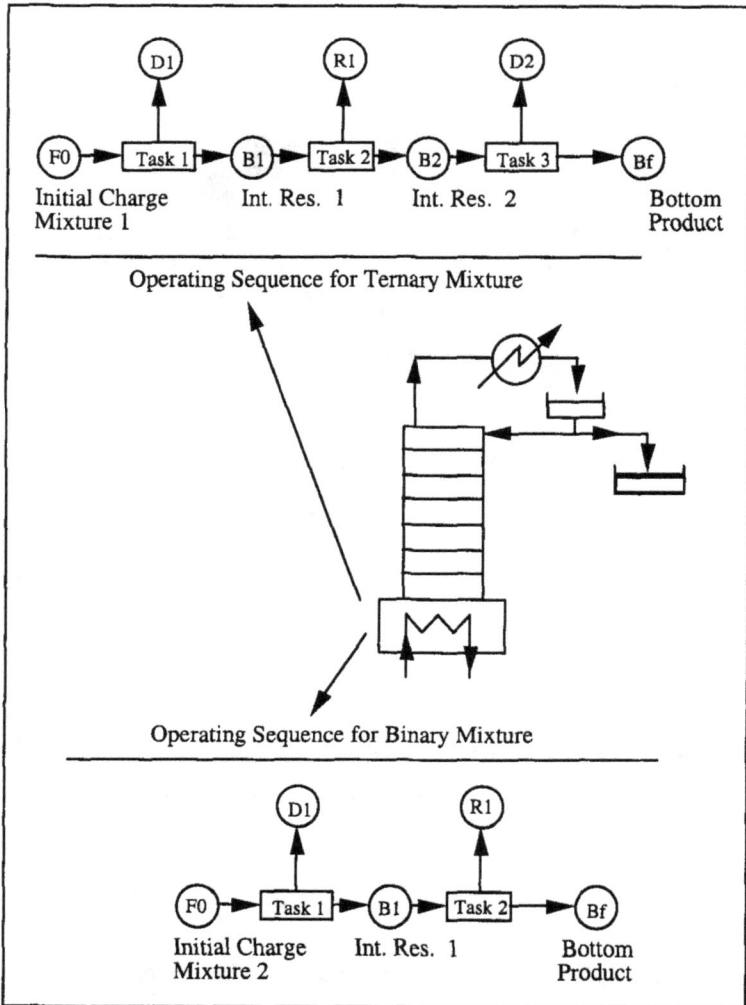

Figure 3.10. STN with Two Separation Duties. [Mujtaba and Macchietto, 1996][b]

[b] Reprinted from *Journal of Process Control*, **6**, Mujtaba, I.M. and Macchietto, S., Simultaneous optimisation of design and operation of multicomponent batch distillation column-single and multiple separation duties, 27-36 , Copyright (1996), with permission from Elsevier Science .

3.2.3. *Constant Distillate Rate*

This mode of operation demands constant rate of distillate throughout. This means that, for constant reflux ratio operation, the vapour load to the condenser is also constant. Boston et al. (1981) and Holland and Liapis (1983) considered this type of operation.

3.2.4. *Constant Reboiler Duty*

In this mode heat input to the reboiler is held constant throughout. In practice it is set to its maximum limit, the value of which depends on the heat exchange system to the reboiler. Domenech and Enjalbert (1974), Greaves et al. (2001), Greaves (2003) used this mode of operation in their laboratory conventional batch column. Cuille and Reklaitis (1986) and Mujtaba (1989) also used this mode of operation in their simulation studies.

3.2.5. *Cyclic Operation*

This is characterized by two modes of operation, called *transient total reflux* and *stripping*. During the total reflux portion of the cycle, liquid reflux is returned to the column, but no product is withdrawn; and during the stripping portion of the cycle, the product is withdrawn; but no reflux is returned to the column (Barb and Holland, 1967). The extreme difficulty of accurate measurement and control of small flow rates in laboratory columns strongly favour cyclic operation (Holland and Liapis, 1983; Sorensen and Skogestad, 1994; Sorensen and Prenzler, 1997; Greaves, 2003).

3.3. Start-up, Production and Shutdown

Operation of batch distillation column can be conveniently described in three parts:

- Start-up period
- Production period
- Shutdown period

Following sections describe briefly each of these operations for a conventional batch distillation column.

3.3.1. Start-up Period

In practice, an empty conventional batch column is started up in the following sequence:

1) The reboiler is charged with the material to be processed and heat is applied to it to bring the material to its boiling point temperature.
2) Depending on the reboiler duty, a part of the material is vaporised and the vapour travels upward both through the plate holes and downcomers and almost instantly reaches the condenser.
3) At this time, the coolant valve is opened and the condensed liquid is stored into a reflux drum. The reflux valve is opened when the liquid fills the condenser holdup tank. At this point some product may also be collected simultaneously.
4) The liquid begins to flow into the top plate and collects on the plate because of the retention made by the vapour flow. When the liquid level passes the weir height (thus filling the holdup), the liquid begins to fall to the plate below and the same phenomenon is repeated until the reboiler is reached. In practice some liquid also trickles down from the plate holes when the flows are initially established.
5) If no product was withdrawn in step 3, the column is now run under total reflux operation until the unit is taken to a steady state or to a state when the distillate composition reaches the desired product purity.

The duration of the first step can usually be considered negligible compared to the overall batch distillation time, whereas the duration of steps 2-5 is important and in some cases it may take a long time to reach a steady state or the desired initial distillate composition (Holland and Liapis, 1983; Nad and Spiegel, 1987; Ruiz, 1988; Logsdon and Biegler, 1993).

3.3.2. Product Period

Generally the product period starts when the removal of distillate from the process is begun. The operation in the product period and its duration depends on the requirements of the product or on the economics of the process. This period can be operated under the following conditions:

1) The start-up period is ended when the desired distillate purity is reached. Product take off is started and the product is collected at constant composition by varying the reflux ratio until a specified amount of distillate has been collected. This type of operation is known as *variable reflux*

operation or *constant distillate composition operation*. In this mode of operation the reflux ratio is such that it always produces on-specification material, stopping the fraction when the reflux ratio has climbed to some value considered to be *uneconomic* (Kerkhof and Vissers, 1978; Logsdon and Biegler, 1993).

2) The total reflux start-up period is ended when the unit reaches its steady state. Product is collected at some constant finite reflux ratio until the accumulated product composition reaches its desired purity. This type of operation is very common in practice and is known as *constant reflux operation*. Under this operation mode the column is operated using a fixed reflux ratio for the whole operation (cut), producing better than specification material at the beginning and below specification material at the end of the fraction (Barolo and Botteon, 1997; Greaves et al., 2001)

The above two types of operations are referred to as *conventional operation* in the literature (Boston et al., 1981; Domenech and Enjalbert, 1974; Holland and Liapis, 1983).

A third type of operation is a trade off between the above two types of operation. Here an optimal reflux policy is chosen so that some objective function is satisfied (*minimum time, maximum product, maximum profit*, etc.), subject to any constraints (product amount and purity) at the end of the process (Coward, 1967; Hansen and Jorgensen, 1986; Diwekar et al., 1987; Mujtaba, 1989; Farhat et al., 1990; Mujtaba and Macchietto, 1992a, 1992b; Logsdon and Biegler, 1993; Mujtaba and Macchietto, 1993, 1996, 1997; Mujtaba, 1997, 1999; Sorensen and Skogestad, 1996). This type of operation is known as *optimal operation*.

3.3.3. Shutdown Period

At the end of production period, a batch distillation column can be shutdown in the following sequence:

1) Heat supply to the column is cut-off.
2) Holdup in the column is collapsed and collected in the reboiler.
3) Condenser holdup may be mixed with the top product or with the reboiler material.

For all other types of column configurations, the operation procedures in start-up, production and shutdown period will be slightly different.

3.3.4. Case Study

Greaves et al. (2001) and Greaves (2003) carried out experimental study using a pilot-plant conventional batch distillation column. The pilot-plant consists of an Aldershaw 35mm column consisting of a 5L reboiler, 40 plates and weir column surround by a pressure jacket, and a total condenser (Figure 3.11). The column is charged initially with the mixture to be separated and there is one outlet for the product to be collected and two sampling points at which the temperature sensors are placed. Methanol-Water system was considered with an initial charge of 900 ml of Methanol and 2100 ml of Water giving a total of 85.04 gmol of the mixture with <0.25, 0.75> mole fractions for Methanol and Water respectively.

The column was operated using *constant reboiler duty* strategy (section 3.2.4) and started up with the procedure outlined in section 3.3.1 with a total reflux for 30 minutes. The reflux in the column is produced by a simple switching mechanism that is controlled by a solenoid in a cyclic pattern (on-off). The valve is open for a fixed period of time (to withdraw distillate) and is closed for a fixed period of time (to return the reflux to the column). In this column the valve is always open (t_{op}) for 2 seconds and then closed (t_{close}) for 2x(R_{exp}) second, where R_{exp} is the reflux setting. Therefore, for a total batch operating time t_{diff}, the total opening time for the valve can be given by,

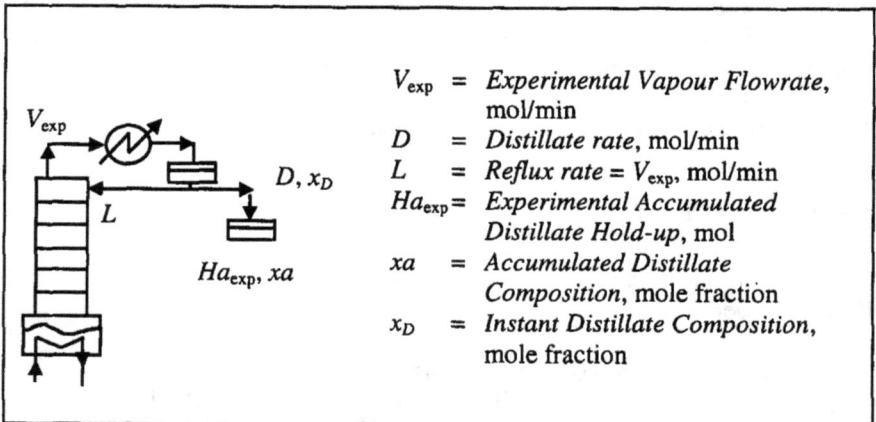

V_{exp}	=	*Experimental Vapour Flowrate,* mol/min
D	=	*Distillate rate,* mol/min
L	=	*Reflux rate* $= V_{exp}$, mol/min
Ha_{exp}	=	*Experimental Accumulated Distillate Hold-up,* mol
xa	=	*Accumulated Distillate Composition,* mole fraction
x_D	=	*Instant Distillate Composition,* mole fraction

Figure 3.11. Schematic of Batch Distillation Column used by Greaves et al.

$$t_{open} = \frac{t_{op}t_{diff}}{t_{op} + t_{close}} = \frac{2}{2\left(1 + R_{exp}\right)}t_{diff} \tag{3.1}$$

If V_{exp} is the vapour rate to the condenser, then at any time the following relation should be satisfied. Note at any time either D or L is equal to zero (due to cyclic nature of the operation).

$$D + L = V_{exp} \tag{3.2a}$$

Therefore, the distillate rate is:

$$D = V_{exp} \text{ (when the valve is open)} \tag{3.2b}$$

and the reflux rate is:

$$L = V_{exp} \text{ (when the valve is closed)} \tag{3.2c}$$

The external reflux ratio (R) defined as ($R = L/D$) will give $R = 0$ or $R = \infty$ at any time for the cyclic operation considered by Greaves et al. (2001).

Assuming continuous flow of reflux and distillate rate over a time period of Δt ($= t_{op} + t_{close}$), the average distillate rate can be expressed by:

$$D_{av} = \frac{Dt_{op}}{\Delta t} = \frac{V_{exp}t_{op}}{\Delta t} \tag{3.2d}$$

and the average reflux rate can be expressed by:

$$L_{av} = \frac{Lt_{close}}{\Delta t} = \frac{V_{exp}t_{close}}{\Delta t} \tag{3.2e}$$

Note that at any time Equation 3.2f (adding Equations 3.2d and 3.3e) will be satisfied.

$$D_{av} + L_{av} = V_{exp} \tag{3.2f}$$

An average external reflux ratio over a period of time Δt can now be defined as:

$$R_{av} = \frac{L_{av}}{D_{av}} \qquad (3.3a)$$

Use of Equations 3.2d, 3.2e and 3.3a will give:

$$R_{av} = \frac{t_{close}}{t_{op}} = \frac{2R_{exp}}{2} = R_{exp} \qquad (3.3b)$$

Note the Equation 3.3b shows that the reflux setting (R_{exp}) in the cyclic operation can be considered as the average external reflux ratio. In other words, the R_{exp} is the ratio of the average reflux rate to the average distillate rate over a period of Δt.

The total amount of distillate collected (Ha_{exp}) over a period of t_{diff} (assuming V_{exp} is constant over that period) is given by,

$$Ha_{exp} = D_{av}\Delta t = Dt_{open} = V_{exp}t_{open} = V_{exp}\frac{1}{\left(1+R_{exp}\right)}t_{diff} \qquad (3.4)$$

As the heat input to the column was fixed, it was not possible to maintain a constant V_{exp} throughout the operation. This results in a dynamic profile for V_{exp} over the operating time t_{diff}, as will be discussed next.

3.3.4.1. Estimation of V_{exp}

For a given R_{exp}, and a small interval of time t_{diff} (t_{diff} greater than Δt or a multiple of Δt), Ha_{exp} is measured and the corresponding V_{exp} can be estimated using Equation 3.4. It is observed that the value of V_{exp} decreases with time. This is due to gradual depletion of the light component from the column leaving behind the heavy component in the column. Since the heat of vaporisation of the heavy component is higher than that of the light component, a fixed heat duty gradually reduces the rate of vapour being produced. It is also observed that at any given time between [0, t_{diff}] the values of V_{exp} are higher for higher R_{exp}. This is due to the fact that the rate of depletion of the lighter component from the column is lower at higher reflux ratio and therefore a fixed heat duty gives a higher vapour rate.

For a given R_{exp}, the vapour rate profile is averaged to obtain an average V_{exp} to be used in the batch distillation model developed by Greaves et al.. Figure 3.12

shows the average V_{exp} vs. R_{exp} curve and Equation 3.5 gives the corresponding relationship.

$$V_{exp} = \left[a\left(\frac{1}{1+R_{exp}} \right)^2 + b\left(\frac{1}{1+R_{exp}} \right) + c \right]^{-1} \qquad (3.5)$$

where a, b and c are constant parameters and are function of the mixture type, column type and the operation strategy.

3.3.4.2. Plant Operation

Using the *constant reflux operation* strategy outlined in section 3.3.2 Greaves et al. (2001) carried out few experiments using different values of R_{exp} and different batch time. The accumulated distillate composition profiles as functions of batch time and distillate holdup are shown in Figures 3.13 and 3.14 respectively. Figure 3.15 shows the instant distillate composition profile for $R_{exp} = 2$.

For a given Ha_{exp} and R_{exp}, average V_{exp} was calculated using Equation 3.5 and batch time t_{diff} was calculated using Equation 3.4 (or Equation 3.6). A rough estimate of batch time thus help designing the number of experiments that can be carried out per day in a pilot-plant. Also it helps planning and implementing shut down procedure (outlined in section 3.3.3) for each set of experiment.

Figure 3.15 can be used to decide the cut time for the production of main-cut and off-cut.

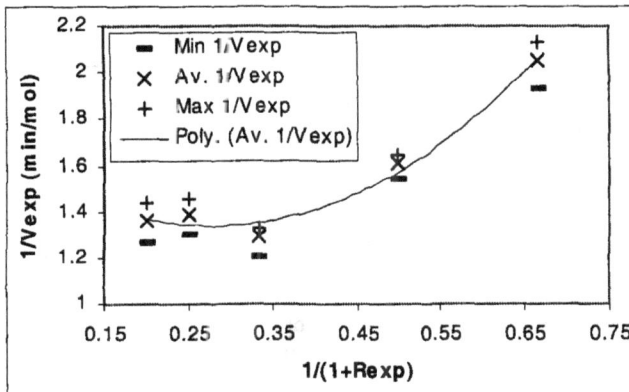

Figure 3.12. Vapour Load vs. Reflux Ratio (Greaves et al., 2001)

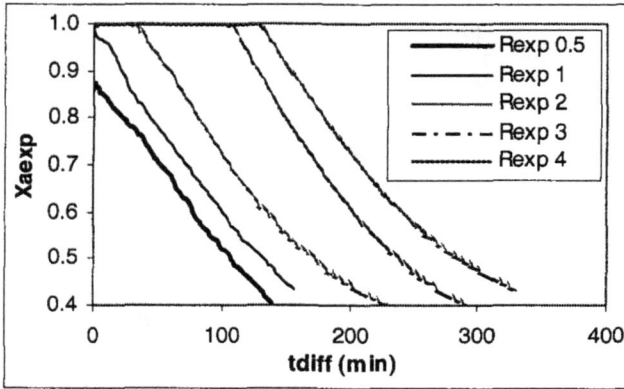

Figure 3.13. Accumulated Distillate Composition vs. Batch Time

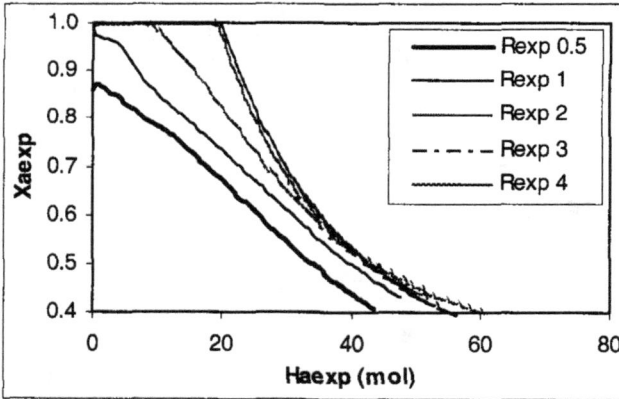

Figure 3.14. Accumulated Distillate Composition vs. Amount of Accumulated Distillate

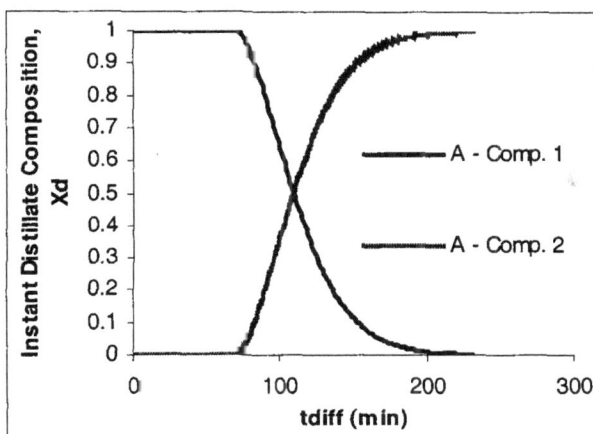

Figure 3.15. Instant Distillate Composition Profile ($R_{exp}= 2$).

3.4. Performance Measure

The performance criteria of a batch distillation column can be measured in terms of *maximum profit, maximum product or minimum time* (Mujtaba, 1999). In distillation, whether batch, continuous or extractive, purity of the main products must be specified as it is driven by the customer demand and product prices. The amount of product and the operation time can be dictated by economics (*maximum profit*) or one of them can be fixed and the other is obtained (*minimum time* with fixed amount of product or *maximum distillate* with fixed operation time). The calculation of each of these will require formulation and solution of optimisation problems. A brief description of these optimisation problems is presented below. Further details will be provided in Chapter 5.

In general, the optimisation problem to optimise the operation of conventional batch distillation can be stated as follows:

given:	the column configuration, the feed mixture, vapour boilup rate,
	a separation task in terms of product purity,
	(+ recovery or amount of product or operation time or none)
determine:	optimal reflux ratio which governs the operation
so as to:	*minimise* the operation time
	or *maximise* the amount of product

or ***maximise*** the profit
subject to: any constraints (e.g. reflux ratio limits)

3.4.1. Case Study- Experiment Based Algorithm for Minimum Time

Greaves et al. (2001) developed an experiment-based algorithm for finding the *minimum batch time* for the column and mixture described in section 3.3.4.

For a given *separation task* [defined in terms of amount and composition of the main-cut (Ha^* , xa^*)] for a binary mixture, Equation 3.4 can be (replacing V_{exp} from Equation 3.5) rearranged as:

$$t_{diff} = \frac{Ha^* (1 + R_{exp})}{f(R_{exp})} = Ha^* g(R_{exp}) \tag{3.6}$$

where f and g are non-linear functions of R_{exp}.

For a given Ha^* , Equation 3.6 shows that the batch time (and so does the distillate composition, xa) increases nonlinearly with the reflux ratio. Figure 3.16 shows these values for $Ha^* = 15$ mol and 40 mol respectively along with the corresponding experimental values. Although each point on any of these curves gives the minimum time for the corresponding (Ha^* , xa), only one point which is the true minimum batch time will correspond to the desired *separation task* (Ha^* , xa^*) and the optimum R_{exp}.

Greaves et al. (2001) proposed an algorithm shown in Figure 3.17 for calculating the optimum reflux ratio and the minimum batch time for a given *separation task*. The algorithm suggests to start with a low value of R_{exp} and gradually increase it and stop at where $xa \sim xa^*$. This approach will require a few iterations to achieve the minimum batch time. Calculations with a large initial value of R_{exp} do not guarantee the optimum reflux ratio and the true minimum batch time at the first point where $xa \sim xa^*$ and therefore may require more iterations. This is explained with reference to two given *separation tasks* as summarised in Table 3.1 and Table 3.2 respectively.

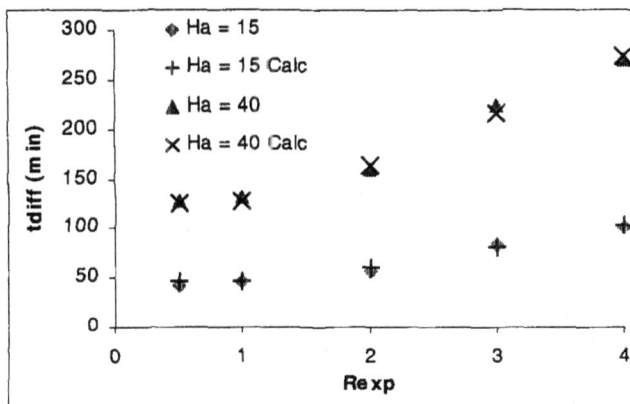

Figure 3.16. Batch Time vs. Reflux Ratio (Equation 3.6)

Table 3.1. Separation Task 1: $Ha^* = 15$, $xa^* = 0.999$

R_{exp}	$1/V_{exp}$	t_{diff}	xa_{exp}
0.5	2.07	46.47	0.731
1	1.58	47.27	0.792
2	1.36	60.93	0.906
3	**1.34**	**80.62**	**0.999**

Table 3.2. Separation Task 2: $Ha^* = 40$, $xa^* = 0.53$

R_{exp}	$1/V_{exp}$	t_{diff}	xa_{exp}
0.5	2.07	123.92	0.439
1	1.58	126.06	0.499
2	**1.36**	**162.49**	**0.529**
3	1.34	214.98	0.531
4	1.32	273.91	0.532

The optimum reflux ratio and the minimum batch time for *separation task* 1 are 3 and 80.62 min (Table 3.1). The *separation task* 2 could be achieved using 3 different reflux ratio (Table 3.2) but however, $R_{exp} = 2$ gives the true minimum batch time which is about 40% lower than the batch time that would be required to achieve the same separation with $R_{exp} = 4$.

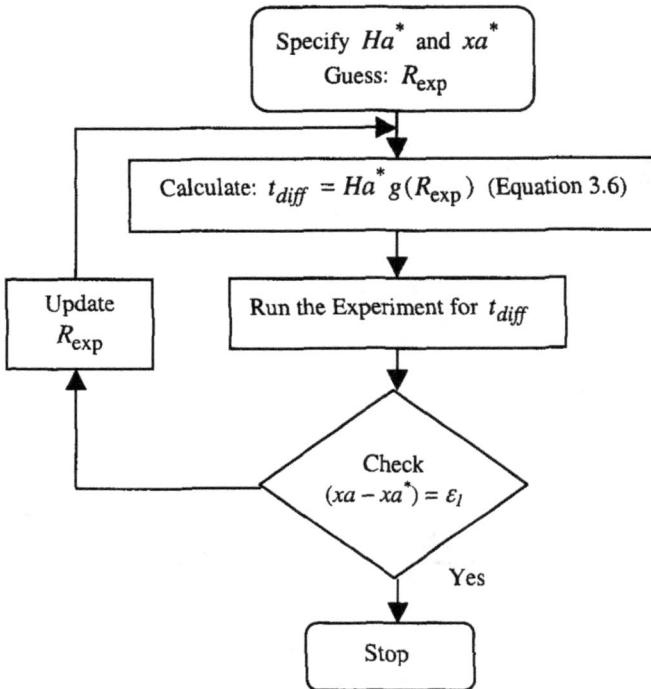

Figure 3.17. Experiment Based Algorithm for Calculating
Minimum Batch Time. ε_l is a small positive number.

The main advantage of the algorithm is that for a given reflux ratio it will estimate the duration of the batch. However, the major disadvantage of this approach is that a time consuming experiment is to be carried out for each new value of the R_{exp} until the given *separation task* can be achieved in minimum time.

Note to avoid abortion of batches during trial experiments (if carried out in an actual plant with large throughput) it is recommended to stop the batch when the distillate composition drops below or the top tray temperature goes above the specification. This will avoid any possible loss of revenue. Also note that, whether it leads to optimal operation or not, the industrial operators decides the cut time based on the above criteria.

3.5. Column Holdup

In a steady state continuous distillation with the assumption of a well mixed liquid and vapour on the plates, the holdup has no effect on the analysis (modelling of such columns does not usually include column holdup) since any quantity of liquid holdup in the system has no effect on the mass flows in the system (Rose, 1985). Batch distillation however is inherently an unsteady state process and the liquid holdup in the system become sinks (accumulators) of material which affect the rate of change of flows and hence the whole dynamic response of the system.

An extensive literature survey indicates that the role of column holdup on the performance of batch distillation has been the subject of some controversy until recently. The following paragraphs outline briefly the investigations carried out on column holdup since 1950. Most of the investigations were restricted to conventional batch distillation columns and binary mixtures. The readers are directed to the original work to develop further understanding of the topic.

Rose et al. (1950) and Rose and O'Brien (1952) studied the effect of holdup for binary and ternary mixtures in a laboratory batch column. They qualitatively defined the term *sharpness of separation* as the sharpness in the break between successive components in the graph of instantaneous distillate composition against percentage distilled. They showed that an increase in column holdup enhanced the sharpness of separation at low reflux ratio but did not have any effect at a very high reflux ratio.

For binary mixtures Converse and Huber (1965) found that in all cases studied, column holdup causes a decrease in the amount of maximum distillate obtained for a fixed time of operation. In another way, for a fixed amount of distillate and purity of the lighter component, higher column holdup increased the batch time. The authors concluded that the presence of significant holdup is bad anyway.

Using binary mixtures, Luyben (1971) studied the effects of holdup, number of plates, relative volatility, etc. on the capacity (total products/hr). For an arbitrarily assumed constant reflux ratio the author observed both positive and negative effects of tray holdup on the capacity for columns with larger number of plates, while only negative effects were observed for columns with smaller number of plates. It is apparent that these observations are related to *the degree of difficulty* of separation.

Mayur and Jackson (1971) simulated the effect of holdup in a three-plate column for a binary mixture, having about 13% of the initial charge distributed as plate holdup and no condenser holdup. They found that for both constant reflux and optimal reflux operation, the batch time was about 15-20% higher for the holdup case compared to the negligible holdup case. Rose (1985) drew similar conclusion about column holdup but mentioned that the adverse effects of column holdup depends entirely on the system, on the performance required (amount of product, purity), and on the amount of holdup. Logsdon (1990) found that column holdup had a small but positive effect on their column operation.

Using binary mixtures, Mujtaba and Macchietto (1998) carried out further investigation to explain the effect of holdup qualitatively or quantitatively and to correlate the facts observed in the past. The goal was to provide a technique for helping in the selection of proper column holdup at the design stage.

The approach used by Mujtaba and Macchietto was to independently (a) characterise the column for a given number plates, mixture and separation, (b) fix the mode of operation which will measure the performance of the column for a given column holdup.

3.5.1. Column Characterisation- the Degree of Difficulty of Separation

Referring to Figure 3.1 for a binary mixture and given:

- the number of plates (N_T) of a batch column
- the type of mixture to be handled (thermodynamic and physical properties)
- an initial charge composition (x_{B0}) and
- a desired distillate product purity (x^*_D)

Christensen and Jorgensen (1987) proposed a measure *q*, *the degree of difficulty* of separation to characterise the column separation task.

$$q = \int_{x_{BF}}^{x_{B0}} \frac{(N_{min} + 1) dx_B}{(x_{B0} - x_{BF})(N_T + 1)} \tag{3.7}$$

Here, x_{BF} is the final bottom product composition and x_B is a reboiler composition intermediate between x_{BF} and x_{B0}.

This measure was based upon the ratio of the minimum necessary number of plates, N_{min} (averaged over the reboiler composition) in a column to the actual number of plates in the given column, N_T. Christensen and Jorgensen assumed that the mixture has a constant relative volatility α and the column operates at total reflux using *constant distillate composition* (x^*_D) strategy (section 3.3.2) and evaluated N_{min} using the Fenske equation:

$$N_{min} + 1 = \frac{\ln\left(\frac{x^*_D}{1 - x^*_D} \frac{1 - x_B}{x_B}\right)}{\ln \alpha} \tag{3.8}$$

Since the relative volatility changes considerably as batch distillation progresses Mujtaba and Macchietto (1998) suggested that α values should be estimated at the top and bottom of the column using rigorous vapour liquid equilibrium model and geometric average α be used in Equation 3.8 as x_B changes.

The measure q reflects in a single number the column used, the type of mixture and the separation requirements. The value of q increases with decreasing relative volatility, increasing distillate purity demands and decreasing number of stages in excess of the minimum. Its value is independent of the amount charged. It ranges from 0 (infinite number of stages) to 1 (minimum number of stages). Typical values are $q < 0.60$ for an easy separation and $q > 0.6$ for a difficult one (Christensen and Jorgensen, 1987). Also this measure is very useful as q does not only measure how close the boiling points of the components are, it also measures the extent a given column is oversized for the purpose.

3.5.2. Performance Measure - Minimum Time

Mujtaba and Macchietto (1998) chose the *minimum time* as the most suitable performance measure, because the separation requirements (e.g. product purities) are fixed and quantitative measure, q could be easily used.

The effects of column holdup can be easily correlated in terms of q and of the minimum batch time required to achieve a given separation task.

3.5.3. Case Study

A summary of several example cases illustrated in Mujtaba and Macchietto (1998) is presented below. Instead of carrying out the investigation in a pilot-plant batch distillation column, a rigorous mathematical model (Chapter 4) for a conventional column was developed and incorporated into the *minimum time* optimisation problem which was numerically solved. Further details on optimisation techniques are presented in later chapters.

Three typical binary mixtures were considered. The mixtures were: 1. Benzene-Toluene, 2. Cyclohexane-Toluene and 3. Butane-Pentane. Some of the results were presented graphically to show the role of holdup in terms of *the degree of difficulty* of separation and *minimum time* operation.

Table 3.3. Summary of Results for Investigation 1. [Adopted from Mujtaba (1989) and Mujtaba and Macchietto (1998)]

Case	Mix	N_T	x^*_D	q	Opt. Reflux Ratio	Opt. Holdup %	T_1 hr	T_2 hr	T_s %
1	1	8	.900	.332	.557	26.0	2.69	2.26	15.98
2	1	6	.900	.427	.601	18.0	2.83	2.51	11.30
3	1	4	.900	.597	.695	10.0	3.38	3.28	2.95
4	1	3	.900	.746	.779	2.0	4.52	4.52	....
5	2	8	.895	.332	.563	28.0	2.72	2.29	15.80
6	2	6	.895	.426	.610	19.0	2.89	2.56	11.40
7	2	4	.895	.597	.699	10.0	3.42	3.32	2.92
8	2	4	.916	.672	.781	2.0	4.56	4.56	....
9	2	3	.895	.746	.782	2.0	4.56	4.56	....
10	3	6	.963	.428	.638	15.0	3.58	2.75	23.18
11	3	4	.963	.598	.764	8.0	4.52	4.24	6.19
12	3	4	.970	.638	.828	4.0	6.11	5.81	4.90
13	3	4	.975	.670	.875	2.0	7.97	7.97	....

Key: = N_T = no of plates, Mix = Mixture
T_1 = minimum batch time for the separation using lowest plate holdup (2%)
T_2 = minimum batch time for the separation using optimal plate holdup
T_s = % time saved = $(T_1 - T_2)$ x $100/T_1$

3.5.3.1. Effect of Plate Holdup on the Column Performance

Column configurations (N_T) and separation requirements (x^*_D) for several cases are presented in Table 3.3. The condenser holdup was fixed to 2% of the total initial charge and the column hold up is varied as a percentage of the total initial charge to the column. The initial charge to the column (B_0) is 5 kmol with a light component mole fraction (x_{B0}) of 0.6. Also the amount of distillate product required (D^*) is set to 3 kmol. The column operates under *constant condenser vapour load* strategy (section 3.2.2) with a vapour load of 3 kmol/hr for all cases.

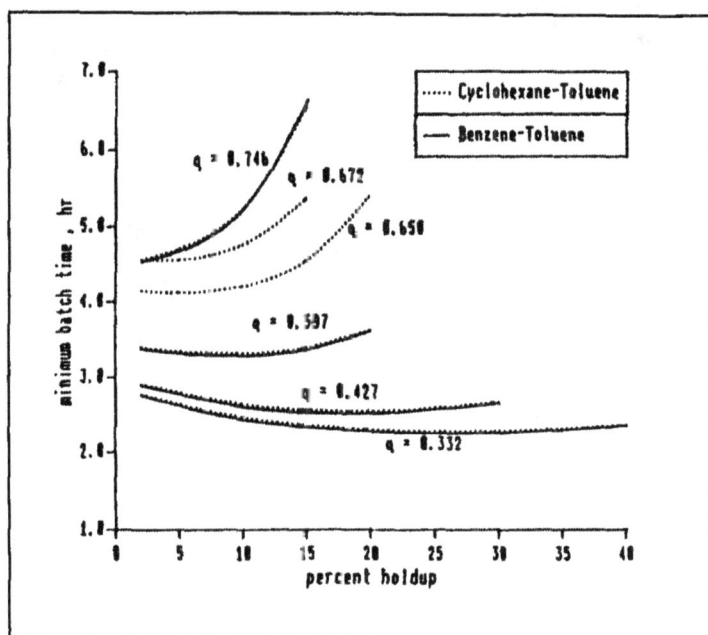

Figure 3.18a. Minimum Batch Time vs Column Holdup at different q^c

A series of *minimum time* problems (Chapter 5) were solved at different values of q with increasing holdup for each case. Figures 3.18a and 3.18b show the minimum time solution vs. percent total holdup in the column for different mixtures at different q and Figures 3.19a and 3.19b show the corresponding optimum reflux ratio (required to get the separation in minimum time) vs. percent total holdup of the column. The results are summarized in Table 3.3 which shows, for each given separation, the optimum value of holdup to achieve the best performance out of the given column. The corresponding best minimum batch time and the optimum reflux ratio to achieve that are also presented in the table for each case.

The last column of Table 3.3 shows the percent reduction in batch time achieved using the optimum plate holdup compared to the batch time with the lowest plate holdup (2%). It clearly shows that the column performance in terms of minimum batch time is improved significantly with increasing plate holdup for easy separation ($q < 0.60$). For one case 23% batch time saving is observed (case 10).

---

c Reprinted from *Chem. Eng. Sci.*, **53**, Mujtaba, I.M. and Macchietto, S., Holdup issues in batch distillation-binary mixture, 2519-2530 , Copyright (1998), with permission from Elsevier Science .

However, for difficult separation ($q > 0.60$) this is reversed. Figures 3.18a and 3.18b clearly show that for $q > 0.60$ the column performance, in terms of minimum batch time, is improved significantly with decreasing plate holdup and suggests that for difficult separations the column holdup should be kept as minimum as possible. This is also clear from the results presented in Table 3.3 which show that for difficult separations optimum column holdup is very close to the minimum (Mujtaba and Macchietto, 1998 used 2% as the minimum column holdup). The minimum batch times for both cases (using minimum and optimum holdup) are almost alike and no time saving could be realized when compared to one another (last column of Table 3.3). The results discussed so far clearly show that holdup may have a dramatic effect on the operation.

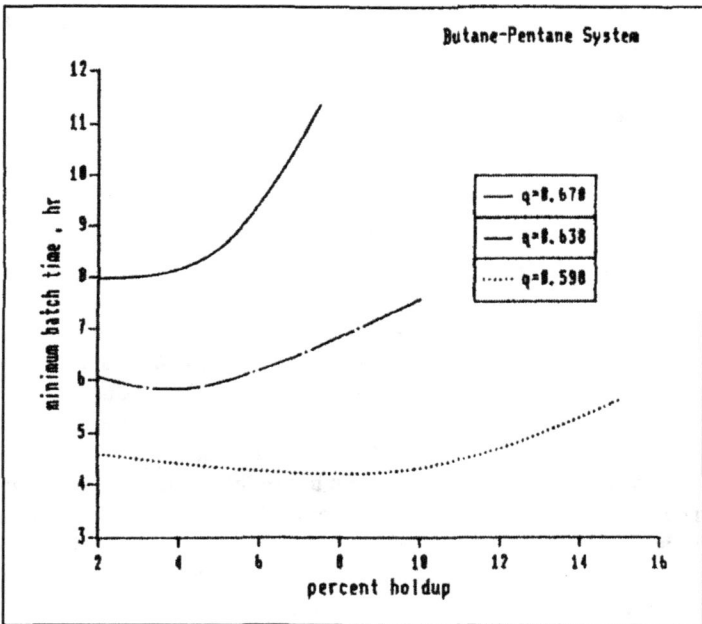

Figure 3.18b. Minimum Batch Time vs Column Holdup at different q^d

---

d   Reprinted from *Chem. Eng. Sci.*, **53**, Mujtaba, I.M. and Macchietto, S., Holdup issues in batch distillation-binary mixture, 2519-2530 , Copyright (1998), with permission from Elsevier Science .

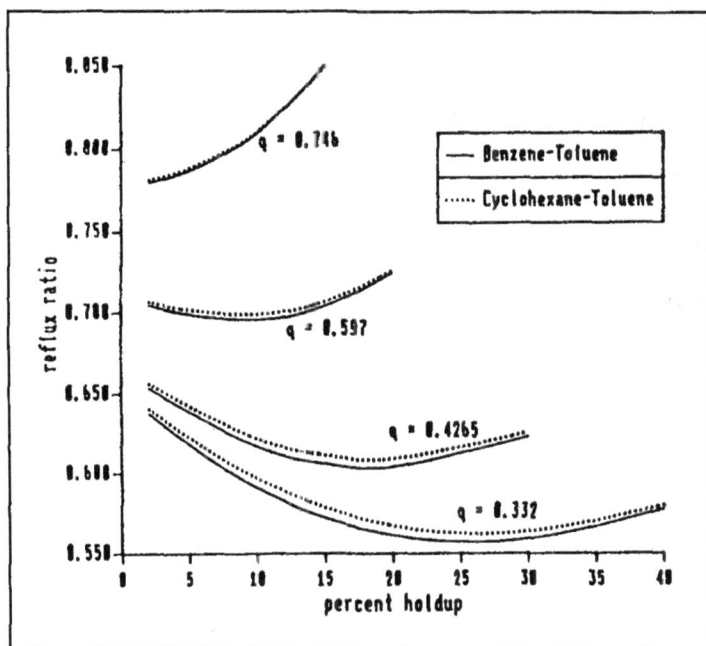

Figure 3.19a. Optimum Reflux Ratio vs Column Holdup at different q^e

Note that q measures how difficult the given separation is in a given column for a particular mixture. However, the batch times are not identical for different mixtures with the same value of q. For the same value of q the batch times can be different for different binary mixtures although the optimum holdup for the binary mixtures are about 10% (Mujtaba and Macchietto, 1998).

The distillate accumulator, plate 1 (top) and the reboiler liquid composition (for benzene) profiles for case 3 are presented in Figures 3.20-3.22. Figure 3.23 presents the plate 1 liquid composition profile for case 4. Using these composition profiles Mujtaba and Macchietto (1998) made the following observations for better understanding of the role of column holdup.

---

[e] Reprinted from *Chem. Eng. Sci.*, **53**, Mujtaba, I.M. and Macchietto, S., Holdup issues in batch distillation-binary mixture, 2519-2530 , Copyright (1998), with permission from Elsevier Science .

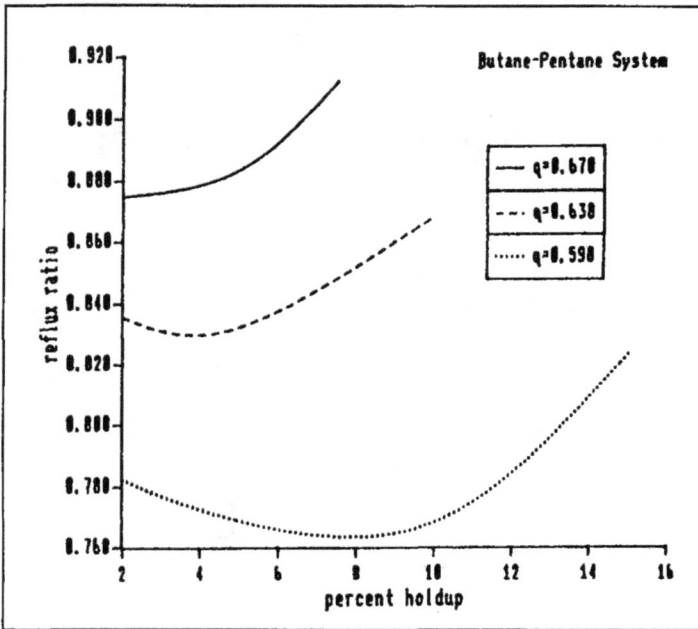

Figure 3.19b. Optimum Reflux Ratio vs Column Holdup at different q^f

As already mentioned, a q value of 0.3-0.4 may be interpreted to mean that 30-40% of the plates available in the column are the required minimum for the given separation and therefore 60-70% of the plates are in excess.

As the holdup on the plates always acts as an accumulator, larger holdup gives slower dynamic responses in the concentration profiles. With very low holdup on the plates, most of the liquid is located in the reboiler. The light component initially moves very quickly up the column, giving an early and high peak in the instant distillate and top plates composition profiles, followed by a fairly quick drop off (Figure 3.20, Holdup=2%). With slightly larger holdup (Figure 3.20, Holdup=10%) the composition profiles become flatter, with a smaller initial overshoot of the distillate composition than before and also a slower change in the composition profiles past the peak. Overall, a lower reflux ratio would be sufficient, resulting in slightly shorter batch times (for the same separation task) than in the previous case (as seen more clearly in Figure 3.21).

---

f   Reprinted from *Chem. Eng. Sci.*, **53**, Mujtaba, I.M. and Macchietto, S., Holdup issues in batch distillation-binary mixture, 2519-2530 , Copyright (1998), with permission from Elsevier Science .

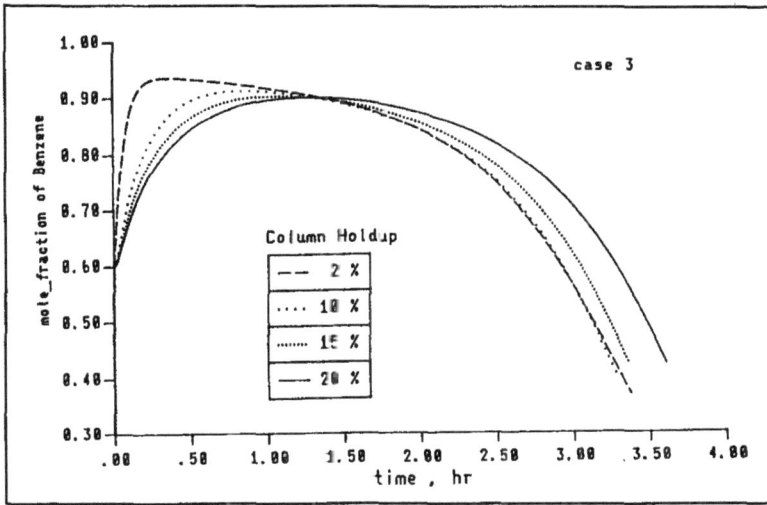

Figure 3.20. Plate 1 Liquid Composition Profiles for Benzene[g]

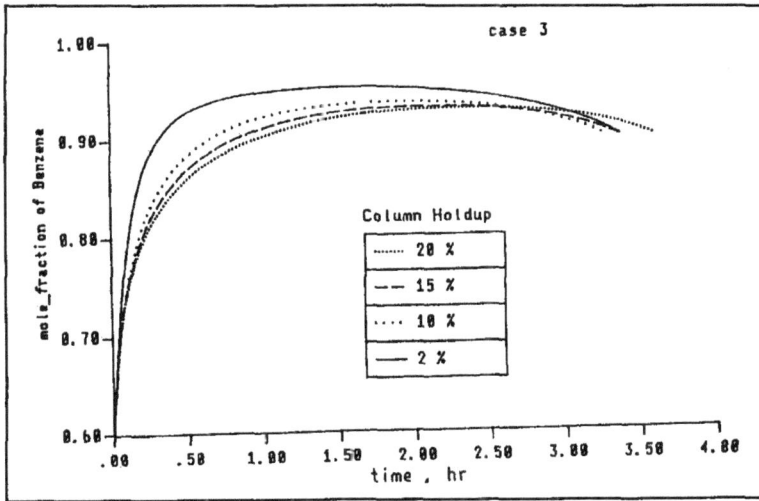

Figure 3.21. Accumulated Distillate Composition Profiles for Benzene[g]

---

[g] Reprinted from *Chem. Eng. Sci.*, 53, Mujtaba, I.M. and Macchietto, S., Holdup issues in batch distillation-binary mixture, 2519-2530 , Copyright (1998), with permission from Elsevier Science .

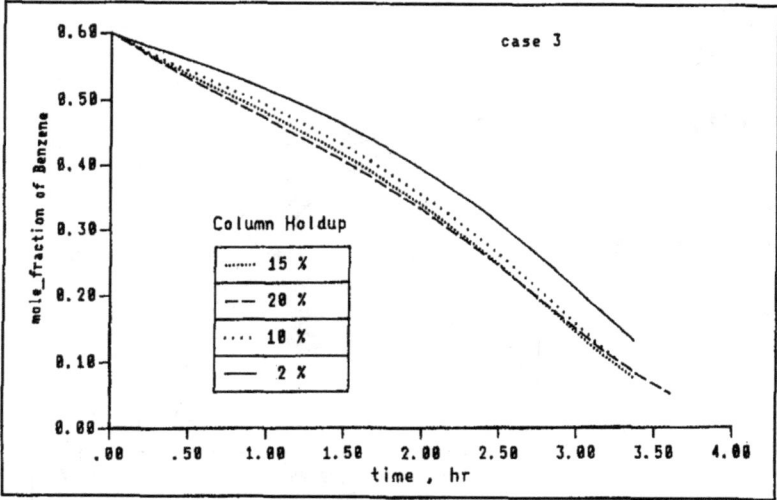

Figure 3.22. Reboiler Liquid Composition Profiles for Benzene[h]

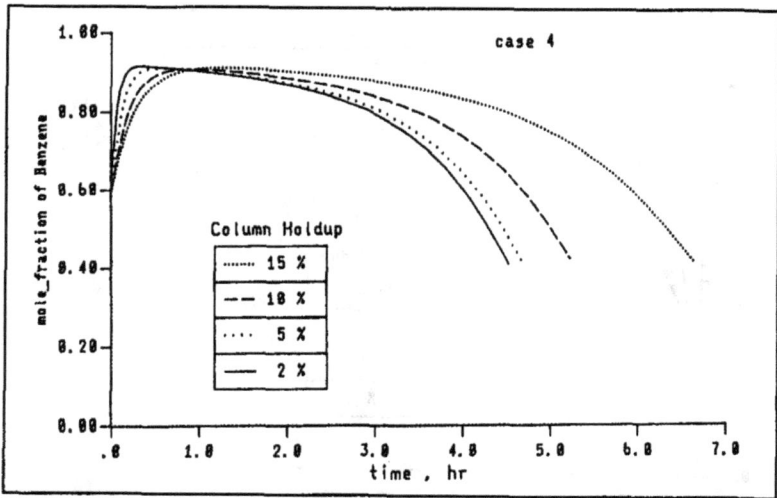

Figure 3.23. Plate 1 Liquid Composition Profiles for Benzene[h]

---

[h]   Reprinted from *Chem. Eng. Sci.*, **53**, Mujtaba, I.M. and Macchietto, S., Holdup issues in batch distillation-binary mixture, 2519-2530 , Copyright (1998), with permission from Elsevier Science .

For a given separation task, further decrease in batch time (due to lower reflux ratio operation) can be obtained by increasing plate holdup until such a stage where any further holdup increases result in the reboiler being quickly depleted of the light component as the distillation progresses at which point the heavy component then begin to rise up the column. At this value of holdup and beyond, low reflux operation is no longer possible. So a high reflux is needed to maintain the product purity which consequently means longer batch time. Figure 3.22 clearly explains the above fact. For the 20% holdup case the column has to operate at a quite high reflux ratio to maintain the product quality because of the significant drop in light component in the reboiler.

With increasing q (more difficult separation) there will be fewer plates in excess of the minimum required and the column must be operated at high reflux ratio to achieve the given separation. The effects of holdup on batch time will be more significant. In case 4 (Table 3.3), for example, the minimum batch time increases from approximately 4.5 hours to about 7 hours as the holdup distributed on the plates increases from 2% of the initial charge to 20% (Figure 3.23). The effects observed at low q with high holdup only are now dominant for any value of the holdup. For the cases studied it has been found that beyond $q = 0.67$ increasing column holdup above zero always increase the minimum batch time (Figure 3.23).

3.5.3.2. Effect of Condenser Holdup on the Column Performance

This study is carried out only for binary mixture 1 with a q value of 0.332. The role of the condenser holdup is examined for three condenser holdup values, 0%, 2% and 5% of the total initial charge. The plate holdup is varied in all cases as a percentage of the total initial charge to the column. There are a total of 8 internal plates with a separation requirement of 90% purity (x^*_D) of benzene.

Figure 3.24 shows the effect of the condenser holdup on the performance of the column. Since, for a total condenser holdup only plays as an accumulator of material but not as a separation stage, larger condenser holdup means longer batch time is always required to achieve a given separation. This is quite clear from Figure 3.24. In practice there must always be a certain amount of condenser holdup to ensure a neat reflux operation, however this should be kept to a minimum. Luyben (1971) also arrived at similar conclusions.

Although the case study presented here identifies an optimum amount of holdup (as a percentage of the total initial charge), in practice the size of the reboiler and the amount of holdup on the plates and in the condenser are fixed for an existing column and are usually dictated by design geometry and pressure drop requirements. Since it is always desirable to charge the reboiler to its full capacity, the holdup to charge ratio will, in general, not be an optimal one for a particular distillation task on hand. In such circumstances it is always possible to trade-off between batch time and recovery of the product to achieve a profitable operation (discussed in detail in Mujtaba and Macchietto, 1993; 1996 and in Chapter 5).

Nevertheless the investigations presented here show that if the holdup effects are understood for the range of separations and mixtures to be handled by a column, it may also be possible to use this information (optimum holdup) at the design stage, and balance the design requirements against additional dynamic effects and column performance variations due to holdup. For further details see Mujtaba and Macchietto (1998).

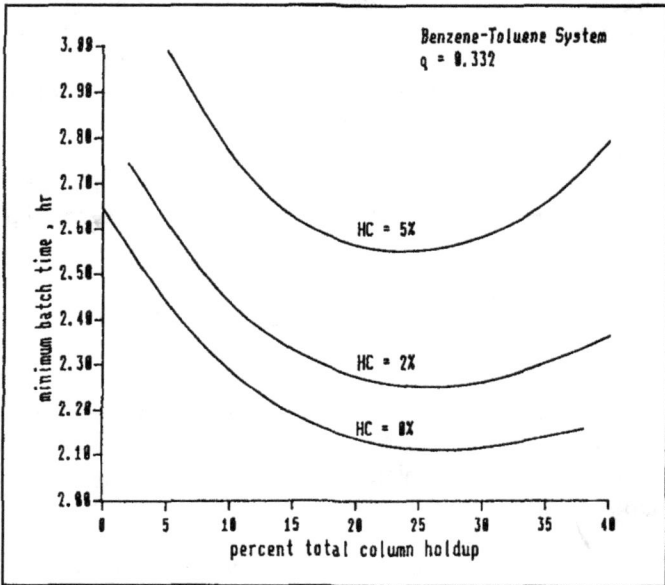

Figure 3.24. Effect of Condenser Holdup[i]

---

[i] Reprinted from *Chem. Eng. Sci.*, **53**, Mujtaba, I.M. and Macchietto, S., Holdup issues in batch distillation-binary mixture, 2519-2530 , Copyright (1998), with permission from Elsevier Science .

3.6. Campaign Operation

More than one batch is considered in a long-term production campaign if the total amount of mixture to be processed is more than the capacity of the column. During processing a particular batch, as the overhead composition varies (Figure 3.25, Mujtaba, 1989), a number of main-cuts and off-cuts are made at the end of various distillation tasks or periods (section 3.1). In campaign mode, each intermediate off-cuts may be collected and stored separately and fed to the reboiler sequentially and reprocessed in subsequent batches (Quintero-Marmol and Luyben, 1990; Mujtaba and Macchietto, 1992a). The other alternative is to collect and store each off-cut separately for sometime and reprocessed when the amount of material of each off-cut reaches to the level of one full batch charge (Mujtaba and Macchietto, 1994).

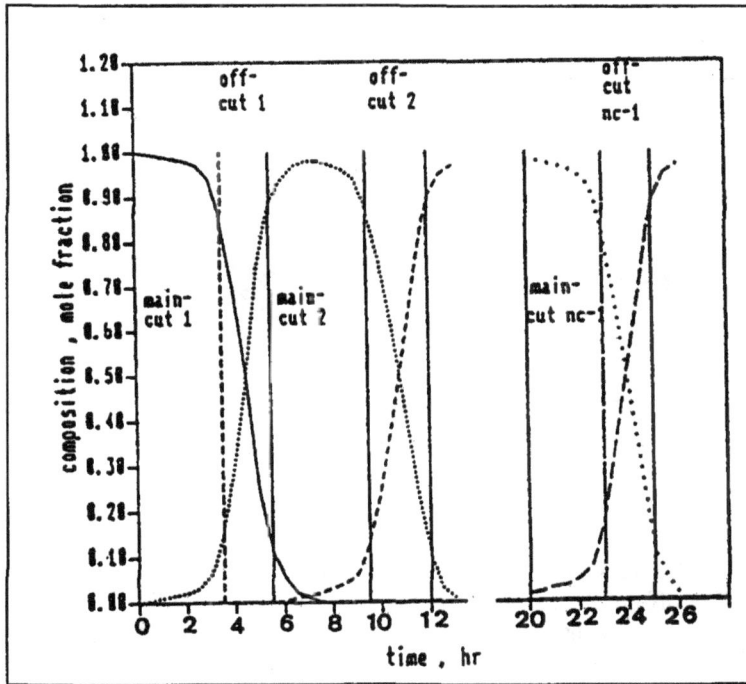

Figure 3.25. Typical Instant Distillate Composition Profile. [Mujtaba, 1989]

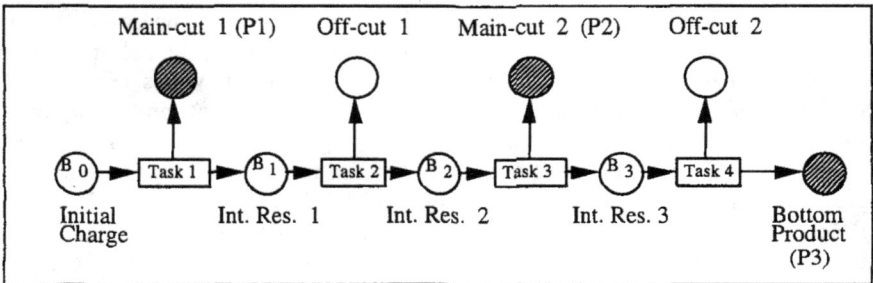

Figure 3.26. Operational Sequence for Fresh Feed Processing
(Product cuts in parenthesis)

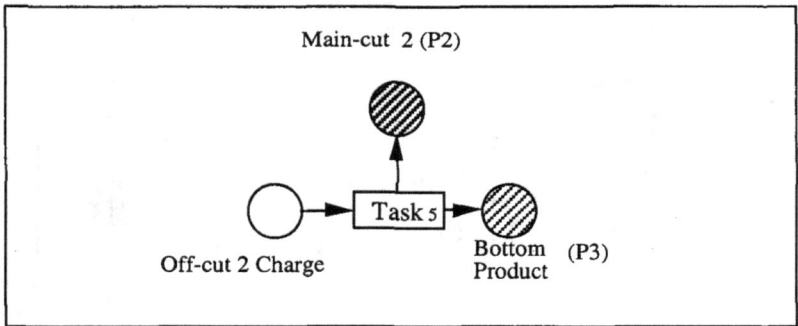

Figure 3.27. Off-cut 2 Reprocessing (Product cuts in parenthesis)

Figure 3.26 shows the operating sequence of processing a fresh ternary feed mixture. The off-cut 2 generated from several batches according to STN of Figure 3.26 can be collected and reprocessed according to the operating sequence shown in Figure 3.27. Figure 3.28 shows the overall campaign mode operation over a time horizon of H (hr). An industrial case study (Mujtaba and Macchietto, 1994) is presented in Chapter 6.

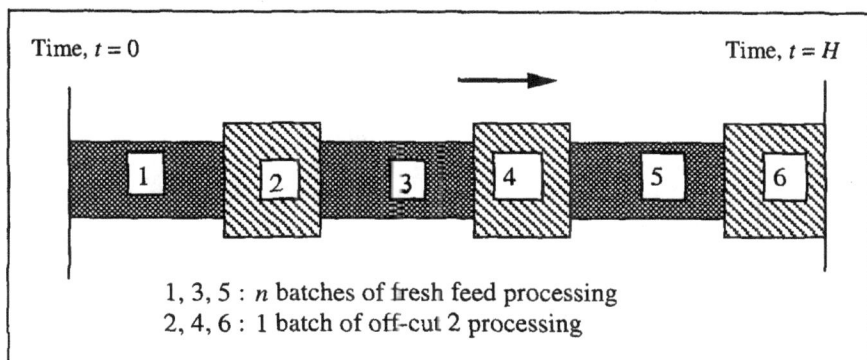

Figure 3.28. Overall Production Campaign (not to scale)

3.7. Recycle Operation

It is well known that the production of off-cuts (Mujtaba and Macchietto, 1993) or production and recycling of off-cuts (Mayur et al., 1970; Christensen and Jorgensen, 1987; Mujtaba and Macchietto, 1988, 1992a) improved the column performance to a great extent depending on *the degree of difficulty* of separation. This type of operation is particularly useful when using an existing column with perhaps a less than optimal number of stages (Mujtaba and Macchietto, 1988). The main advantages with recycling of off-cuts are a potential reduction in distillation time and also recovery of valuable products (Christensen and Jorgensen, 1987; Mujtaba and Macchietto, 1988b, 1992b).

3.7.1. Binary Mixture

For binary mixtures there is usually one main-cut and one off-cut (Figure 3.4) and therefore the operational strategy with off-cut recycle is straight forward. Typical choices to be taken are whether to recycle or not the off-cut fraction, the amount and composition of this cut and the reflux ratio profile to be adopted during production of both cuts. As the batch cycle is repeated (under campaign mode operation discussed in section 3.6) with a similar operation policy with recycling, a *quasi-steady state* mode of operation is attained which is characterised by the fact that the amount and composition of the off-cut fraction recycled from the previous batch and the off-cut fraction from the current batch become identical (Figure 3.29). Luyben

(1988) indicates that in practice such a *quasi-steady state* mode is approximately achieved after three or four cycles.

Mujtaba and Macchietto (1988) used the measure, q, *the degree of difficulty* of separation (section 3.5.1) to correlate the extent of benefit of production and recycling of off-cuts for binary mixtures. It is reported that for some separations more than 70% batch time savings were possible when compared to no off-cut production and no recycle cases. Using the measure, q, the authors were also able to explain whether and when an initial total reflux operation was required in batch distillation. Further details are presented in Chapter 8.

3.7.2. Multicomponent Mixture

For multicomponent mixtures of n_c components there are n_c -1 possible off-cuts (Figure 3.25). There are a number of operational choices regarding the way the off-cuts are recycled (whether and how to store them, when and how to recycle them) and determining the best strategy is more complex. On the other hand the recycle policy may have a strong effect on the overall economy of the process, and selecting the best one is very important.

The collection of intermediate off-cuts permits achieving more easily the high purity requirements of the main distillate (and bottoms) products, at the expense of some loss of valuable material in the off-cuts. Collecting the off-cuts and reprocessing them in subsequent batches can increase recoveries. The most common recycling strategy (Luyben, 1988) consists of collecting all the off-cuts produced during a batch cycle in a single vessel. At the end of the cycle the contents of the vessel are recycled to the still, and mixed with fresh feed to complete a new charge. This is shown by simple diagram in Figure 3.30.

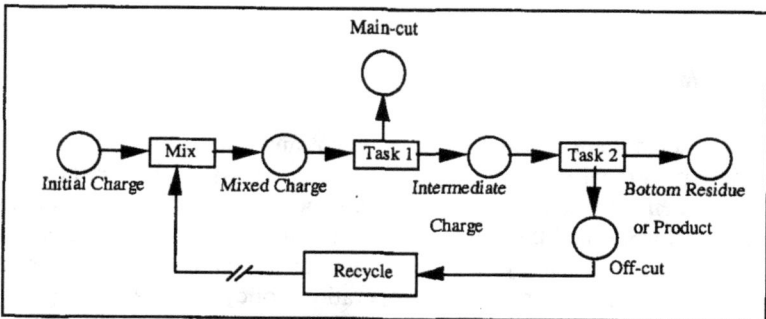

Figure 3.29. Quasi-steady State Off-cut Recycle Strategy in Binary Conventional Batch Distillation

Mujtaba (1989) suggested an alternative strategy to collect and store the off-cuts separately and recycle each of them to the reboiler in sequential order (Figure 3.31). Quintero-Marmol and Luyben (1990) discussed a similar recycling strategy. They also considered other alternative off-cuts handling strategies for multicomponent mixtures.

Bonny et al. (1996) developed a further strategy for off-cut recycling. Further details are discussed in Chapter 8.

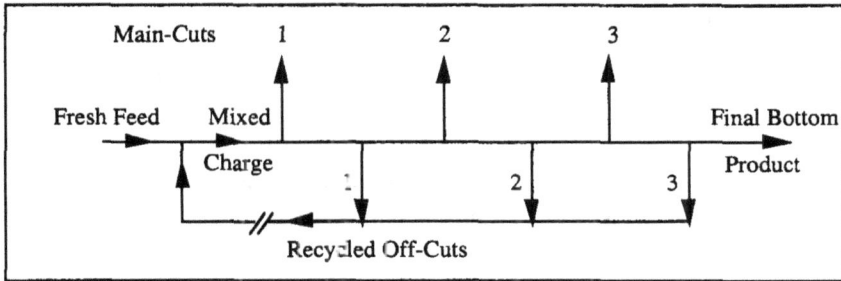

Figure 3.30. Multicomponent Batch Operation with Recycled Off-Cuts
(Luyben, 1988)

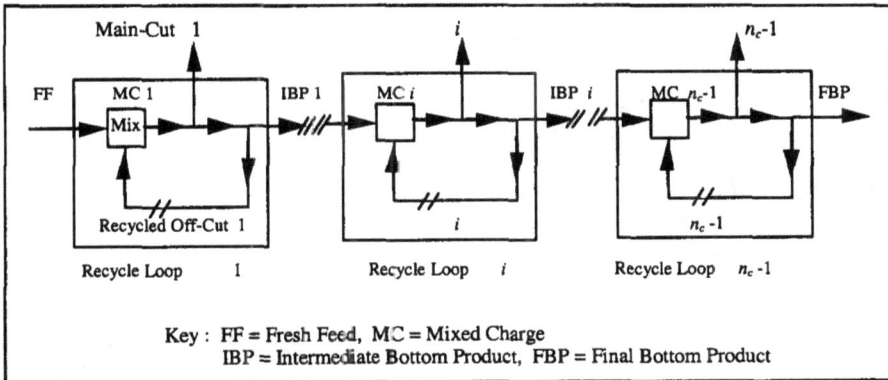

Figure 3.31. Multicomponent Batch Operation with Off-Cut Recycle
(Mujtaba, 1989)

References

Barb, D.K. and Holland, C.D., *In Proceedings of the 7th World Petroleum Congress*, **4** (1967), 31.

Barolo, M. and Botteon, F., *AIChE J.* **43** (1997), 2601.

Boston, J.F., Britt, H.I., Jirapongphan, S. and Shah, V.B.,1980, *In Foundation of Computer Aided Chemical Process Design*, (Engineering Foundation, New York), **II** (1980), 203.

Converse, A.O. and Huber, C.I., *IEC Fund.* **4** (1965), 475.

Coward, I., *Chem. Eng. Sci.* **22** (1967), 503.

Christensen, F.M. and Jorgensen, S.B., *Chem. Eng. J.* **34** (1987), 57.

Cuille, P.E. and Reklaitis, G.V., *Comput. chem. Engng.* **10** (1986), 389.

Diwekar, U.M., Malik, R.K. and Madhavan, K.P., *Comput. chem. Engng.* **11** (1987), 629.

Domenech, S. and Enjalbert, M., *Chem. Eng. Sci.* **29** (1974),1529

Domenech, S. and Enjalbert, M., *Comput. chem. Engng.* **5** (1981), 181.

Farhat, S., Czernicki, M., Pibouleau, L. and Domenech, S., *AIChE J.* **36** (1990), 1349.

Greaves, M. A., *Hybrid Modelling, Simulation and Optimisation of Batch Distillation Using Neural Network Techniques*. Ph.D. Thesis, (University of Bradford, Bradford, UK, 2003).

Greaves, M.A., Mujtaba, I.M. and Hussain, M.A., In *Application of Neural Network and Other Learning Technologies in Process Engineering*, eds. Mujtaba, I.M. and Hussain, M.A. (Imperial College Press, London. 2001), 149.

Hansen, T.T. and Jorgensen, S.B., *Chem. Eng. J.* **33** (1986), 151.

Holland, C.D. and Liapis, A.I., *Computer methods for solving dynamic separation problems* (McGraw-Hill Book Company, New York, 1983).

Kerkhof, L.H. and Visseres, H.J.M., *Chem. Eng. Sci.* **33** (1978), 961.

Kondili, E., Pantelides, C.C. and Sargent, R.W.H. *In Proceedings of 3rd International Symposium on Process Systems Engineering*, (1988) 62.

Logsdon, J.S. and Biegler, L.T., *IEC Res.* **32** (1993), 700.

Logsdon, J.S., *Efficient determination of optimal control profiles for differential and algebraic systems*, PhD Thesis, (Carnegie Mellon University, USA, 1990).

Luyben, W.L., *IEC Proc. Des. Dev.* **10** (1971), 54.

Luyben, W.L., *IEC Res.* **27** (1988), 642.

Macchietto, S and Mujtaba, I.M., In *Batch Processing Systems Engineering: Fundamentals and Applications for Chemical Engineering*, eds. G.V. Reklaitis et al., Series F: Computer and Systems Sciences, (Springer Verlag, Berlin). **143**, (1996), 174.

Mayur, D.N. and Jackson, R., *Chem. Eng. J.* **2** (1971), 150.

Mayur, D.N., May, R.A., and Jackson, R., *Chem. Eng. J.* **1** (1970), 15.

Mujtaba, I.M., *Optimal Operational Policies in Batch Distillation*. PhD Thesis, (Imperial College, University of London, 1989).

Mujtaba, I.M., *Trans. IChemE*, **75A** (1997), 609.

Mujtaba, I.M., *Trans. IChemE*, **77A** (1999), 588.

Mujtaba, I.M. and Macchietto, S. In *Recent Progres en Ginie de Procedes*, eds. Domenech, S. et al. (Lavoisier Techniche et Documentation, Paris, 1988) **2**, 191.

Mujtaba, I.M. and Macchietto, S., In *IMACS Annals on Computing and Applied Mathematics, Vol 4: Computing and Computers for Control Systems*, eds. Borne, P. et al. (J.C. Baltzer AG, Scientific Publishing Co., Basel, Switzerland. 1989), 55.

Mujtaba, I.M. and Macchietto, S., *Comput. chem. Engng.* **16S** (1992a), S273.

Mujtaba, I.M. and Macchietto, S., Optimal operation of multicomponent batch distillation, *Presented at the AIChE Spring National Meeting*, New Orleans, USA, March 29 - April 2, Paper no. 61c, 1992b.

Mujtaba, I.M. and Macchietto, S., Optimal operation of reactive batch distillation. *Presented at the AIChE Annual Meeting*, Miami Beach, USA, Nov. 1-6, paper no. 135g, 1992c.

Mujtaba, I.M. and Macchietto, S., *Comput. chem. Engng.* **17** (1993), 1191.

Mujtaba, I.M. and Macchietto, S., *In Proceedings of the 1994 IChemE Research Event*, IChemE, UK, **2** (1994) 911.

Mujtaba, I.M., Stuart, G. and Macchietto, S., *In Proceedigns of DYCORD'95*, Ed. J.B. Rawlings, (Danish Automation Society, Denmark. 1995), 135.

Mujtaba, I.M. and Macchietto, S., *J. Proc. Control.* **6** (1996), 27.

Mujtaba, I.M. and Macchietto, S., *IEC Res.* **36** (1997), 2287.

Mujtaba, I.M. and Macchietto, S., *Chem. Eng. Sci.* **53** (1998), 2519.

Nad, M. and Spiegel, L., *In Proceedings CEF 87*, Sicily, Italy, April (1987), 737.

Quintero-Marmol, E. and Luyben, W.L., *IEC Res.* **29** (1990), 1915.

Robinson, E.R., *Chem Eng. Sci.* **25** (1970), 921.

Robinson, C.S. and Gilliland, E.R., *Elements of Fractional Distillation*, (4th ed., McGraw Hill, New York, 1950).

Rose, L.M., *Distillation Design in Practice* (Elsevier, New York, 1995).

Rose, A. and O'Brien, V.J. (Jr), *Ind. Eng. Chem.* **44** (1952), 1480.

Rose, A., William, T.J. and Prevost, C., *Ind. Eng. Chem.* **42** (1950), 1876.

Ruiz, C.A., *In Proceedings of CHEMDATA 88*, 13-15 June, Sweden, (1988), 330.

Sorensen, E. and Prenzler, M., *Comput. chem. Engng.* **21S** (1997), S1215.

Sorensen, E. and Skogested, S. *In Proceedings of the 5th International Symposium on Process Systems Engineering- PSE'94*, Korea (1994), 449.

Sorensen, E. and Skogested, S. *Chem. Eng. Sci.* **51** (1996), 4949.

Wajge, R.M. and Reklaitis, G.V., *Chem. Eng. J.* **75** (1999), 57.

Walsh, S. Mujtaba, I.M. and Macchietto, S., *Acta Chimica Slovenica*, **42** (1995), 69.

CHAPTER 4

4. MODELLING AND SIMULATION

4.1. Introduction

Simulating the actual operation (both start-up and product period) of conventional columns has been the subject of much research for more than half a century. The main interest was usually to develop a model (consisting of mass and energy balances, hydraulic model, physical properties, etc.) that could best predict the operation of the column.

Modelling of batch distillation begins with the well-known and simple Rayleigh Model (Rayleigh, 1902). With the development of high speed digital computers the main issues in modelling were whether and how to include energy balances, column holdup, plate hydraulics, accurate physical properties, etc. to simulate the actual operation of batch columns. In many cases it was found that the models had to be simplified because of several reasons: size and complexity of the problem, capabilities of the computer, availability of suitable numerical methods to integrate the model equations, gain in accuracy in the prediction of real operation vs computation time, etc.

4.1.1. Simulation of Start-up Period

The simulation of actual start-up operation is very difficult unless there is a detailed rigorous model including plate hydraulics. However, the simulation of start-up operation from step 5 (section 3.3.1) is fairly easy and can be done without considering a detailed hydraulic model. To do this the filling of the holdup in steps 3 and 4 (section 3.3.1) can be achieved in different ways:

(a) directly with still pot liquid at the boiling point temperature (Converse and Huber, 1965; Hitch and Rousseau, 1988; Mujtaba and Macchietto, 1988, 1998; Logsdon and Biegler, 1993, Hasebe et al., 1992). This means that the plate and condenser compositions in the model equations are initialised to the fresh feed composition for simulation;

56

(b) operating the unit without reflux i.e. with only one rectification theoretical stage (using Rayleigh model). Vapours from the reboiler are condensed and stored in the overhead equipment until the liquid fills the condenser and the column holdups. This mode was suggested by Luyben (1971) and was used by Gonzalez-Velasco et al. (1987).

Step 5 (section 3.3.1) now simply requires to run the column (i.e. simulate the column operation using a model) at total reflux until the unit reaches the steady state (Holland and Liapis, 1983; Nad and Spiegel, 1987) or until the instant distillate composition reaches the product composition so that the product can be collected at constant distillate composition (Coward, 1967; Kerkhof and Vissers, 1978; Logsdon and Biegler, 1993). A variation of this total reflux operation concept can also be found in the literature and in practice. In this variation only a part of the condensed liquid is returned to the column and the rest is taken out as product (product period starts from step 3 of section 3.3.1) (Converse and Huber, 1965; Mayur and Jackson, 1971; Mujtaba, 1989; Mujtaba and Macchietto, 1992). Mujtaba and Macchietto (1988) and Mujtaba (1989) clearly explain whether and when an initial total reflux operation is required.

So far, only Ruiz (1988) considered the actual start-up procedure in his simulation by incorporating a very detailed model including plate hydraulics, weeping, channelling of vapour through downcomers, the actual liquid sealing of downcomers, etc. The author employed a generalised dynamic model (DYNAM), developed for continuous distillation by Gani et al. (1986), in batch distillation. Most recently van Lith et al. (2003) considered simulation of start-up period using fuzzy logic based model.

4.1.2. Simulation of Product Period

In comparison, the simulation of product period is fairly easy and was considered by several authors in the past with different types of models and for conventional columns (Huckaba and Danly, 1960; Meadow, 1963; Domenech and Enjalbert, 1981; Cuille and Reklaitis, 1986; Diwekar and Madhavan, 1986; Hitch and Rousseau, 1988; Ruiz, 1988; Galindez and Fredenslaund, 1988; Mujtaba, 1989; Mujtaba and Macchietto, 1998; Diwekar and Madhavan, 1991a,b; Sundaram and Evans, 1993a,b). Some experimental simulations of the product period were also reported with modelling (Domenech and Enjalbert, 1974; Nad and Spiegel, 1987).

Robinson and Gilliland (1950), Abrams et al. (1987) and Hasebe et al. (1992, 1995), Sorensen and Skogestad (1996a), Barolo et al. (1998), Furlonge et al. (1999) reported simulation results using inverted, batch distillation columns with middle vessel and multi-effect batch distillation, respectively. Mujtaba (1997) simulated batch distillation operation using a continuous distillation model.

To the best of author's knowledge no work has so far been reported on the simulation of actual shutdown operation in batch distillation.

4.2. Models for Conventional Batch Distillation

4.2.1. *Rayleigh Model - Model Type I*

The Rayleigh model was developed for a single stage batch distillation· where a liquid mixture is charged in a still and a vapour is produced by heating the liquid. At any time, the vapour on top of the liquid is in equilibrium with the liquid left in the still. The vapour is removed as soon as it is produced but no part of the vapour is returned as reflux to the still after condensation.

Starting with an initial amount of liquid (B_0) with composition x_{B0} charged in the still, the amount (B_1) of liquid left in the still and its composition (x_{B1}) at any time can be calculated using the Rayleigh model (Equation I.1).

$$\ln \frac{B_0}{B_1} = \int_{x_{B1}}^{x_{B0}} \frac{dx_B}{y - x_B} \tag{I.1}$$

where y is the vapour composition at any time and is in equilibrium with x_B.

The amount of distillate at any given time can be calculated using:

$$H_a = B_0 - B_1 \tag{I.2}$$

For a binary mixture and given that the relative volatility (α) between the components is constant, instantaneous distillate composition y can be calculated using:

$$y = \frac{\alpha x_B}{1 + (\alpha - 1) x_B} \tag{I.3}$$

4.2.1.1. Example

A liquid mixture of benzene and toluene with $B_0 = 1$ kmol and $x_{B0} = <0.5, 0.5>$ molefraction is subject to batch distillation. The relative volatility of the mixture over the operating temperature range is assumed constant with a value of $(\alpha =)$ 2.44. Table 4.1 shows the distillation profiles obtained using the Rayleigh model.

4.2.2. Short-cut Model- Model Type II

The shortcut model is developed based on the assumption that batch distillation operation can be represented by a series of continuous distillation operation of short duration and employs modified Fenske-Underwood-Gilliland (FUG) shortcut model of continuous distillation (Diwekar and Madhavan, 1991a,b; Sundaram and Evans, 1993a,b). Starting with an initial charge (B_0, x_{B0}) at time $t=t_0$ and for a small interval of time $\Delta t = t_1 - t_0$, the batch distillation column conditions at t_0 and t_1 is schematically shown in Figure 4.1 (Galindez and Fredenslund, 1988).

For a given vapour rate, the distillate rate can be calculated by

$$D = \frac{V}{1+R} \tag{II.1}$$

where R is the reflux ratio (external) defined as $R = \dfrac{L}{D}$.

The accumulated distillate over $\Delta t = t_1 - t_0$ can be obtained by

$$H_a = D\Delta t \tag{II.2}$$

Overall mass balance:

$$B_0 = B_1 + H_a$$

Table 4.1. Distillation Profile Using The Rayleigh Model

x_{B1}	B_1	H_a	y
0.5 (x_{B0})	1.0 (B_0)	0.0	0.71
0.43	0.70	0.3	0.64
0.37	0.54	0.46	0.59
0.21	0.24	0.76	0.40
0.0	0.0	1.0	0.00

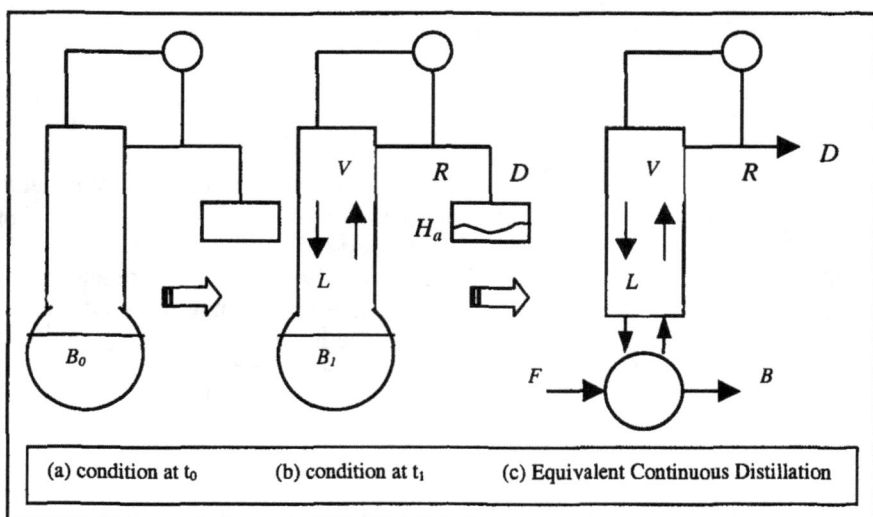

Figure 4.1. Batch Distillation as a Series of Continuous Distillation

or,

$$B_0 / \Delta t = B_1 / \Delta t + H_a / \Delta t \qquad \text{(II.3)}$$

Let, $F = B_0 / \Delta t; B = B_1 / \Delta t$; Therefore,

$$F = B + D \qquad \text{(II.4)}$$

The equivalent continuous distillation operation for a short interval of Δt can now be shown in Figure 4.1.(c).

The component balance:

$$Fx_F^i = Bx_{B_{new}}^i + Dx_D^i$$

or,

$$x_{B_{new}}^i = \frac{Fx_F^i - Dx_D^i}{B} \qquad \text{(II.5)}$$

Note that $x^i_{B_{new}}$ from the current distillation step becomes x^i_F for the next distillation step. At time, $t = 0$, $x^i_F = x^i_{B0}$. The distillate compositions x^i_D are estimated using FUG method. Also note that Diwekar (1992, 1995) and Sundaram and Evans (1993a) developed alternative expressions for $x^i_{B_{new}}$. Diwekar (1992, 1995) describes the method for calculating the top composition (x^i_D) using modified FUG method.

Fenske Equation:

$$N_{min} = \frac{\ln\left(\frac{x^i_D \; x^k_B}{x^k_D \; x^i_B}\right)}{\ln \alpha_i} \tag{II.6}$$

$$C_1 \approx N_{min} \tag{II.7}$$

Hengestebeck-Geddes' Equation:

$$x^i_D = \left(\frac{\alpha_i}{\alpha_1}\right)^{C_1} \frac{x^1_D}{x^1_B} x^i_B \qquad i = 2, 3, \ldots, n_c \tag{II.8}$$

Summation of fractions and x^1_D estimation:

$$\sum_{i=1}^{n} x^i_D = 1 \tag{II.9}$$

$$x^1_D = \frac{1}{\sum_{i=1}^{n}\left(\frac{\alpha_i}{\alpha_1}\right)^{C_1} \frac{x^i_B}{x^1_B}} \tag{II.10}$$

Underwood Equations:

$$\sum \frac{\alpha_i x^i_B}{\alpha_i - \phi} = 0 \tag{II.11}$$

$$R_{min,u} + 1 = \sum_{i=1}^{n} \frac{\alpha_i x_D^i}{\alpha_i - \phi} \qquad (\text{II}.12)$$

Gilliland Correlation:

$$X = \frac{R - R_{min,u}}{R + 1} \qquad (\text{II}.13)$$

$$Y = \frac{N - N_{min}}{N + 1} \qquad (\text{II}.14)$$

$$Y = 1 - \exp\left[\frac{(1 + 54.4X)(X - 1)}{(11 + 117.2X)\sqrt{X}}\right] \qquad (\text{II}.15)$$

$$R_{min,g} = R - X(R + 1) \qquad (\text{II}.16)$$

Condition for C_1 estimation:

$$G_c = \frac{R_{min,u}}{R} - \frac{R_{min,g}}{R} \approx 0 \qquad (\text{II}.17)$$

Equations II.6-II.17 are iterative to estimate the right value of C_1 and therefore to calculate x_D^i.

Note that Sundaram and Evans (1993a,b) used FUG method of continuous distillation directly and developed time explicit model, while Diwekar (1992) developed modified FUG method as described above and time implicit model for batch distillation. Sundaram and Evans used time as an independent variable of the model while Diwekar (1992) used reboiler composition as independent variable. Both models are based on zero column holdup and does not include plate-to-plate calculations. See the original references for further details.

4.2.2.1. Example

Seader and Henley (1998) considered the separation of a ternary mixture in a batch distillation column with $B_0 = 100$ moles, $x_{B0} = $ <A, B, C> = <0.33, 0.33, 0.34> molefraction, relative volatility $\alpha = $ <2.0, 1.5, 1.0>, theoretical plates $N = 3$, reflux ratio $R = 10$ and vapour boilup ratio $V = 110$ kmol/hr. The column operation was simulated using the short-cut model of Sundaram and Evans (1993a). The results in terms of reboiler holdup (B_l), reboiler composition profile (x_{Bl}), accumulated distillate composition profile (x_a), minimum number of plates (N_{min}) and minimum reflux ratio (R_{min}) are summarised in Table 4.2. At any time, the amount of distillate (H_a) can be calculated using Equation II.3. The values of x_a are calculated using:

$$x_a = \frac{B_0 x_{B0} - B_1 x_{B1}}{H_a}.$$

4.2.3. Simple Model- Model Type III

Referring to Figure 4.2 the model is developed based on the assumptions of constant relative volatility and equimolal overflow and include detailed plate-to-plate calculations. Further assumptions are listed below:

i)   constant molar holdup for condenser and internal plates.
ii)   total condensation without sub-cooling.
iii)   negligible vapour holdup.
iv)   perfect mixing of liquid and vapour on the plates.
v)   negligible heat losses.
vi)   theoretical plates.
vii)   feed mixture at its bubble point.
viii)   adiabatic column.
ix)   negligible pressure drop across the column.

Table 4.2. Simulation by Shortcut Model of Sundaram and Evans (1993a)

Time	B_1	x_{B1} (reboiler)			N_{min}	R_{min}
hr	kmol	A	B	C		
0.0	100 (B_0)	0.3300	0.3300	0.3400 (x_{B0})	-	-
0.5	95	0.3143	0.3329	0.3528	2.6294	1.2829
1.0	90	0.2995	0.3348	0.3657	2.6249	1.3092
1.5	85	0.2839	0.3365	0.3796	2.6199	1.3385
2.0	80	0.2675	0.3378	0.3947	2.6143	1.3709

Time	x_D (Instant Distillate)			x_a (Accumulated Distillate)		
hr	A	B	C	A	B	C
0.0	0.6449	0.2720	0.0831	-	-	-
0.5	0.5957	0.2962	0.1081	0.6283	0.2749	0.0968
1.0	0.5803	0.3048	0.1149	0.6045	0.2868	0.1087
1.5	0.5633	0.3142	0.1225	0.5912	0.2932	0.1156
2.0	0.5446	0.3242	0.1312	0.5800	0.2988	0.1212

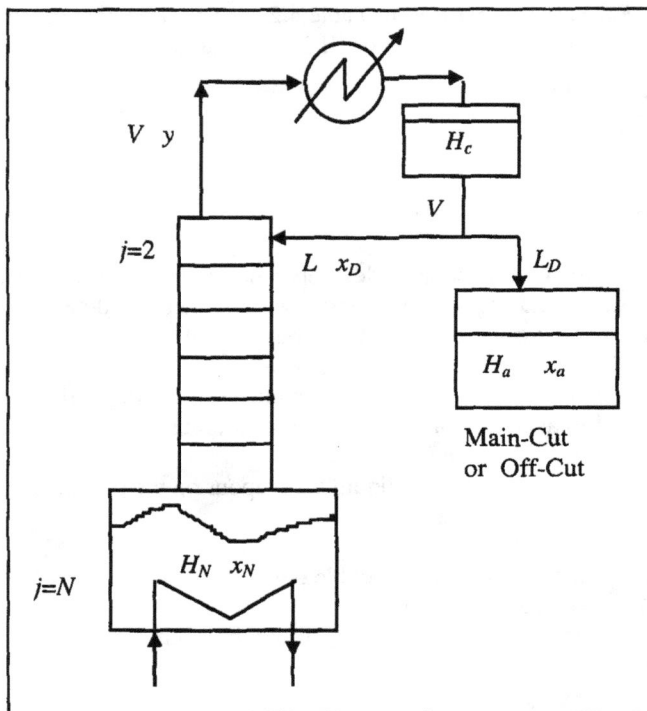

Figure 4.2. CBD with Notations for Simple Model (Type III)

First the equations for the condenser are presented. Then the equations for the accumulator, followed by the equations for the plates in the column and the reboiler are presented. The trays are counted from the top to the bottom. j refers to plates and i refers to components.

Condenser and Accumulator, $j=1$; $i = 1$ to n_c

The amount of product in the accumulator, H_a, changes from time to time due to incoming liquid, L_D, from the condenser according to the overall mass balance equation:

$$\frac{dH_a}{dt} = L_D \tag{III.1}$$

The component mass balance in the accumulator is:

$$\frac{dx_{ai}}{dt} = \frac{L_D}{H_a}\left(x_{1i} - x_{ai}\right)$$ (III.2)

The condenser characterises the first tray and the holdup tank contains an amount of liquid H_C which is kept constant at all times. The component balance for the holdup tank is:

$$\frac{dx_{1i}}{dt} = \frac{V}{H_c}\left(y_{2i} - x_{1i}\right)$$ (III.3)

The reflux ratio (internal) is defined as:

$$L = rV$$ (III.4)

The distillate-rate to the accumulator or product tank is therefore:

$$L_D = V(1-r)$$ (III.5)

Internal Plates, $j=2$ to $(N-1)$; $i = 1$ to n_c

Like the condenser, the molar plate holdup remains at a constant value. The component balance on tray j is:

$$\frac{dx_{ji}}{dt} = \frac{V}{H_j}\left(y_{j-1,i} - y_{j,i}\right) + \frac{L}{H_j}\left(x_{j-1,i} - x_{j,i}\right)$$ (III.6)

The vapour liquid equilibrium relationship is written as:

$$y_{j,i} = \frac{\alpha_i x_{j,i}}{\sum_{k=1}^{n_c}\alpha_k x_{j,k}}$$ (III.7)

Reboiler, $j=N$; $i = 1$ to n_c

The amount of mixture left in the reboiler depends on the liquid and the vapour flow rate through the column. The total mass balance is written as:

$$\frac{dH_N}{dt} = L - V \tag{III.8}$$

The component mass balance is:

$$H_N \frac{dx_{N,i}}{dt} = L(x_{N-1,i} - x_{N,i}) - V(y_N - x_{N,i}) \tag{III.9}$$

The vapour liquid equilibrium relationship is same as the equation (III.7) with $j = N$.

Converse and Huber (1965), Robinson (1970), Mayur and Jackson (1971), Luyben (1988) and Mujtaba (1997) used this model for simulation and optimisation of conventional batch distillation. Domenech and Enjalbert (1981) used similar model in their simulation study with the exception that they used temperature dependent phase equilibria instead of constant relative volatility. Christiansen et al. (1995) used this model (excluding column holdup) to study parametric sensitivity of ideal binary columns.

4.2.3.1. Example

A liquid binary mixture with $B_0 = 10$ kmol and x_{B0} = <0.6, 0.4> molefraction is subject to conventional batch distillation shown in Figure 4.3. The relative volatility of the mixture over the operating temperature range is assumed constant with a value of ($\alpha =$) 2. The total number of plates is, $N = 20$. The vapour boilup rate is, $V = 5.0$ kmol/hr and the reflux ratio is, $r = 0.75$. The condenser and total plate holdups are 0.2 and 0.2 kmol respectively.

Figure 4.3 shows the accumulated distillate amount, the bottom product amount and the compositions at the end of 5 hr operation, obtained using the simple model outlined in section 4.2.3. Figure 4.4 shows the instant distillate composition profile.

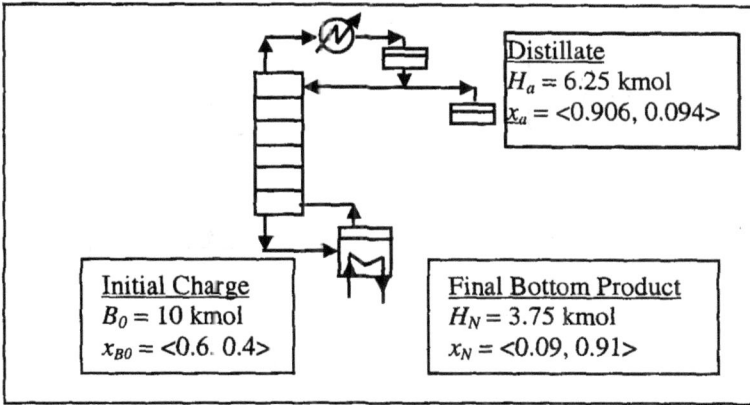

Figure 4.3. Batch Distillation using Simple Model

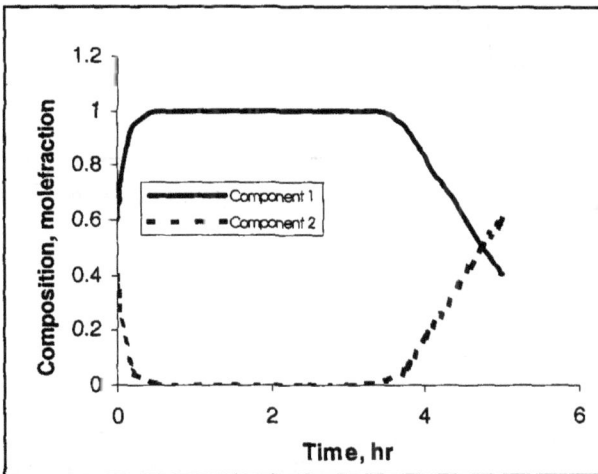

Figure 4.4. Instant Distillate Composition Profile (Example 4.2.3.1)

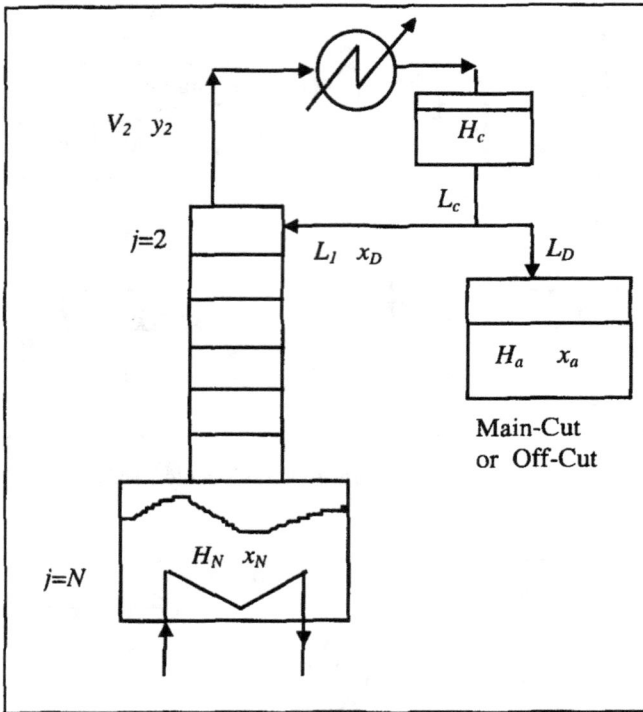

Figure 4.5. CBD with Notations for Detailed Model (Type IV-CVH)

4.2.4. *Rigorous Model - Model Type IV*

4.2.4.1. Constant Volume Holdup (CVH) Model

The model equations are presented in the following section, with reference to the column configuration shown in Figure 4.5. The main assumptions are listed below:

i)	constant volume holdup for reboiler, condenser and internal plates.
ii)	total condensation without sub-cooling.
iii)	negligible vapour holdup.
iv)	perfect mixing of liquid and vapour on the plates.
v)	negligible heat losses.
vi)	theoretical plates

vii)   feed mixture at its bubble point
viii)   adiabatic column

Condenser and Accumulator, $j=1$; $i = 1$ to n_c

Accumulator total mass balance:

$$\frac{dH_a}{dt} = L_D \qquad\qquad (IV.1)$$

Component bass balance:
   a)   Accumulator

$$\frac{d(H_a x_{ai})}{dt} = L_D x_{Di} \qquad\qquad (IV.2)$$

   b)   Condenser Holdup Tank

$$\frac{d(H_c x_{Di})}{dt} = V_2 y_{2i} - L_c x_{Di} \qquad\qquad (IV.3)$$

Energy balance:

$$\frac{d(H_c h_1^L)}{dt} = V_2\, h_2^V - L_c h_1^L - Q_C \qquad\qquad (IV.4)$$

Physical properties and other equations:

$$L_1 = r L_c \qquad\qquad (IV.5)$$
$$L_D = L_c\,(1-r) \qquad\qquad (IV.6)$$
$$T_1 = T_1(\mathbf{x}_D, P) \qquad\qquad (IV.7)$$
$$h_1^L = h_1^L(\mathbf{x}_D, T_1, P) \qquad\qquad (IV.8)$$
$$H_c = V_c \rho_c \qquad\qquad (IV.9)$$
$$\rho_c = \rho_c(\mathbf{x}_D, T_1, P) \qquad\qquad (IV.10)$$

Internal Plates, $j=2$ to $(N-1)$; $i = 1$ to n_c

Total mass balance:

$$\frac{dH_j}{dt}=L_{j-1}+V_{j+1}-L_j-V_j \qquad \text{(IV.11)}$$

Component mass balance:

$$\frac{d(H_j x_{ji})}{dt}=L_{j-1}x_{j-1,i}+V_{j+1}y_{j+l,i}-L_j x_{ji}-V_j y_{ji} \qquad \text{(IV.12)}$$

Energy balance:

$$\frac{d(H_j h_j^L)}{dt}=L_{j-1}h_{j-1}^L+V_{j+1}h_{j+1}^V-L_j h_j^L-V_j h_j^V \qquad \text{(IV.13)}$$

Equilibrium:

$$y_{ji}=k_{ji}x_{ji} \qquad \text{(IV.14)}$$

Restrictions:

$$\sum y_{ji}=1 \qquad \text{(IV.15)}$$

Physical properties:

$$k_{ji}=k_{ji}(\mathbf{y}_j,\mathbf{x}_j,T_j,P) \qquad \text{(IV.16)}$$

$$h_j^L=h_j^L(\mathbf{x}_j,T_j,P) \qquad \text{(IV.17)}$$

$$h_j^V=h_j^V(\mathbf{y}_j,T_j,P) \qquad \text{(IV.18)}$$

$$H_j=V_j\rho_j \qquad \text{(IV.19)}$$

$$\rho_j=\rho_j(\mathbf{x}_j,T_j,P) \qquad \text{(IV.20)}$$

<u>Reboiler, *j=N*; *i* = 1 to *n*$_c$</u>

Total mass balance:

$$\frac{dH_N}{dt}=L_{N-1}-V_N \qquad \text{(IV.21)}$$

Component mass balance:

$$\frac{d(H_N x_{Ni})}{dt} = L_{N-1} x_{N-1,i} - V_N y_{Ni} \tag{IV.22}$$

Energy balance:

$$\frac{d(H_N h_N^L)}{dt} = L_{N-1} h_{N-1}^L - V_N h_N^V + Q_R \tag{IV.23}$$

The other equations for the reboiler are same as Equations IV.14-IV.20 where j is replaced by N.

Meadow (1963), Distefano (1968), Boston et al. (1980), Bosley and Edgar (1994a,b), Mori et al. (1995), Mujtaba and Macchietto (1998) used similar to this type of model in their simulation and optimisation studies.

4.2.4.1.1. Example - CVH Model

Boston et al. (1980) considered a complex separation problem, involving a quaternary mixture, 5 operational steps (3 main-cuts), with the addition of a secondary charge after CUT1. The first two operational steps form CUT1, where the most volatile component, propane is removed from the system. Then the main accumulator is dumped and a secondary charge containing 40% butane and 60% hexane is added to the reboiler instantaneously, before beginning the third step. The third step (CUT 2) is a production step which produces butane. The accumulator is dumped again before the beginning of the next two steps (CUT 3) when propane is removed from the system, leaving hexane as the final bottom product. The problem is defined in detail in Table 4.3 and the results obtained by Boston et al. are presented in Table 4.4.

4.2.4.2. Constant Molar Holdup (CMH) Model

A subset of the type IV model equations can be obtained using the assumption of constant molar holdup in the condenser and in intermediate plates and fast energy dynamics. These assumptions will make the resulting set of DAEs an index 1 rather than index 2 system. In the literature this type of model is referred to as the rigorous constant molar holdup model. Galindez and Fredenslund (1988), Mujtaba and Macchietto (1988, 1992, 1993, 1996, 1998) used this type of model in their studies. Refer to Mujtaba and Macchietto (1998) for the model equations.

Table 4.3. Problem Specifications in Boston et al. (1980)

<u>**Column Specification**</u>

No. of internal stages (N)	= 8
Internal stage holdup, ft^3	= 0.01
Condenser holdup, ft^3	= 0.10
Pressure, psia	= 14.7

Initial charge, 1bmol	= 100.0
Components:	Propane, n- Butane, n- Pentane, n- Hexane
Composition	= <0.1, 0.3, 0.1, 0.5> molefraction

Intermediate charge, lbmol	= 20
Composition:	= <0.0, 0.4, 0.6, 0.0> molefraction

<u>**Physical Property Models**</u>

K-values	Constant relative to the K-value of pentane. Pentane K-value by Raoult's law with extended Antione for vapour pressure.
Enthalpies	Ideal gas for vapour, Watson equation for liquid.
Densities	Rackett equation.

4.2.4.2.1. Example 1 - CMH Model

Domenech and Enjalbert (1974) carried out a series of experimental tests in a laboratory batch distillation column. A binary mixture of Cyclohexane and Toluene was considered for the purpose. The experimental equipment used was a perforated plate column, with 4 trays and a 60 litre reboiler heated with a heat transfer coefficient of 3 kw. The experimental results obtained by Domenech and Enjalbert together with column input data are presented in Table 4.5.

Mujtaba (1989) used CMH model to simulate the operations considered by Domenech and Enjalbert (1974). Since the overall stage efficiency in the experimental column was 75%, the number of theoretical plates used by Mujtaba was 3. The column was initialised at its total reflux steady state values. Soave-Redlich-Kwong (SRK) model was used for the VLE property calculations. Vapour phase enthalpies were calculated using ideal gas heat capacity values and the liquid phase enthalpies were calculated by subtracting heat of vaporisation from the vapour enthalpies. The input data required to evaluate these thermodynamic properties were taken from Reid et al. (1977).

The comparison between the experimental data of Domenech and Enjalbert (1974) and those obtained by Mujtaba (1989) is shown in Figure 4.6 which shows a good match between the experimental data and the model predictions.

Table 4.4. Summary of the Results of Boston et al (1980)

Operations step	1	2	3	4	5
Purpose	C3	Removal	C4 Production	C5	Removal
Reflux ratio	5	20	25	15	25
Vapour flow rate	12	42	52	32	52
Distillate rate	2	2	2	2	2
Time (h)	4.07	1.81	18.27	4.31	1.78
Present distillate					
Propane	0.80000	0.01500	-	-	-
Butane	0.20000	0.98500	0.16440	-	-
Pentane	-	-	0.83560	0.80000	0.01610
Hexane	-	-	-	0.20000	0.98390
Accumulated Distillate					
Propane	0.98880	0.84970	0.00020	-	-
Butane	0.01120	0.15030	0.99000	0.00630	0.00450
Pentane	-	-	0.00980	0.98750	0.79140
Hexane	-	-	-	0.00610	0.20410
(1bmol)	8.1321	11.7600	36.5430	8.6129	12.1670
Still pot					
Propane	0.02060	-	-	-	-
Butane	0.32540	0.31910	0.00060	-	-
Pentane	0.10900	0.11350	0.13350	0.01730	0.00020
Hexane	0.54500	0.56740	0.86580	0.98270	0.99980
(1bmol)	91.737	88.124	71.600	62.998	59.398

Table 4.5. Experimental Data from Domenech and Enjalbert (1974)

Case 1		Case 2		Case 3		Case 4	
R = 3		R = 4		R = 6		R = 9	
$x_{B0} = 0.62$		$x_{B0} = 0.55$		$x_{B0} = 0.42$		$x_{B0} = 0.30$	
H_a	x_D	H_a	x_D	H_a	x_D	H_a	x_D
0.0	0.96	0.0	0.945	0.0	0.92	0.0	0.90
21.0	0.94	10.0	0.93	10.0	0.90	5.0	0.86
29.0	0.93	20.0	0.91	25.0	0.875	12.5	0.85
50.0	0.92	29.0	0.905	39.0	0.865	23.0	0.84
60.0	0.91	40.0	0.90	60.0	0.83	36.0	0.80
75.0	0.895	50.0	0.895	80.0	0.73	43.0	0.77
90.0	0.875	60.0	0.885	89.0	0.63	51.0	0.69
111.0	0.825	81.0	0.85	9.0	0.475	63.0	0.56
129.0	0.70	93.0	0.83			75.0	0.40
140.0	0.55	104.0	0.77			0.83	0.30
		118.0	0.65				
		123.0	0.53				
		131.0	0.37				

Binary mixture of cyclohexane and toluene		
Rectification column with sieve trays	4	trays
Operating pressure	1.0	atm
Total condenser		
Initial charge in the still pot	200.0	mol
Liquid holdup in each tray and condenser	2.5	mol
Reboiler heat duty	3	KW
Stage efficiency	75.0	%

Note: Reflux ratio R is defined as external Reflux ratio while the model uses internal reflux ratio

4.2.4.2.2. Example 2 - CMH Model

Mujtaba (1989) simulated the example considered by Boston et al. (1980) presented in section 4.2.4.1.1 using CMH model. The volume holdups used by Boston et al. were converted to molar holdups at the initial conditions. These were 0.00493 lbmol for each internal plates and 0.0493 lbmol for the condenser. Equilibrium k values were calculated using Antoine's vapour pressure correlation and enthalpies by the same procedure outlined in section 4.2.4.2.1. The simulation results are presented in Table 4.6. Note the slight differences in predictions (Table 4.4 and 4.6) are due to the use of different types of models (CVH and CMH) and thermodynamic property calculations.

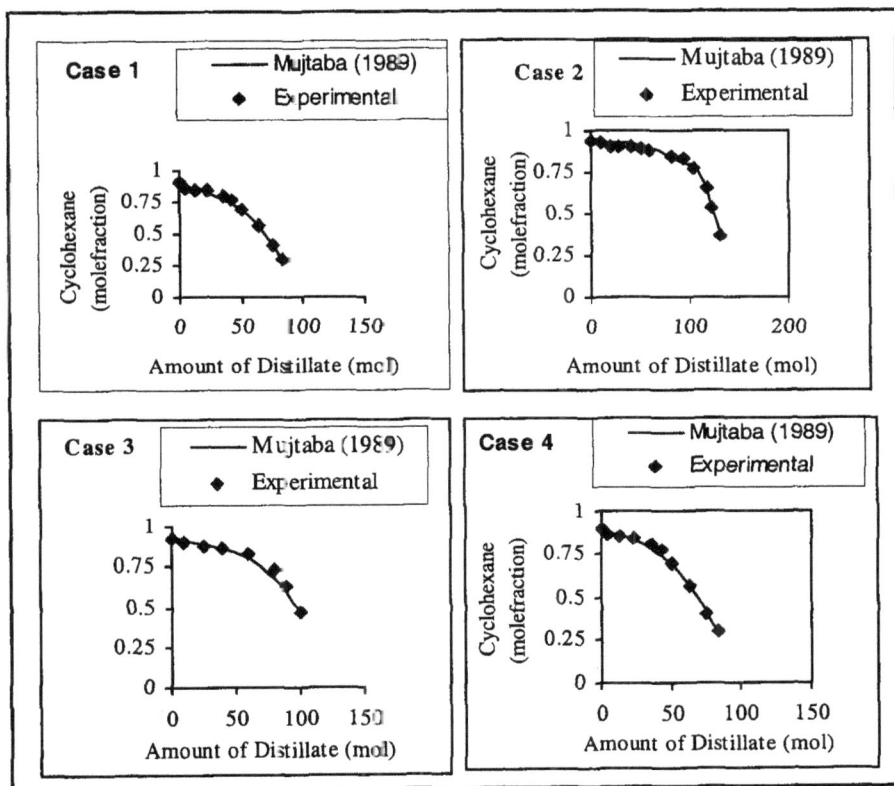

Figure 4.6. Simulation Results using Rigorous CMH Model and Comparison with Experiments.

Table 4.6. Simulation Results using CMH Model by Mujtaba (1989)

Operation Step	1	2	3	4	5
Purpose	C3	Removal	C4 Production	C5	Removal
Reflux Ratio	5	20	25	15	25
Vapour Flow rate	12	42	52	32	52
Distillate Rate	2	2	2	2	2
Time (Hr)	4.07	1.81	18.27	4.31	1.78
Present Dist.					
Propane	0.754	0.031	-	-	-
Butane	0.246	0.969	0.254	-	-
Pentane	-	-	0.745	0.613	0.091
Hexane	-	-	-	0.387	0.909
Accumulated					
Distillate					
	0.981	0.850	-	-	-
Propane	0.019	0.150	0.988	0.017	0.012
Butane	-	-	0.012	0.940	0.778
Pentane	-	-	-	0.043	0.210
Hexane					
(lb Mol)	8.139	11.760	36.548	8.619	12.180
Still Pot					
Propane	0.021	-	-	-	-
Butane	0.325	0.319	0.001	-	-
Pentane	0.109	0.113	0.133	0.023	0.002
Hexane	0.545	0.567	0.866	0.977	0.998
(lb Mol)	91.86	88.240	71.680	63.061	59.380

4.2.4.2.3. Example 3 - CMH Model

Nad and Spiegel (1987) carried out experimentation in a conventional packed batch distillation column using a cyclohexane-heptane-toluene mixture. The column consists of 20 theoretical stages (equivalent) including the condenser and reboiler. The feed to the column was 2.93 kmol of which 1.9% was total column holdup and 1.2% was condenser holdup. The column underwent an initial total reflux operation for about 2.54 hr before any product was collected. After then the mixture was separated into 3 main-cuts with 2 off-cuts in between, leaving a final product in the reboiler.

Nad and Spiegel also used the computer programme DISBATCH of Galindez and Fredenslund (1988) to reproduce the experimental data. The authors obtained reasonably good agreement in their calculations. However, to do so, they had to use a reflux profile considerably different from that reported as the experimentally measured one. They investigated the influence of a number of input variables on the simulated behaviour of the column and tried with different reflux ratio profiles until the best agreement with the experiment was obtained.

Mujtaba (1989) simulated the same example for the first product cut using a reflux ratio profile very close to that used by Nad and Spiegel in their own simulation and a nonideal phase equilibrium model (SRK). The purpose of this was to show that a better model (model type IV) and better integration method achieves even a better fit to their experimental data. Also the problem was simulated using an ideal phase equilibrium model (Antoine's equation) and the computational details were presented. The input data to the problem are given in Table 4.7.

Vapour phase enthalpies were calculated using ideal gas heat capacity values and the liquid phase enthalpies were calculated by subtracting heat of vaporisation from the vapour enthalpies. The input data required to evaluate these thermodynamic properties were taken from Reid et al. (1977). Initialisation of the plate and condenser compositions (differential variables) was done using the fresh feed composition according to the policy described in section 4.1.1.(a). The simulation results are presented in Table 4.8. It shows that the product composition obtained by both ideal and nonideal phase equilibrium models are very close those obtained experimentally. However, the computation times for the two cases are considerably different. As can be seen from Table 4.8 about 67% time saving (compared to nonideal case) is possible when ideal equilibrium is used.

Figure 4.7 shows the simulated instant distillate composition profiles by Mujtaba and by that of Nad and Spiegel using nonideal phase equilibrium models. The figure also includes experimentally obtained instant distillate composition data and the adjusted reflux ratio profiles used by Nad and Spiegel and Mujtaba.

Table 4.7. Problem Definition (Example 3)

No. of Internal Plates	$= 18$
Total Condenser	$= 1$
Partial Reboiler	$= 1$
Total N	$= 20$
Internal Plate Holdup	$= 3.093$ E-3 kmol
Condenser Holdup	$= 35.160$ E-3 kmol
Vapor Load to the condenser	$= 2.75$ kmol/hr
Column Pressure	$= 1.013$ bar
Reboiler Feed, B_0	$= 2.93$ kmol
Components:	Cyclohexane, n-Heptane, Toluene
Composition, x_{B0}	$= <0.407, 0.394, 0.199>$, molefraction

Table 4.8. Experimental and Simulated Results of Example 3

Experimental Results of Nad and Spiegel (1987)	
Accumulated distillate	= 1.16 kmol
Cyclohexane composition	= 0.895 molefraction
Operation time (including total reflux period)	= 5.16 hr
Simulation Results by Mujtaba (1989)	
1. Nonideal case	
Accumulated distillate	= 1.16 kmol
Cyclohexane composition	= 0.891 molefraction
Operation time (including total reflux period)	= 5.16 hr
Total simulation time (approx. using IBM 370)	= 4.9 min
Total time in physical properties calculations	= 3.3 min
% of total time spent in physical properties	= 67.0 %
2. Ideal Case	
Accumulated distillate	= 1.16 kmol
Cyclohexane composition	= 0.906 molefraction
Operation time (including total reflux period)	= 5.16 hr
Total simulation time (approx).	= 1.16 min
Total time in physical properties calculations	= 0.4 min
% of total time spent in physical properties	= 24.0 %
Time saved by using ideal physical properties	= 67.0 %

Key: Time saving is based on the time required by non-ideal case.

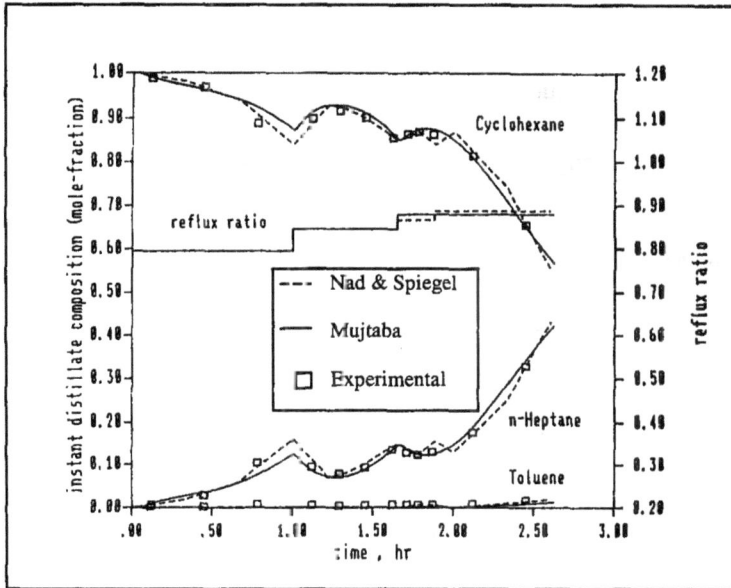

Figure 4.7. Simulation Results by Mujtaba (1989) and Comparison with Experiment and Simulation of Nad and Spiegel (1987)

4.2.5. Rigorous Model with Chemical Reactions – Model Type V

Referring to Figure 4.5 and the assumptions listed in section 4.2.4.1 the rigorous model for batch distillation with chemical reactions is presented below. Further assumptions are: no chemical reactions in the vapour phase and in the condenser accumulator.

<u>Condenser and Accumulator, $j=1$: $i = 1$ to n_c</u>

Accumulator total mass balance:

$$\frac{dH_a}{dt} = L_D \qquad\qquad (V.1)$$

Component mass balance:
  a) Accumulator

$$\frac{d(H_a x_{ai})}{dt} = L_D x_{Di} \tag{V.2}$$

b) Condenser Holdup Tank

$$\frac{d(H_c x_{Di})}{dt} = V_2 y_{2i} + r_{1i} H_c - L_c x_{Di} \tag{V.3}$$

Energy balance:

$$\frac{d(H_c h_1^L)}{dt} = V_2 h_2^V - L_c h_1^L - Q_C \tag{V.4}$$

Physical properties and other equations:

$$L_1 = r L_c \tag{V.5}$$
$$L_D = L_c (1 - r) \tag{V.6}$$
$$T_1 = T_1 (x_D, P) \tag{V.7}$$
$$h_1^L = h_1^L (x_D, T_1, P) \tag{V.8}$$
$$H_c = V_c \rho_c \tag{V.9}$$
$$\rho_c = \rho_c (x_D, T_1, P) \tag{V.10}$$
$$r_{1i} = r_{1i} (k_r, x_D) \tag{V.11}$$
$$\Delta n_1 = \sum r_{1i} \tag{V.12}$$

Internal Plates, j=2 to (N-1); i = 1 to n_c

Total mass balance:

$$\frac{dH_j}{dt} = L_{j-1} + V_{j+1} - L_j - V_j + \Delta n_j H_j \tag{V.13}$$

Component mass balance:

$$\frac{d(H_j x_{ji})}{dt} = L_{j-1} x_{j-1,i} + V_{j+1} y_{j+l,i} - L_j x_{ji} - V_j y_{ji} + r_{ji} H_j \tag{V.14}$$

Energy balance:

$$\frac{d(H_j h_j^L)}{dt} = L_{j-1} h_{j-1}^L + V_{j+1} h_{j+1}^V - L_j h_j^L - V_j h_j^V \tag{V.15}$$

Equilibrium:

$$y_{,i} = k_{ji} x_{ji} \tag{V.16}$$

Restrictions:

$$\sum y_{ji} = 1 \tag{V.17}$$

Physical properties:

$$k_{ji} = k_{ji}(\mathbf{y}_j, \mathbf{x}_j, T_j, P) \tag{V.18}$$

$$h_j^L = h_j^L(\mathbf{x}_j, T_j, P) \tag{V.19}$$

$$h_j^V = h_j^V(\mathbf{y}_j, T_j, P) \tag{V.20}$$

$$H_j = V_j \rho_j \tag{V.21}$$

$$\rho_j = \rho_j(\mathbf{x}_j, T_j, P) \tag{V.22}$$

$$r_{ji} = r_{ji}(k_r, x_D) \tag{V.23}$$

$$\Delta n_j = \sum r_{ji} \tag{V.24}$$

Reboiler, $j=N$; $i = 1$ to n_c

Total mass balance:

$$\frac{dH_N}{dt} = L_{N-1} - V_N + \Delta n_N H_N \tag{V.25}$$

Component mass balance:

$$\frac{d(H_N x_{Ni})}{dt} = L_{N-1} x_{N-1,i} - V_N y_{Ni} + r_{Ni} H_N \tag{V.26}$$

Energy balance:

$$\frac{d(H_N h_N^L)}{dt} = L_{N-1} h_{N-1}^L - V_N h_N^V + Q_R \qquad (V.27)$$

The other equations for the reboiler are same as equations (V.16-V.24) where j is replaced by N.

Cuille and Reklaitis (1986) and Albet et al. (1991) used similar model to simulate batch reactive distillation process. Egly et al. (1979), Reuter et al. (1989), Mujtaba (1989) and Mujtaba and Macchietto (1992, 1997) used a modified version of this model based on constant molar holdup in their studies. Sorensen and Skogestad (1996c), Sorensen et al. (1996b), Balasubramhanya and Doyle III (2000) used simple models for studying control strategies in batch reactive distillation.

4.2.5.1. Example

Mujtaba and Macchietto (1997) and Greaves (2003) simulated the esterification of ethanol and acetic acid using a Type V-CMH model. The reaction products are ethyl acetate (main product) and water. The reversible reaction scheme together with the boiling point temperatures are shown below:

Acetic Acid + Ethanol <=> Ethyl Acetate + Water			
(1)	(2)	(3)	(4)

	(1)	(2)	(3)	(4)
Boiling Points, K	391.1	351.5	350.3	373.2

Ethyl acetate, the main product, has the lowest boiling temperature in the mixture and therefore has the highest volatility. The number of plates (defining the column configuration), feed, feed composition, column holdup, etc. for the problem are given in Table 4.9. Four percent of the total feed charge is considered as the total column holdup in this example. Fifty percent of this holdup is taken as the condenser holdup and the rest is equally divided for the plate holdup. Plate compositions, product accumulator compositions and reflux drum compositions are the differential variables of the model equations. These variables are initialised to the feed compositions at time $t = 0$ which ensure consistent initialisation of the DAE system used in this work (see Pantelides, 1988 for detail). The vapour liquid equilibrium data and the kinetic data are taken from Simandl and Svrcek (1991) and Bogacki et al. (1989) respectively and are shown in Table 4.10. The vapour and liquid enthalpies are calculated using data from Reid et al. (1977). It is to be emphasised that these data do not account for detailed VLE calculations and for any azeotropic formed. Typical plots of accumulated distillate and reboiler composition profiles with reflux ratio, $r = 0.952$ are shown in Figures 4.8 and 4.9.

Table 4.9. Input Data for Ethanol Esterification.
[Adopted from Mujtaba and Macchietto, 1997]

No. of Ideal Separation Stages
(including a reboiler and a total condenser)   = 10
Total Fresh Feed, B_0, kmol   = 5.0
Feed Composition x_{B0}, molefraction =
<Acetic Acid, Ethanol, Ethyl Acetate, Water> = <0.45, 0.45, 0.0, 0.1>

Column Holdup, kmol:
    Condenser   = 0.1
    Internal Plates   = 0.0125
Condenser Vapour Load, kmol/hr   = 2.5
Column Pressure, bar   = 1.013

Table 4.10. VLE and Kinetic Data for Ethanol Esterification
[Adopted from Mujtaba and Macchietto, 1997]

<u>Vapour liquid equilibrium:</u>

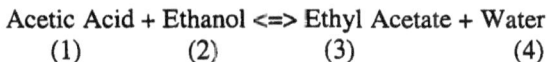

  Acetic Acid + Ethanol <=> Ethyl Acetate + Water
    (1)   (2)   (3)   (4)

  $K_1 = 2.25 \times 10^{-2} \, T - 7.812,$ $T > 347.6$ Kelvin
  $K_1 = 0.001$ $T \le 347.6$
  $\log K_2 = -2.3 \times 10^3 \, /T + 6.588$
  $\log K_3 = -2.3 \times 10^3 \, /T + 6.742$
  $\log K_4 = -2.3 \times 10^3 \, /T + 6.484$

<u>Kinetic Data:</u>

Rate of reaction, gmols/(litre.min); $r = k_1 \, C_1 \, C_2 - k_2 \, C_3 \, C_4$
where, rate constants are $k_1 = 4.76 \times 10^{-4}$ and $k_2 = 1.63 \times 10^{-4}$
and C_i stands for concentration in gmols/litre for the ith component

---

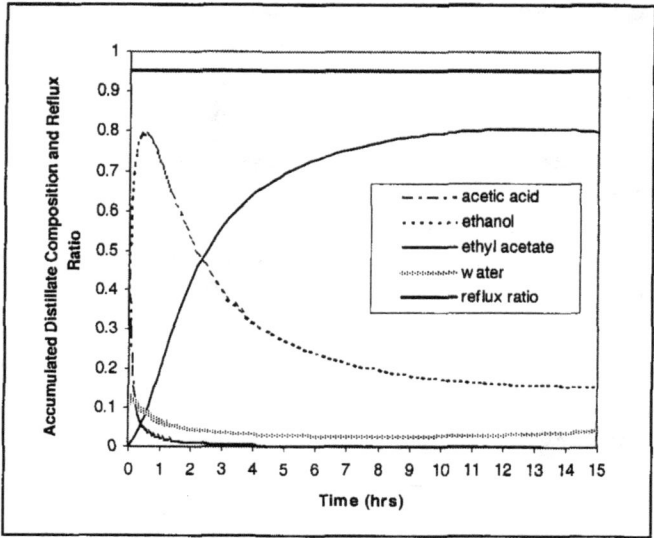

Figure 4.8. Distillate Composition Profile for Ethanol Esterification

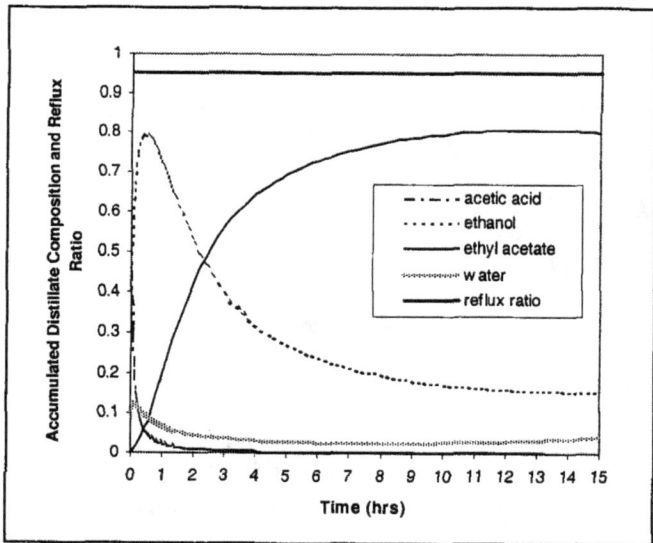

Figure 4.9. Reboiler Composition Profile for Ethanol Esterification

4.3. Models for Unconventional Batch Distillation

4.3.1. Continuous Column for Batch Distillation

Referring to Figure 4.10 of a continuous distillation column the model is developed based on the assumptions of constant relative volatility and equimolal overflow and include detailed plate-to-plate calculations. Further assumptions are listed below:

i)   constant molar holdup for condenser and internal plates.
ii)   total condensation without sub-cooling.
iii)   negligible vapour holdup.
iv)   perfect mixing of liquid and vapour on the plates.
v)   negligible heat losses.
vi)   theoretical plates
vii)   feed mixture at its bubble point.

First the equations for the condenser will be presented. Then the equations for the accumulator, followed by the equations for the plates in the column and the reboiler are presented. The trays are counted from the top to the bottom.

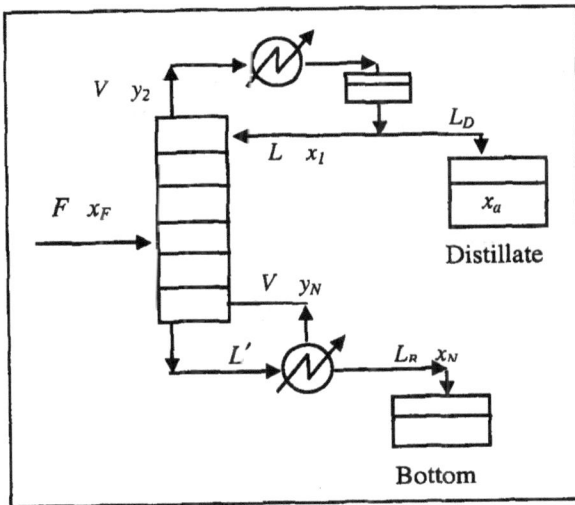

Figure 4.10. Continuous Distillation with Notations

<u>Condenser and Accumulator, $j=1$; $i = 1$ to n_c</u>

The composition of the distillate and of the liquid reflux is same as that of the vapour going into the condenser. Therefore:

$$y_{2i} - x_{1i} = 0 \qquad\qquad (SS.1)$$

$$x_{ai} - x_{1i} = 0 \qquad\qquad (SS.2)$$

The reflux ratio is defined as:

$$L = rV \qquad\qquad (SS.3)$$

The distillate rate to the accumulator or product tank is therefore:

$$L_D = V(1-r) \qquad\qquad (SS.4)$$

<u>Internal Plates, $j=2$ to $(N-1)$; $i = 1$ to n_c</u>

There will be no effect of plate holdup on the steady state mass balance. The component balance on tray j is:

above feed plate: $L(x_{j-1,i} - x_{ji}) + V(y_{j+1,i} - y_{ji}) = 0$

feed plate: $Lx_{j-1,i} - L'x_{ji} + V(y_{j+1,i} - y_{ji}) + Fx_F = 0$

below feed plate: $L'(x_{j-1,i} - x_{ji}) + V(y_{j+1,i} - y_{ji}) = 0 \qquad (SS.5)$

Feed plate total mass balance:

$$F + L - L' = 0 \qquad\qquad (SS.6)$$

Equilibrium:

$$y_{ji} = \frac{\alpha_i x_{ji}}{\sum_{k=1}^{n_c} \alpha_k x_{jk}} \qquad\qquad (SS.7)$$

<u>Reboiler, $j=N$; $i = 1$ to n_c</u>

Total mass balance:

$$L' - L_B - V = 0 \qquad \text{(SS.8)}$$

Component balance:

$$L' x_{N-1,i} - L_B x_{Ni} - V y_{Ni} = 0 \qquad \text{(SS.9)}$$

The vapour liquid equilibrium relationship is same as Equation SS.7 with $j = N$.

Attarwala and Abrams (1974) and Mujtaba (1997) used the above model for batch distillation task using a continuous column (see section 2.2.4).

4.3.1.1. Example

A liquid binary mixture with $B_0 = 10$ kmol and $x_{B0} = <0.6, 0.4>$ molefraction is subject to batch distillation using a continuous column shown in Figure 4.11. The relative volatility of the mixture over the operating temperature range is assumed constant with a value of ($\alpha =$) 2. The total number of plates is, $N = 20$ and the feed is introduced at the bottom plate of the column. The vapour boilup rate is, $V = 5.0$ kmol/hr. The column is operated with a continuous feed rate, $F = 10$ kmol/hr, therefore the column runs for 1 hr to process the total amount of feed (B_0). The reflux ratio, $r = 0.509$ ensures distillate at the desired purity of $x_D^1 = 0.90$ molefraction. Figure 4.11 shows the accumulated distillate amount, the bottom product amount and composition at the end of 1 hr of operation, obtained using the continuous column model outlined in section 4.3.1.

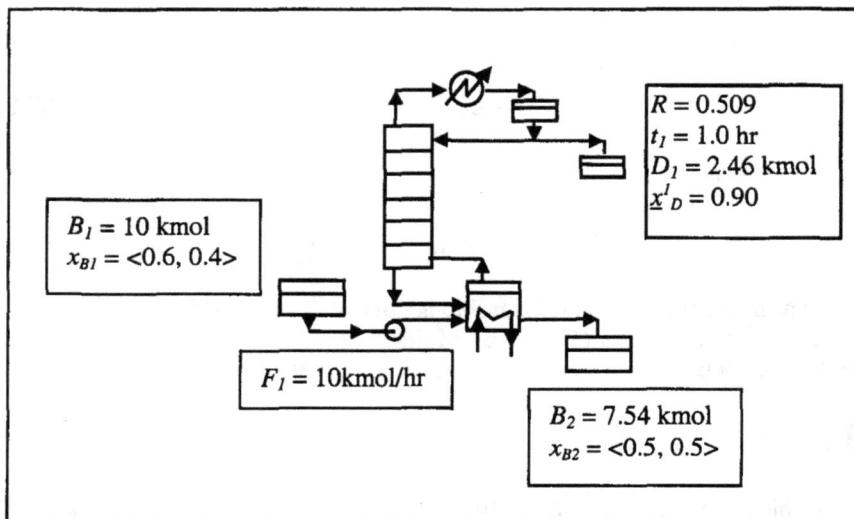

Figure 4.11. Batch Distillation Task in a Continuous Column

4.3.2. *Inverted Batch Distillation (IBD) Column*

Referring to Figure 4.12 for inverted batch distillation column, the intermediate plate equations in model types III, IV and V presented in section 4.2 will remain the same. The model equations for the condenser and for the reboiler for types III, IV and V models are presented below.

4.3.2.1. Type III - Simple model

Note, for Type III model $V_2 = V_N = V$ and $L_1 = L_{N-1} = L$.

Condenser: $j=1; i = 1$ to n_c

Component balance:

$$H_c \frac{dx_{1i}}{dt} = Vy_{2i} - Lx_{1i} \qquad \text{(III.a1)}$$

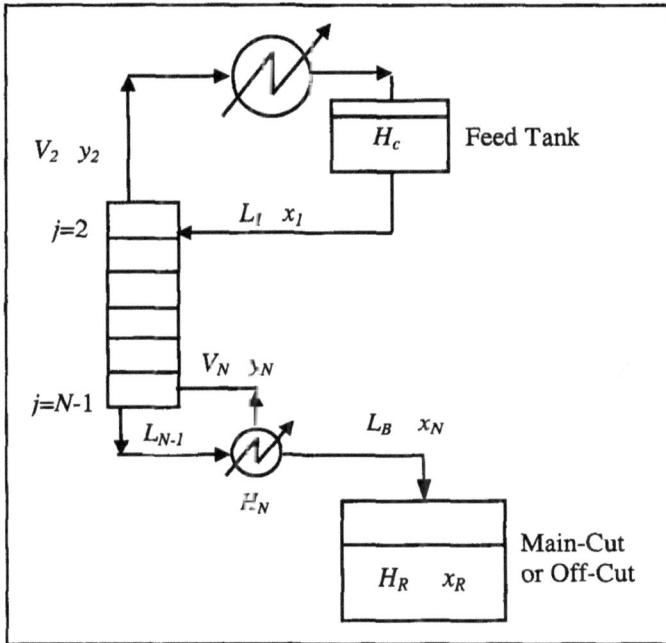

Figure 4.12. IBD Column with Notations

<u>Reboiler Holdup and Product Tanks; *j=N*; *i* = 1 to n_c</u>

Product Tank total mass balance:

$$\frac{dH_R}{dt} = L_B \qquad\qquad \text{(III.a2)}$$

Component mass balance:

a)   Product Tank

$$H_R \frac{dx_{Ri}}{dt} = L_B (x_{Ni} - x_{Ri}) \qquad\qquad \text{(III.a3)}$$

b)   Reboiler Holdup Tank

$$H_N \frac{dx_{Ni}}{dt} = L x_{N-1,i} - L_B x_{Ni} - V y_{Ni} \tag{III.a4}$$

Reboil ratio definition:

$$r_b = V / L \tag{III.a5}$$

4.3.2.2. Type IV - Rigorous Model - CVH

Condenser: $j=1$; $i = 1$ to n_c

Total mass balance:

$$\frac{dH_c}{dt} = V_2 - L_1 \tag{IV.a1}$$

Component mass balance:

$$\frac{d(H_c x_{1i})}{dt} = V_2 y_{2i} - L_1 x_{1i} \tag{IV.a2}$$

Energy balance:

$$\frac{d(H_c h_1^L)}{dt} = V_2 h_2^V - L_1 h_1^L - Q_C \tag{IV.a3}$$

Reboiler Holdup and Product Tanks: $j=N$; $i = 1$ to n_c

Total mass balance:

a)   Product Tank

$$\frac{dH_R}{dt} = L_B \tag{IV.a4}$$

b)   Reboiler Holdup Tank

$$\frac{dH_N}{dt} = L_{N-1} - L_B - V_N \qquad\qquad \text{(IV.a5)}$$

Component mass balance:

a) Product Tank

$$\frac{d(H_R x_{Ri})}{dt} = L_B (x_{Ni} - x_{Ri}) \qquad\qquad \text{(IV.a6)}$$

b) Reboiler Holdup Tank

$$\frac{d(H_N x_{Ni})}{dt} = L_{N-1} x_{N-1,i} - L_B x_{Ni} - V_N y_{Ni} \qquad\qquad \text{(IV.a7)}$$

Energy balance:

$$\frac{d(H_N h_N^L)}{dt} = L_{N-1} h_{N-1}^L - L_B h_N^l - V_N h_N^V + Q_R \qquad\qquad \text{(IV.a8)}$$

Reboil ratio definition:

$$r_{\jmath} = V_N / L_{N-1} \qquad\qquad \text{(IV.a9)}$$

Constant Molar Holdup (CMH) model can be easily derived from the above set of equations.

4.3.2.3. Type V - Rigorous Model with Chemical Reaction

Condenser: $j=1; i = 1$ to n_c

Total mass balance:

$$\frac{dH_c}{dt} = V_2 - L_1 + \Delta n_1 H_c \qquad\qquad \text{(V.a1)}$$

Component mass balance:

$$\frac{d(H_c x_{ci})}{dt} = V_2 \; y_{2i} - L_1 \; x_{1i} + r_{ci} H_c \tag{V.a2}$$

Energy balance:

$$\frac{d(H_c h_1^L)}{dt} = V_2 h_2^V - L_1 h_1^L - Q_C \tag{V.a3}$$

<u>Reboiler Holdup and Product Tanks: $j=N$; $i = 1$ to n_c</u>

Total mass balance:

a)   Product Tank

$$\frac{dH_R}{dt} = L_B \tag{V.a4}$$

b)   Reboiler Holdup Tank

$$\frac{dH_N}{dt} = L_{N-1} - L_B - V_N + \Delta n_N H_N \tag{V.a5}$$

Component mass balance:

a)   Product Tank

$$\frac{d(H_R x_{Ri})}{dt} = L_B (x_{N,i} - x_{Ri}) \tag{V.a6}$$

b)   Reboiler Holdup Tank

$$\frac{d(H_N x_{Ni})}{dt} = L_{N-1} x_{N-1,i} - L_B x_{Ni} - V_N y_{Ni} + r_{Ni} H_N \tag{V.a7}$$

Energy balance:

$$\frac{d(H_N h_N^L)}{dt} = L_{N-1} h_{N-1}^L - L_B h_N^l - V_N h_N^V + Q_R \qquad (V.a8)$$

Reboil ratio definition:

$$r_b = V_N / L_{N-1} \qquad (V.a9)$$

Constant Molar Holdup (CMH) model can be easily derived from the above set of equations.

Note in all three types of model thermodynamic, kinetic and other physical properties can be calculated using the appropriate equations defined in section 4.2.

4.3.2.4. Example

A liquid binary mixture with B_0 = 10 kmol (H_c) and x_{B0} = <0.6, 0.4> (x_i) molefraction is subject to inverted batch distillation shown in Figure 4.12. The relative volatility of the mixture over the operating temperature range is assumed constant with a value of (α =) 2. The number of plates is, N = 10. The vapour boilup rate is, V = 10.0 kmol/hr. The total plate holdup is 0.3 kmol and the reboiler holdup is 0.1 kmol. The total batch time of operation is 4 hr with two time intervals. The first interval is of duration 1 hr and the column is operated with a reboil ratio of 0.8. The second interval is of duration 3 hrs when the column is operated with a reboil ratio of 0.9. The column operation is simulated with the type III model (section 4.3.2.1).

The results in terms of the amount and composition in the condenser holdup tank and in the bottom product accumulator are presented in Tables 4.11 and 4.12 respectively. The condenser holdup tank and bottom product accumulator composition and reboil ratio profiles are shown in Figures 4.13 and 4.14 respectively.

Table 4.11. Condenser Holdup Tank Profile (IBD Column)

Time, hr	H_c kmol	$x_{1,1}$ molefraction	$x_{1,2}$ molefraction
0 (initial charge)	10.0	0.500	0.500
0.5	8.74	0.573	0.426
1.0	7.50	0.650	0.349
1.5	6.93	0.709	0.291
2.0	6.37	0.764	0.236
2.5	5.81	0.822	0.177
3.0	5.24	0.880	0.119
3.5	4.70	0.924	0.076
4.0	4.17	0.954	0.046

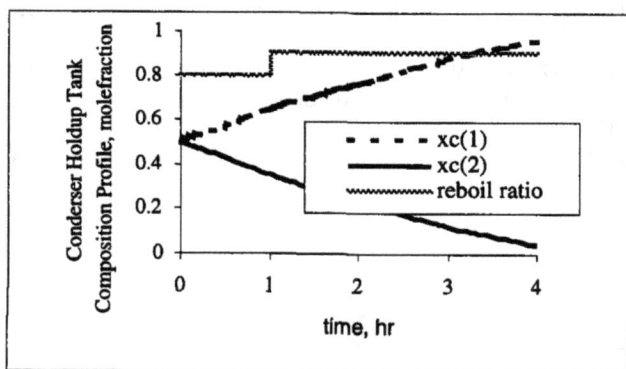

Figure 4.13. Condenser Holdup Tank Composition Profile (IBD Column)

Table 4.12. Bottom Product Accumulator Profile (IBD Column)

Time, hr	H_R kmol	$x_{R,1}$ molefraction	$x_{R,2}$ molefraction
0	0.0	0.0	0.0
0.5	1.26	0.078	0.922
1.0	2.50	0.072	0.928
1.5	3.07	0.064	0.936
2.0	3.63	0.059	0.941
2.5	4.19	0.059	0.940
3.0	4.76	0.072	0.928
3.5	5.30	0.105	0.895
4.0	5.83	0.150	0.850

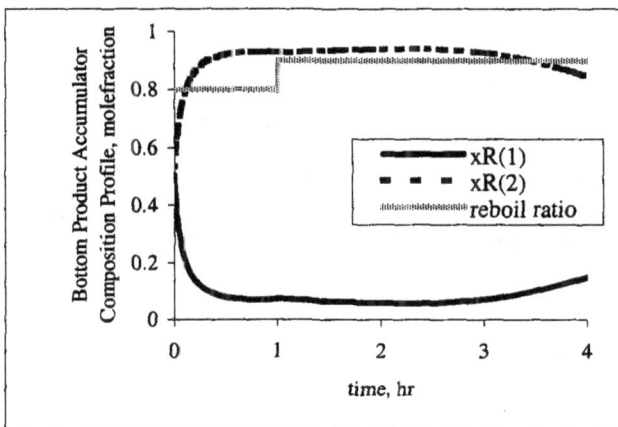

Figure 4.14. Bottom Product Accumulator Composition Profile (IBD Column)

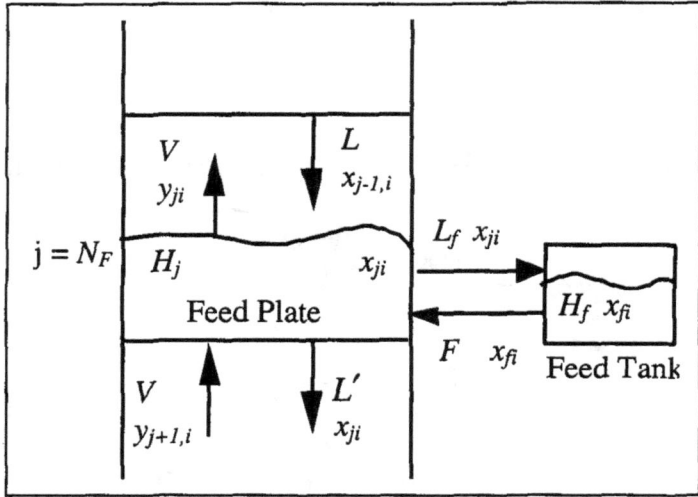

Figure 4.15. Schematic Diagram of the Feed Plate and Feed Tank for
MVC Column

4.3.3. Middle Vessel Batch Distillation Column (MVC)

Referring to Figure 2.2 for MVC column configuration, the model equations for the rectifying section are the same (except the reboiler equations) as those presented for conventional batch distillation column (Type III, IV, V in section 4.2). The model equations for the stripping section are the same (except the condenser equations) as those presented for inverted batch distillation column (Type III, IV, V in section 4.3.2). However, note that the vapour and liquid flow rates in the rectifying and stripping sections will not be same because of the introduction of the feed plate.

Here, the model equations for the feed plate (Figure 4.15) will be presented for model type III, IV and V.

4.3.3.1. Simple Model - Type III

<u>Feed Tank and Feed Plate: $j = N_F$; $i = 1$ to n_c</u>

    a)   Feed Tank

Total mass balance:

$$\frac{dH_f}{dt} = L_f - F \qquad \text{(III.m1)}$$

Component balance:

$$H_f \frac{dx_{\bar{f}i}}{dt} = L_f (x_{ji} - x_{fi}) \qquad \text{(III.m2)}$$

b) Feed Plate

Total mass balance:

$$L - L' + F - L_f = 0 \qquad \text{(III.m3)}$$

Component mass balance:

$$H_j \frac{dx_{ji}}{dt} = Lx_{j-1,i} - L'x_{ji} + V(y_{j+1,i} - y_{ji}) + Fx_{fi} - L_f x_{ji} \qquad \text{(III.m4)}$$

4.3.3.2. Rigorous Model (CVH)- Type IV

Note that liquid and vapour flow rates vary throughout the column (unlike that is shown in Figure 4.15)

<u>Feed Tank and Feed Plate: j = N_F; i = 1 to n_c</u>

a) Feed Tank

Total mass balance:

$$\frac{dH_f}{dt} = L_f - F \qquad \text{(IV.m1)}$$

Component balance:

$$\frac{d(H_f x_{fi})}{dt} = L_f x_{ji} - F x_{fi} \qquad \text{(IV.m2)}$$

b) Feed Plate

Total mass balance:

$$\frac{dH_j}{dt} = L_{j-1} + V_{j+1} + F - L_j - V_j - L_f \qquad \text{(IV.m3)}$$

Component mass balance:

$$\frac{d(H_j x_{ji})}{dt} = L_{j-1} x_{j-1,i} + V_{j+1} y_{j+1,i} + F x_{fi} - V_j y_{ji} - L_j x_{ji} - L_f x_{ji} \text{ (IV.m4)}$$

Energy balance:

$$\frac{d(H_j h_j^L)}{dt} = L_{j-1} h_{j-1}^L + V_{j+1} h_{j+1}^V + F h_F^L - L_j h_j^L - V_j h_j^V - L_f h_j^L \qquad \text{(IV.m5)}$$

4.3.3.3. Rigorous Model with Chemical Reaction- Type V

Feed Tank and Feed Plate: $j = N_F$; $i = 1$ to n_c

a) Feed Tank

Total mass balance:

$$\frac{dH_f}{dt} = L_f - F + \Delta n_f H_f \qquad \text{(V.m1)}$$

Component balance:

$$\frac{d(H_f x_{fi})}{dt} = L_f x_{ji} - F x_{fi} + r_{fi} H_f \qquad \text{(V.m2)}$$

b) Feed Plate

Total mass balance:

$$\frac{dH_j}{dt} = L_{j-1} + V_{j+1} + F - L_j - V_j - L_f + \Delta n_j H_j \qquad \text{(V.m3)}$$

Component mass balance:

$$\frac{d(H_j x_{ji})}{dt} = L_{j-1} x_{j-1,i} + V_{j+1} y_{j+1,i} + F x_{fi} - V_j y_{ji} - L_j x_{ji} - L_f x_{ji} + r_{ji} H_j$$

$$\text{(V.m4)}$$

Energy balance:

$$\frac{d(H_j h_j^L)}{dt} = L_{j-1} h_{j-1}^L + V_{j+1} h_{j+1}^V + F h_F^L - L_j h_j^L - V_j h_j^V - L_f h_j^L \qquad \text{(V.m5)}$$

4.3.3.4. Example 1

Barolo et al. (1998) developed a mathematical model of a pilot-plant MVC column. The model was validated using experimental data on a highly non-ideal mixture (ethanol-water). The pilot plant and some of the operating constraints are described in Table 4.13. The column is equipped with a steam-heated thermosiphon reboiler, and a water-cooled total condenser (with subcooling of the condensate). Electro-pneumatic valves are installed in the process and steam lines. All flows are measured on a volumetric basis; the steam flow measurement is pressure- and temperature-compensated, so that a mass flow measurement is available indirectly. Temperature measurements from several trays along the column are also available. The plant is interfaced to a personal computer, which performs data acquisition and logging, control routine calculation, and direct valve manipulation.

The model consists of dynamic mass and algebraic energy balances (the model type is somewhat in between type III and IV). Internal plates, condenser, reboiler, middle vessel are assumed to be well mixed. A linear pressure profile is also assumed. The internal liquid flows are calculated using the Francis Weir formula. The start-up period i.e. the filling of the condenser hold up tank and of the internal plates (from top to bottom) is also modelled. The integration of the differential equations is performed using Runge-Kutta-Fehlberg algorithm.

Table 4.13. Column Configuration and Operating Constraints

Column Configuration

Total number of plates $= 30$
Plate type = Sieve.   Column diameter $= 0.3$ m, Column height $= 9.9$ m.
Reboiler type = Vertical steam-heated thermosiphon
Maximum reboiler capacity = 90 L
Condenser type = Horizontal water-cooled shell-and-tube
Reflux drum capacity = 40 L (open to atmosphere)
Maximum capacity of the middle vessel = 500 L

Operating Constraints

Subcooled reflux to maintain low ethanol vapour pressure
Subcooled liquid from tray to middle vessel

Three parameters were identified and adjusted to validate the model against the experiments. The parameters are: the heat losses, the nominal tray holdup and the Murphree tray efficiency (E_M). Figure 4.16 shows how E_M is adjusted to match the dynamic model prediction and experimental temperature profile measured on Plate #12. Figure 4.17 shows the comparison between the experimental and model prediction of ethanol composition in the reflux drum, middle vessel and in the bottom of the column. Figures 4.16-17 show a good match between the model prediction and experiments.

Using the model, Barolo et al. (1998) further studied the dynamic behaviour of the column and interactions of different design and operating parameters on the column operability and productivity were established. See the original reference for further details. The optimisation study carried out by Greaves et al (2003) using a Neural Network based dynamic model is presented in Chapter 12.

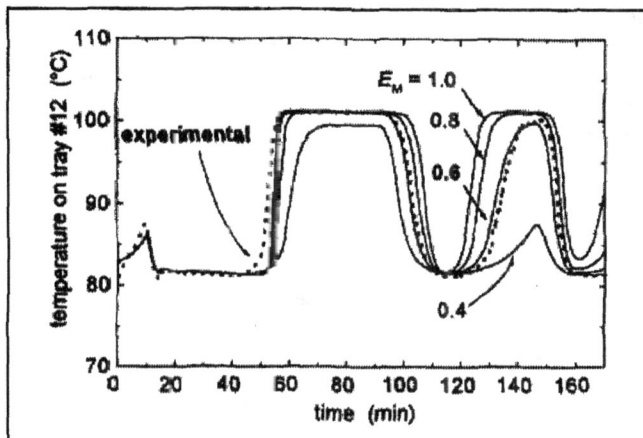

Figure 4.16. Adjustment of Murphree Tray Efficiency. Dotted line: Experiment. Full lines: Model Predictions. [Barolo et al., 1998][a]

4.3.3.5. Example 2

Using the model type III but with chemical reaction Mujtaba and Macchietto (1994) considered simulation of simultaneous chemical reaction and distillation in an MVC column, with the following reversible reaction scheme.

$$A + B \iff C + D$$
$$(1) \quad (2) \quad \quad (3) \quad (4)$$

The reaction products are C (main product) and D, the latter being the most volatile component and C being the least volatile component in the reaction mixture. Separation of the product by distillation permits increasing conversion, while at the same time yielding the product in concentrated form.

The input data defining the column configurations, feed, feed composition, column holdup, etc. are given in Table 4.14. The reaction is modelled by simple rate equations (Table 4.14). The feed tank location was $N_F = 7$ (stages numbered from the top down). The given batch time is 12 hrs. Conversion to product C was 70%.

---

[a] Reprinted from *Computers & Chemical Engineering.* **22**, Barolo, M., Guarise, G. B., Rienzi S. A. and Trotta, A., Understanding the dynamics of a batch distillation column with a middle vessel, S37-40, Copyright (1998) with permission from Elsevier Science

Also 1.52 kmol of product C with an accumulated composition of 0.70 molefraction was obtained in the distillate tank. Figure 4.18 shows the accumulated distillate, feed tank and bottom product composition profiles for the operation and Figure 4.19 shows the holdup profiles in the distillate accumulator, feed tank and bottom product accumulator.

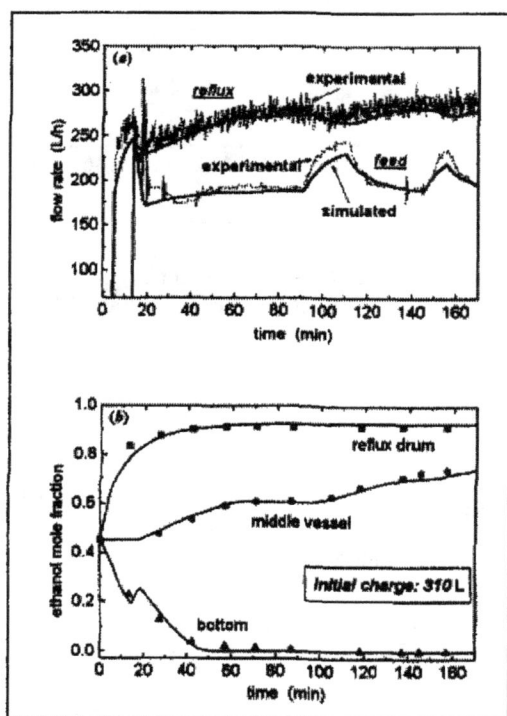

Figure 4.17. Experimental (symbol and dotted lines) and Calculated (full lines) Profiles of (a) Manipulated Flow rates and (b) Compositions in the Vessels[b].

---

[b] Reprinted from *Computers & Chemical Engineering*. **22**, Barolo, M., Guarise, G. B., Rienzi S. A. and Trotta, A., Understanding the dynamics of a batch distillation column with a middle vessel, S37-40, Copyright (1998) with permission from Elsevier Science.

Table 4.14. Input Data for Example 2

No. of Plates	
(including reboiler, total condenser)	= 13
Total fresh feed in the feed tank, B_0, kmol	= 5.0
Component	<A, B, C, D>
Feed composition, x_{B0}, mole fraction:	= <0.5, 0.5, 0.0, 0.0>
Column holdup, kmol:	
Condenser = 0.07 Reboiler	= 0.07
Internal plates, total	= 0.11
Vapour load, kmol/hr	= 2.5
Flow rate from feed tank to the feed plate, kmol/hr = 5.0	
Relative volatility, α	= <0.56, 0.50, 0.44, 1.0>
Rate of reaction, 1/hr: $r = k_i\, x_1\, x_2 - k_2\, x_3\, x_4$	
(with $k_1 = 2k_2$; x_i is the mole fraction of the ith component)	

4.3.4. Multivessel Batch Distillation Column (MultiBD)

Referring to Figure 2.3 of multivessel batch distillation (MultiBD) column, the model equations for condenser, reboiler and internal plates are the same as those presented for conventional batch distillation column (section 4.2). The model equations for the vessels are the same as those presented for feed tank of the MVC column (section 4.3.3). Note however, that there are no feed plate model equations as in the case of an MVC column.

4.3.4.1. Example

Skogestad et al. (1997) simulated a MultiBD column with a similar to type III model using a total reflux operation. The input data for the column is presented in Table 4.15.

The column was operated until composition in each vessel (including the condenser and the reboiler) reaches steady state values (more than 3 hr of operation). The final product compositions in each vessel are shown in Figure 4.20.

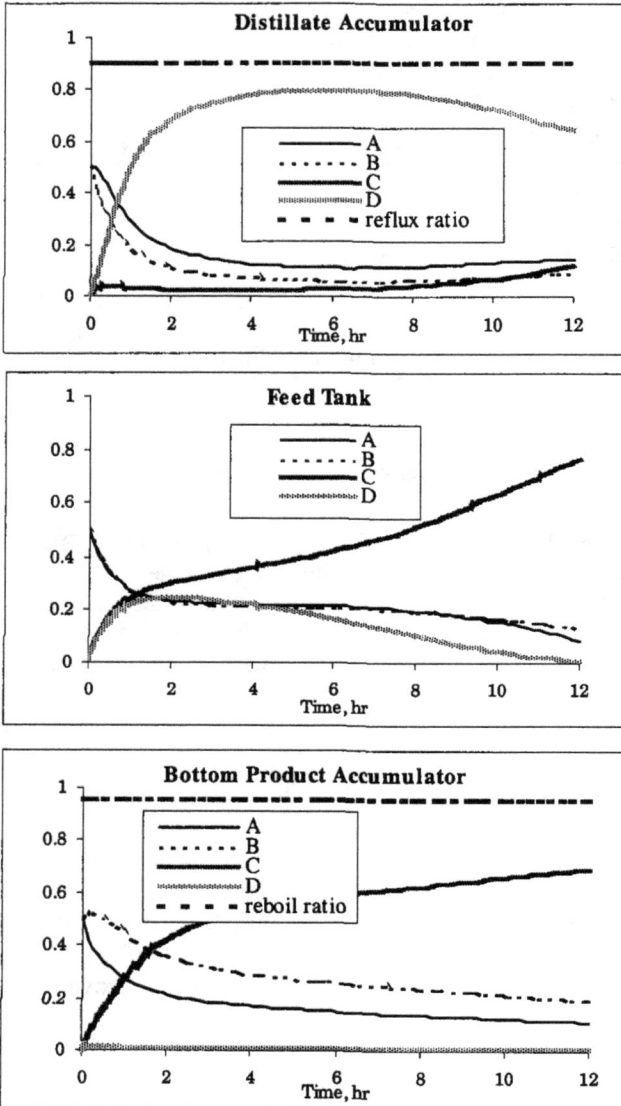

Figure 4.18. MVC with Chemical Reaction - Composition Profiles

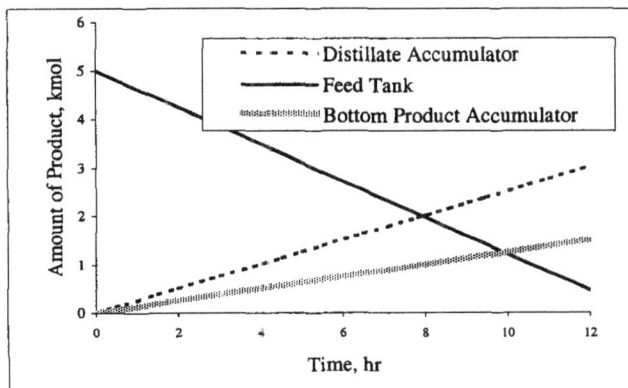

Figure 4.19. MVC with Chemical Reaction - Holdup Profiles

Table 4.15. Input Data for MultiBD Column

No. of column sections		= 3
No. of plates in each section,	N_k	= 11
Molar holdup in each vessel, (including condenser and reboiler)	$B_{0,m}$	= 2.5 kmol
Plate constant molar holdup,	$H_{j,k}$	= 0.01 kmol (per plate)
Liquid flow,	L	= 10 kmol/hr
Vapour flow,	V	= 10 kmol/hr
No. of components,	n_c	= 4
Initial charge composition,	x_{B0}	= <0.25, 0.25, 0.25, 0.25>
Relative volatility,	α	= <10.2, 4.5, 2.3, 1.0>

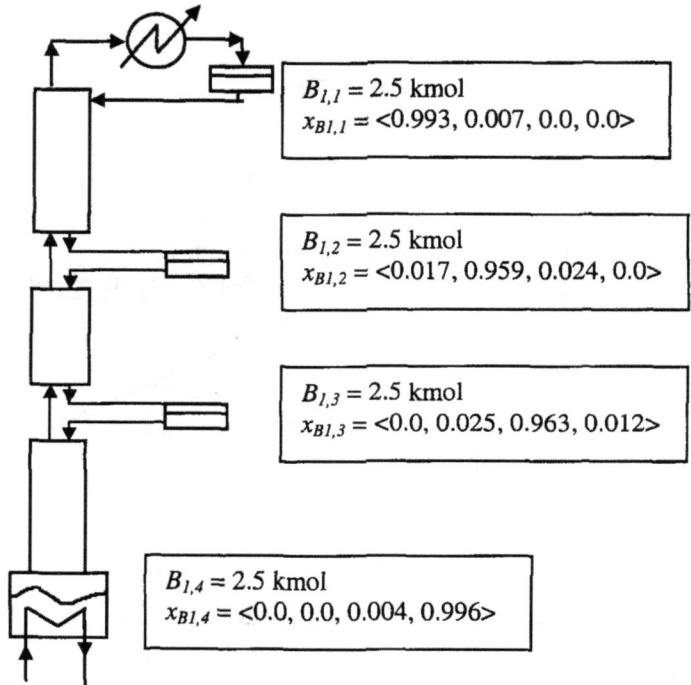

Figure 4.20. MultiBD Column - Product Amounts and Compositions

4.4. Packed Batch Distillation and Model

Extensive literature survey shows that only a few considered modelling, simulation and optimal control of such column (Hitch and Rousseau, 1988; Hansen and Jorgensen, 1986). Nad and Spiegel (1987) simulated their experimental packed column using the methods developed for tray columns by estimating the number of trays representative of the packed height. The reasons for overlooking this area are deficiencies remain evident in several key areas such as (a) the inability of current numerical scheme to solve the complicated systems of nonlinear model equations (partial differential and algebraic equations, PDAEs) required for a rigorous treatment of simultaneous mass and heat transfer phenomena, (b) lack of flexibility of most current packed column algorithms as they are typically process dependent with narrow operating regions. In addition, inherent dynamic nature of batch distillation makes PDAEs more complicated.

Wajge et al. (1997) attempted to develop rigorous PDAE model for packed batch distillation with and without chemical reaction and used finite difference and orthogonal collocation techniques to solve such model. The main purpose of the study was to investigate the efficiencies of the numerical methods employed. The authors observed that the collocation techniques are computationally more efficient compared to the finite difference method, however the order of approximating polynomial needs to be carefully chosen to achieve a right balance between accuracy and efficiency. See the original reference for further details.

4.5. Numerical Issues

Unlike continuous distillation, batch distillation is inherently an unsteady state process. Dynamics in continuous distillation are usually in the form of relatively small upsets from steady state operation, whereas in batch distillation individual species can completely disappear from the column, first from the reboiler (in the case of CBD columns) and then from the entire column. Therefore the model describing a batch column is always dynamic in nature and results in a system of Ordinary Differential Equations (ODEs) or a coupled system of Differential and Algebraic Equations (DAEs) (model types III, IV and V).

4.5.1. Classification of ODEs and DAEs

System of ODEs and DAEs can be classified according to their index (Morison, 1984). The model equations presented in the previous sections mostly constitute a coupled system of DAEs of index one (type III) or two (Type IV and V). Index is simply defined by the maximum number of differentiations required to reduce a DAE system to an ODE system. DAE system of index exceeding unity occurs in many areas of chemical engineering modelling. For example if the assumption of total condensation in the condenser is removed and material and energy balances are written for a fixed condenser volume, the resulting DAEs will be of index two (type IV and V - CVH models). This was shown in detail by Pantelides et al. (1988). Solution of such DAE systems is sometimes difficult. Classification of DAE system according to their index is given in some detail in Morison (1984).

Systems of index zero or one are ODE systems or simple DAE systems, and should cause no problem when integrated by well known existing methods. Integration of higher index DAEs requires special treatment (Gritsis, 1990; Bosley and Edgar, 1994a,b). See the original references for further details.

4.5.2. Integration Methods

Depending on the numerical techniques available for integration of the model equations, model reformulation or simplified version of the original model has always been the first step. In the Sixties and Seventies simplified models as sets of ordinary differential equations (ODEs) were developed. Explicit Euler method or explicit Runge-Kutta method (Huckaba and Danly, 1960; Domenech and Enjalbert, 1981; Coward, 1967; Robinson, 1969, 1970; etc) were used to integrate such model equations. The ODE models ignored column holdup and therefore non-stiff integration techniques were suitable for those models.

However, in batch distillation, the system is frequently very stiff, owing either to wide ranges in relative volatilities or large differences in tray and reboiler holdups. Therefore, if methods for non-stiff problems are applied to stiff problems (ODE models but having column holdup and/or energy balances), a very small integration step must be used to ensure that the solution remains stable (Meadow, 1963; Distefano, 1968; Boston et al., 1980; Holland and Liapis; 1983, etc.).

In the Eighties and Nineties ODE models were discretized as algebraic equations using orthogonal collocation on finite elements techniques and solved (Hansen and Jorgensen, 1986; Christensen and Jorgensen, 1987; Logsdon and Biegler, 1993; Li et al., 1998). Diwekar and co-workers (1986, 1987, 1991a, 1991b, 1992, 1995) and Sundaram and Evans (1993a,b) assumed that the batch distillation column could be considered as continuous column with changing feed. That is, for small interval of time the batch column behaviour is analogous to a continuous column and they employed widely used Fenske-Underwood-Gilliland shortcut method to integrate the model equations. Galindez and Fredenslund (1988) also considered a similar approach but employed rigorous continuous distillation model and modified Naphtali-Sandholm method (Christiansen et al., 1979; Naphtali and Sandholm, 1971) to integrate the model equations. Naphtali and Sandholm method requires grouping of the model equations by stages (number of plates) and the use of Newton-Raphson method. Mori et al. (1995) reformulated DAEs model of Distefano (1968) as ODEs and used the generalized implicit Euler's method to integrate the model equations.

Rigorous and stiff batch distillation models considering mass and energy balances, column holdup and physical properties result in a coupled system of DAEs. Solution of such model equations without any reformulation was developed by Gear (1971) and Hindmarsh (1980) based on Backward Differentiation Formula (BDF). BDF methods are basically predictor-corrector methods. At each step a prediction is made of the differential variable at the next point in time. A correction procedure corrects the prediction. If the difference between the predicted and corrected states is less than the required local error, the step is accepted. Otherwise the step length is reduced and another attempt is made. The step length may also be increased if possible and the order of prediction is changed when this seems useful.

Cuille and Reklaitis (1986), Ruiz (1988), and Mujtaba and co-workers (1989-2003) used this method in batch distillation studies.

Bosley and Edgar (1994a,b) developed techniques to calculate directly dynamic distillation (DDD) derivatives (non iterative), which were then employed to integrate rigorous DAEs (index 2) based batch distillation model.

Seader and Henley (1998) provided a detailed account of the merits and demerits of different integration methods with typical conventional batch distillation examples. They have noted three particular issues with the integration methods applied in batch distillation calculations that require attention. These are the *truncation error*, *stability* and *stiffness ratio*. These will be briefly discussed in the following.

4.5.2.1. Truncation Error

Local truncation errors are the result of approximating the functions on the right hand side of the ODEs at each time step. Although locally small, these errors can propagate as calculation proceeds to subsequent time steps and can result to large global truncation errors. The net effect is the gradual but significant loss of accuracy in the computed dependent variables. Use of smaller time step will reduce truncation errors.

Consider the following equation (y is dependent variable, x is the independent variable, h is the integration step size),

$$\frac{dy}{dx} = f(x, y) \tag{4.1}$$

If explicit Euler's approximation is implemented then,

$$y_{i+1} = y_i + f(x_i, y_i)h \tag{4.2}$$

The full representation however is given by,

$$y_{i+1} = y_i + f(x_i, y_i)h - \frac{f'(x_i, y_i)}{2!}h^2 + ... + \frac{f^{(n-1)}(x_i, y_i)^{(n)}}{n!}h^n \tag{4.3}$$

Hence the truncation error of the approximation is,

$$Error = \frac{f'(x_i, y_i)}{2!}h^2 + ... + \frac{f^{(n-1)}(x_i, y_i)^{(n)}}{n!}h^n \tag{4.4}$$

Alternatively the truncation error is approximated to the second order expression,

$$Error \approx \frac{f'(x_i, y_i)}{2!} h^2 \tag{4.5}$$

Conveniently this approximation actually over estimates the truncation error, as the other order terms will only reduce its magnitude.

4.5.2.2. Stability

Each numerical method has its own criteria for selecting the maximum time step. If this is not satisfied instability in the calculations may arise which makes the computed dependant variables totally inaccurate. In addition, to avoid oscillations in the dependant variables, a further reduction of the maximum time step may be necessary.

Seader and Henley (1998) noted the following criteria for explicit Euler's method to prevent the instability and oscillations respectively.

$$h_{max} = \frac{2}{|\lambda|_{max}} ; \qquad \text{[for instability]} \tag{4.6}$$

$$h_{max} = \frac{1}{|\lambda|_{max}} , \qquad \text{[for oscillations]} \tag{4.7}$$

where, $|\lambda|_{max}$ is the maximum absolute eigenvalue of the Jacobian matrix of the ODEs.

With an example of batch distillation, Seader and Henley showed that the time step needed in implicit Euler's method was 200 times of that needed for explicit Euler's method.

4.5.2.3. Stiffness Ratio

Quite often chemical engineering systems are encountered with widely different time constants, which give rise to both long-term and short-term effects. The corresponding ordinary differential equations have widely different eigenvalues. Differential equations of this type are known as stiff systems. Seader and Henley (1988) derived the expressions for maximum and minimum eigenvalues for the differential component mass balance equations related to intermediate plates and reboiler respectively.

$$\left|\lambda\right|_{max} \approx 2\left[\frac{L_j + k_{ij}V_j}{H_j}\right] \quad \text{for intermediate plates} \tag{4.8}$$

$$\left|\lambda\right|_{min} \approx \left[\frac{L_{N-1} + k_{iN}V_N}{H_N}\right] \quad \text{for reboiler} \tag{4.9}$$

The *stiffness ratio*, *SR*, is defined as:

$$SR = \frac{\left|\lambda\right|_{max}}{\left|\lambda\right|_{min}} \tag{4.10}$$

Seader and Henley (1998) noted that for $SR = 20$, the system is not stiff, $SR = 1000$, the system is stiff and $SR = 1,000,000$, the system is very stiff.

4.5.3. Initialisation of the DAE System

The initialisation of variables in a system of equations is important. While, in systems of ODEs all of the state variables must be initialised, in DAE systems only some of the variables need to be initialised, which is the same as the number of differential variables for index one system. The other variables can be determined using the algebraic equations. It is inconvenient for the user to be required to initialise all of the variables as this might require the solution of a set of nonlinear algebraic equations. Pantelides (1988) developed a procedure for consistent initialisation of DAE systems. Readers are directed to this reference for further details.

Nomenclature

D, L_D	distillate flow rate (kmol/hr)
F	feed flow rate in a continuous column or flow rate out of feed tank in MVC columns (kmol/hr)
H_a, H_c	accumulator and condenser holdup respectively (kmol)
H_j, H_N, H_f	plate, reboiler and feed tank holdup respectively (kmol)
h^L, h^V	Liquid, vapour enthalpy (kJ/kmol)
L_f	liquid flow rate into the feed tank in MVC columns (kmol/hr)
L, L', V	liquid, vapour flow rate, vapour boilup rate (kmol/hr)
L_B	bottom product flow rate (kmol/hr)
N, N_F, N_{min}	number of plates, feed tank location, minimum number of plates

n_c	no of components
Q_C, Q_R	condenser or reboiler duty (kJ/hr)
T, P	temperature (K), pressure (bar)
K, k	vapour-liquid equilibrium constant
k_r	array of reaction rate constants
r	internal reflux ratio [0, 1]
r_b	internal reboil ratio [0, 1]
R, R_{min}	external reflux ratio, minimum reflux ratio
$R_{min,u}$	minimum reflux ratio (Underwood)
$R_{min,g}$	minimum reflux ratio (Gilliland)
t	batch time (hr)
x, y	liquid or vapour composition (molefraction)
x, y	array of liquid or vapour composition (molefraction)
x_a	accumulated distillate composition (molefraction)
x_D	instant distillate composition (molefraction)
r	reaction rate (1/hr)
Δn	change in moles due to chemical reaction

Superscripts and subscripts

i	component number $(1, n_c)$
j	stage number $(1, N)$
k	column section in MultiBD columns
m	vessel no. in MultiBD columns

Greek letters

α	relative volatility
ϕ	common root

References

Abrams, H.J., Miladi, M. and Attarwala, F.T., Preferable alternatives to conventional batch distillation, *Presented at the IChemE Symposium Series no. 104*, Brighton, 7-9 September (1987).

Albet, J., Le Lann, J.M., Joulia, X. and Koehret, B., *In Proceedings Computer Oriented Process Engineering* (Elsevier Science Publishers B.V., Amsterdam, 1991), 75.

Attarwala, F.T., and Abrams, H.J., Optimisation techniques in binary batch distillation. *IChemE Annual Research Meeting*, London, 1974.

Balasubramhanya, L.S. and Doyle III, F.J., *J. Proc. Cont.* **10** (2000), 209.

Barolo, M., Guarise, G. B., Rienzi S. A. and Trotta, A., *Comput. chem. Engng.* **22** (1998), S37.

Bogacki, M.B.; Alejski, K. and Szymanowski, J., *Comput. chem. Engng.* **13** (1989), 1081.

Bosley, J.R. Jr. and Edgar, T.F. *In Proceedings of 5th International Seminar on Process Systems Engineering, Kyongju, Korea, 30 May-3 June*, **1** (1994a), 477.

Bosley, J.R. Jr. and Edgar, T.F., *J. Proc. Control.* **4** (1994b), 195.

Boston, J.F., Britt, H.I., Jirapongphan, S. and Shah, V.B., *In Foundation of Computer Aided Chemical Process Design*, (Engineering Foundation, New York), **II** (1980), 203.

Converse, A.O. and Huber, C.I., *IEC Fund.* **4** (1965), 475.

Coward, I., *Chem. Eng. Sci.* **22** (1967), 503.

Christensen, F.M. and Jorgensen, S.B., *Chem. Eng. J.* **34** (1987), 57.

Christiansen, A.C., Jacobsen, E.W., Perkins. J.D. and Skogestad, S., On the dynamics of batch distillation: A study of parametric sensitivity in ideal binary columns. *Presented at the AIChE Annual Meeting*, Miami, USA, November, paper no. 184d, 1995.

Christiansen, L.J., Michelsen, M.L and Fredenslund, A., *Comput. chem. Engng.* **3** (1979), 535.

Cuille, P.E. and Reklaitis, G.V., *Comput. chem. Engng.* **10** (1986), 389.

Distefano, G.P., *AIChE. J.* **14** (1968), 190

Diwekar, U.M., *Batch Distillation: Simulation, Optimal Design and Control* (Taylor and Francis, Washington, D.C., 1995).

Diwekar, U.M. and Madhavan, K.P., *In Proceedings of World Congress III of Chemical Engineering.*, Tokyo. **4** (1986) 719.

Diwekar, U.M., Malik, R.K. and Madhavan, K.P., *Comput. chem. Engng.* **11** (1987), 629.

Diwekar, U.M. and Madhavan, K.P., *IEC. Res.* **30** (1991a), 713.

Diwekar, U.M. and Madhavan, K.P., *Comput. chem. Engng.* **15** (1991b), 833.

Diwekar, U.M., *AIChE J.* **38** (1992), 1571.

Domenech, S. and Enjalbert, M., *Chem. Eng. Sci.* **29** (1974), 1529.

Domenech, S. and Enjalbert, M., *Comput. chem. Engng.* **5** (1981), 181.

Egly, H. Ruby, V. and Seid, B., *Comput. chem. Engng.* **3** (1979), 169.

Furlonge, H.I., Pantelides, C.C. and Sorensen, E., *AIChE J.* **45** (1999), 781.

Galindez, H. and Fredenslund, A., *Comput. chem. Engng.* **12** (1988), 281.

Gani, R., Ruiz, C.A. and Cameron, I.T., *Comput. chem. Engng.* **10** (1986), 181.

Gear, C.W., *IEEE Trans. Circuit Theory*, **CT-18** (1971), 89.

Gonzalez-Velasco, J.R. and et al., *IEC Res.* **26** (1987), 745.

Greaves, M. A., *Hybrid Modelling, Simulation and Optimisation of Batch Distillation Using Neural Network Techniques*. Ph.D. Thesis, (University of Bradford, Bradford, UK, 2003).

Greaves, M.A., Mujtaba, I. M., Barolo, M., Trotta, A. and Hussain, M. A., *Trans. IChemE*, **81A** (2003), 393.

Gritsis, D., *The Dynamic Simulation and Optimal Control of Systems Described by Index Two Differential-algebraic Equations*. PhD. Thesis, (Imperial College, University of London, 1990).

Hansen, T.T. and Jorgensen, S.B., *Chem. Eng. J.* **33** (1986), 151.

Hasebe, S., Abdul Aziz, B.B., Hashimoto, I. and Watanabe, T., In *Proceedings IFAC Workshop*, London 7-8 September (1992), 177.

Hasebe, S., Kurooka, T. and Hashimoto, I., *In Proceeding of IFAC Symposium DYCORD'95*, Danish Automation Society, Denmark (1995), 249.

Hindmarsh, A.C., *Ass. Comput. Mach., Signam Newsl.* **15** (1980), 10.

Hitch, D.M. and Rousseau, R.W., *IEC Res.* **27** (1988), 1466.

Holland, C.D. and Liapis, A.I., *Computer Methods for Solving Dynamic Separation Problems* (McGraw-Hill Book Company, New York, 1983).

Huckaba, C.E. and Danly, D.E., *AIChE. J.* **6**(1960), 335.

Kerkhof, L.H. and Vissers, H.J.M., *Chem. Eng. Sci.* **33** (1978), 961.

Li, P., Hoo, H.P. and Wozny, G., *Chem. Eng. Tech.* **21** (1998), 853.

Logsdon, J.S. and Biegler, L.T., *IEC Res.* **32** (1993), 700.

Luyben, W.L., *IEC Proc. Des. Dev.* **10** (1971), 54.

Luyben, W.L., *IEC Res.* **27** (1988), 642.

Mayur, D.N. and Jackson, R., *Chem. Eng. J.* **2** (1971), 150.

Meadow, E.L., *Chem. Eng. Prog. Symp. Ser.* **59** (1963), 48.

Mori, H., Goto, H., Yang, Z.C., Aragaki, T. and Koh, S., *J. Chem. Eng. Japan.* **28** (1995), 576.

Morison, K.R., *Optimal Control of Processes Described by Systems of Differential and Algebraic Equations*. PhD. Thesis, (Imperial College, University of London, 1984).

Mujtaba, I.M., *Optimal Operational Policies in Batch Distillation*. PhD Thesis, (Imperial College, University of London, 1989).

Mujtaba, I.M., *Trans. IChemE*, **75A** (1997), 609.

Mujtaba, I.M. and Macchietto, S. In *Recent Progres en Ginie de Procedes*, eds. Domenech, S. et al. (Lavoisier Techniche et Documentation, Paris, 1988) **2**, 191.

Mujtaba, I.M. and Macchietto, S., *Comput. chem. Engng.* **16S** (1992), S273.

Mujtaba, I.M. and Macchietto, S., *Comput. chem. Engng.* **17** (1993), 1191.

Mujtaba, I.M. and Macchietto, S. *In Proceedings ADCHEM'94- IFAC Symposium on Advanced Control of Chemical Processes*, Kyoto, Japan, 25-27 May, (1994), 415.

Mujtaba, I.M. and Macchietto, S., *J. Proc. Control.* **6** (1996), 27.

Mujtaba, I.M. and Macchietto, S., *IEC Res.* **36** (1997), 2287.

Mujtaba, I.M. and Macchietto, S., *Chem. Eng. Sci.* **53** (1998), 2519.

Nad, M. and Spiegel, L., *In Proceedings CEF 87* , Sicily, Italy, April (1987), 737.

Naphtali, L.M. and Sandholm, D.P., *AIChE J.* **17** (1971), 148.

Pantelides, C.C., *SIAM J Sci. Stat. Comput.* **9** (1988), 213.

Pantelides, C.C., Gritsis, D., Morison, K.R. and Sargent, R.W.H., *Comput. chem. Engng.* **12** (1988), 449.

Rayleigh, L., *Phil. Mag.* **4** (1902), 527.

Reid, R.C., Prausnitz, J.M. and Sherwood, T.K., *The Properties of Gases and Liquids* (3rd ed., Mc-Graw Hill Book Company, 1977).

Reuter, E., Wozny, G. and Jeromin, L., *Comput. chem. Engng.* **13** (1989), 499.

Ruiz, C.A., *In Proceedings of CHEMDATA 88*, 13-15 June, Sweden, (1988), 330.

Robinson, C.S. and Gilliland, E.R., *Elements of Fractional Distillation*, (4th ed., McGraw Hill, New York, 1950).

Robinson, E.R., *Chem Eng. Sci.* **24** (1969), 1661.

Robinson, E.R., *Chem Eng. Sci.* **25** (1970), 921.

Seader, J.D. and Henley, E.J., *Separation Process Principles* (John Wiley & Sons, Inc., 1998).

Simandl, J. and Svrcek, W.Y., *Comput. chem. Engng.* **15** (1991), 337.

Skogestad, S., Wittgens, B., Litto, R. and Sorensen, E., *AIChE J.* **43** (1997), 971.

Sorensen, E., Macchietto, S., Start, G., and Skogestad, S., *Comput. chem. Engng.* **20** (1996b), 1491.

Sorensen, E. and Skogestad, S., *Chem. Eng. Sci.* **51** (1996a), 4949.

Sorensen, E., and Skogestad, S., *J. Proc. Cont.* **4** (1996c), 205.

Sundaram, S. and Evans, L.B., *IEC Res,* **32** (1993a), 511.

Sundaram, S. and Evans, L.B., *IEC Res,* **32** (1993b), 500.

van Lith, P.F., Betlem, B.H.L., Roffel, B., *Comput. chem. Engng.* **27** (2003), 1021.

Wajge, R.M., Wilson, J.M., Pekny, J.F. and Reklaitis, G.V., *IEC Res.* **36** (1997), 1738.

CHAPTER 5

5. DYNAMIC OPTIMISATION

5.1. Optimisation

"Of Two Evils, Always Choose the Lesser" was written centuries ago on the walls of an ancient Roman bathhouse in connection with a choice between two aspirants for the emperor of Rome. This describes one of the most well known approaches to *the principle of optimisation* (Edgar and Himmelblau, 1988).

Typical chemical engineering problems (such as process design or plant operation) have many, and possibly an infinite number of solutions. The term optimisation is freely used to describe the complete spectrum of techniques from the basic multiple run approach of trial and error to highly complex numerical strategies. This assortment stems from the fact that optimisation is not idyllic in the real world but there are a lot of issues that require a practical approach. However, the potential benefit is huge and hence it is the next logical step after developing a model. So to avoid the numerous pitfalls it must not be flippantly treated or downsized in complexity!

5.1.1. Essential Features of Optimisation Problems

Every optimisation problem will have:

1. At least one **Objective Function** to be optimised (e.g. profit function, cost function, etc.) often called the *economic model*
2. One or more **Equality Constraints** (e.g. model equations)
3. One or more **Inequality Constraints** (e.g. lower and uppers bounds of operating variables, such as temperature in a reactor, reflux ratio in a distillation column)

A set of variables that satisfy the items 2 and 3 precisely will provide a *feasible solution* of the optimisation problem.

116

A set of variables that satisfy the items 2 and 3 and also provide an optimal value for the function in item 1 will provide an *optimal solution* of the optimisation problem.

Two types of optimisation problems are often encountered:

(a) *Linear Optimisation:* Objective function, constraints are linear.
(b) *Non-linear Optimisation:* Objective function, constraints are non-linear or combination of linear and non-linear systems.

Reklaitis et al. (1983), Edgar and Himmelblau (1988) have discussed several solution methods for solving linear and nonlinear optimisation problems. Here, some of the optimisation techniques used in batch distillation will be discussed.

5.2. Dynamic Optimisation (Optimal Control) of Batch Distillation

Batch distillation is inherently a dynamic process and thus results to optimal control or dynamic optimisation problems (unless batch distillation task is carried out in a continuous distillation column).

The optimal control of a process can be defined as a control sequence in time, which when applied to the process over a specified control interval, will cause it to operate in some optimal manner. The criterion for optimality is defined in terms of an objective function and constraints and the process is characterised by a dynamic model. The optimality criterion in batch distillation may have a number of forms, maximising a profit function, maximising the amount of product, minimising the batch time, etc. subject to any constraints on the system. The most common constraints in batch distillation are on the amount and on the purity of the product at the end of the process or at some intermediate point in time. The most common control variable of the process is the reflux ratio for a conventional column and reboil ratio for an inverted column and both for an MVC column.

The optimal operation of a batch column depends of course on the objectives one wishes to achieve at the end of the process. Depending on the objective function and any associated constraints, a variety of dynamic optimisation problems were defined in the past for conventional batch distillation column. Brief formulations of these optimisation problems are presented in the following subsections. Situations in which each formulation can be applied are discussed.

All the formulations are presented with reference to batch distillation operation schematically represented by a STN. Figure 5.1 shows a single distillation task i producing a main-cut j (D_j, x_D^j) and a bottom residue product (B_i, x_B^i) from an initial charge (B_{i-1}, x_B^{i-1}) charged to the reboiler at the beginning of the task (t_0^i).

The batch time for the operation is given by $t_C^i = t_F^i - t_0^i$, where t_F^i is the end time of distillation task i. Note that if $t_0^i = 0$, $(B_{i-1},\ x_B^{i-1})$ becomes $(B_0,\ x_{B0}^0)$ and t_C^i becomes equal to t_F^i. The batch time, here, is simply the time required to complete a distillation task. It excludes any charging, cleaning and transfer time before or at the end of the cut. For simplicity, we drop the use of superscript i in the following sections.

In general, the optimisation problem to optimise the operation of a CBD column can be stated as follows:

> *given:*   the column configuration, the feed mixture, vapour boilup rate
>   a separation task in terms of product purity
>   (+ recovery or amount of product or operation time or none)
> *determine:* optimal reflux ratio which governs the operation
> *so as to:* *minimise* the operation time
>   or *maximise* the amount of product
>   or *maximise* the profit
> *subject to:* equality and inequality constraints (e.g. model equations)

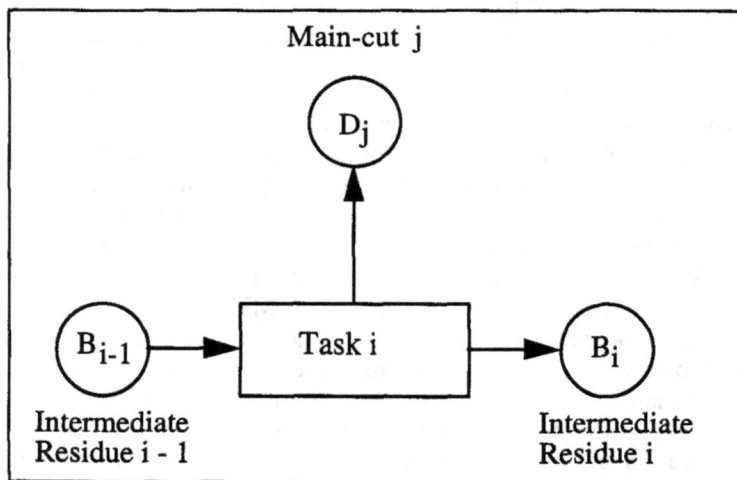

Figure 5.1. State Task Network for Batch Distillation

Mathematically it can be represented as:

P0   Minimise (or Maximise) J
        $u(t)$

subject to   Equality Constraints (Model)
              Inequality Constraints

where $u(t)$ denotes all the optimisation variables [e.g. controls (reflux ratio) and its switching times and or the final time]. Inequality constraints refer to simple bounds on $u(t)$ and final time constraints to the amount and/or purity of top or bottom product. Different types of dynamic optimisation problem formulations are now presented in the following sub-sections.

5.2.1. Minimum Time Problem

The minimum time problem is also known as the time optimal control problem. Coward (1967), Hansen and Jorgensen (1986), Robinson (1970), Mayur et al. (1970), Mayur and Jackson (1971), Mujtaba (1989) and Mujtaba and Macchietto (1992, 1993, 1996, 1998) all minimised the batch time to yield a given amount and composition of distillate using conventional batch distillation columns. The time optimal operation is often desirable when the amount of product and its purity are specified a priori and a reduction in batch time can produce either savings in the operating costs of the column itself or permit improved scheduling of other batch operations elsewhere in a process. Mathematically the problem can be written as:

P1   Min t_F
        $r(t)$

s.t.   Model Equations   (equality constraints)

        $x_D \geq x_D^*$ (inequality constraints)

        $D \geq D^*$ (inequality constraints)
        Linear bounds on reflux ratio (inequality constraints)

where D, x_D are the amount of distillate and its composition at the final time t_F, D^*, x_D^* are the specified amount of distillate product and its purity. $r(t)$ is the reflux ratio profile which is optimised. Note here that unless other constraints become limiting, at the optimum point we expect (D, x_D) to be equal or very close to (D^*, x_D^*) respectively (no product give away).

In the past researchers, using simple models, reported timesaving for the time optimal solutions of more than 10% compared to constant reflux operations. Mujtaba (1989) reported a batch timesaving of 30 - 45% compared to constant reflux operations for a number of example problems using detailed dynamic column models (Type IV-CMH).

5.2.2. Maximum Distillate Problem

Converse and Gross (1963), Converse and Huber (1965), Murty et al. (1980), Diwekar et al. (1987), Mujtaba (1989), Logsdon (1990) and Logsdon et al. (1990) considered an optimisation problem which maximises the amount of distillate product (of specified purity) for a given time of operation t_F^*. This type of operation is often useful when a fixed period is assigned to a particular batch unit for a particular job. Mathematically the optimisation problem can be written as:

P2   Max $J = D$
 $r(t)$

 s.t.   Model Equations   (equality constraints)

 $x_D \geq x_D^*$ (inequality constraints)

 $t = t_F^*$

 Linear bounds on reflux ratio (inequality constraints)

where D, x_D and x_D^* are same as defined in section 5.2.1. Note here that in this formulation final time t_F^* is not included in $r(t)$. It is reported that optimal reflux ratio policies obtained by solving this optimisation problem resulted in product yields up to 3-7% using simplified models and up to 10-13% using detailed models over those obtained with constant reflux ratio policies. Logsdon and Biegler (1993) however solved this maximum distillate problem with path constraint. Instead of satisfying the product purity at the end of the batch time, the product was collected all through at constant composition. The maximum amount of product achieved using this approach was reduced by about 8% compared to that obtained by solving problem **P2**.

5.2.3. Maximum Profit/Productivity Problem

Kerkhof and Vissers (1978) combined the *minimum time* and the *maximum distillate* problems into an economic profit function P to be maximised.

$$P = \frac{\text{added value to product}}{\text{time}} - \text{operating cost} = \frac{C_1 D - C_2 B_0}{t_F + t_s} - C_3 \qquad (5.1)$$

where, P is the profit ($/hr), C_1 is the price of the distillate product ($/kmol), C_2 is the raw material cost ($/kmol), C_3 is the operating cost (energy, wages, depreciation, etc.) ($/hr), t_s is the set up time (charging and cleaning time between batches).

Both the amount of distillate and the time of operation are to be established. The only constraint arises from the required purity of the distillate product. Mathematically, the problem can be written as:

P3   Max $J = P$
          $R(t)$

      s.t.   Model Equations   (equality constraints)

          $x_D \geq x_D^*$ (inequality constraints)

          Linear bounds on reflux ratio   (inequality constraints)

Kerkhof and Vissers and Mujtaba (1989) considered the operating cost C_3 as constant, but in practice it may vary, say, with different boilup rates. More general expressions (Logsdon et al., 1990, Mujtaba and Macchietto, 1996) could also be used to evaluate the operating cost per hour. When $C_1 = 1.0$, $C_2 = C_3 = t_s = 0$, the *maximum profit* problem results in a *maximum productivity* problem (Mujtaba, 1989, 1999).

Kerkhof and Vissers showed that for difficult separations an optimal reflux control policy yields up to 5% more distillate, corresponding to 20-40% higher profit, than either constant distillate composition or constant reflux ratio policies.

All the optimisation problem formulations presented above were aimed to achieve optimal operation policies for a variety of objective functions but for a single period operation (i.e. single distillation task). In single period operation only one product cut is made from both binary and multicomponent mixtures and optimal operation policy is restricted only to that period.

However, in batch distillation, as the overhead composition varies during the operation, a number of main-cuts and off-cuts are made at the end of various distillation tasks or periods (see Chapter 3). Purities of the main products are usually determined by market or downstream process requirements but the amounts recovered must be selected based on the economic trade off between longer distillation times (hence productivity), reflux ratio levels (hence energy costs), product values, etc. Increasing the recovery of a particular species in a particular cut may have strong effects on the recovery of other species in subsequent cuts or, in fact, on the ability to achieve at all the required purity specifications in subsequent cuts. The profitable operation of such processes therefore requires consideration of the whole (multiperiod) operation. The optimisation problem formulations of such problems are presented in Chapter 6.

5.3. Summary of the Past Work on Dynamic Optimisation

Table 5.1 summarises the past work on dynamic optimisation of batch distillation using different types of column configurations.

As can be seen from Table 5.1, most of the authors used very simplified model based on constant relative volatility, equimolal overflow and no holdup assumptions. Although the simple/shortcut model/solution methods have advantages over the rigorous model/methods in terms of computation time, their use is very much restricted to ideal systems and to an initial trial design or to the finding initial trial operating policies (Robinson, 1970; Mayur and Jackson, 1971; Diwekar et al., 1987; Diwekar, 1992). These must still be verified and refined using rigorous dynamic column models, accurate physical property data and rigorous integration methods, in particular with nonideal mixtures and when holdups are significant or by experiments.

Table 5.1. Summary of the Past Work on Optimisation of Batch Distillation

Reference	Column Type	Model Type	Mixture	Operation	Optimisation Problem
Converse & Huber (1965)	CBD	III	Binary	Normal	P2
Coward (1967)	CBD	III[1]	Binary	Normal	P1
Robinson (1969)	CBD	III[1]	Binary & Multicomponent	Normal	P2
Robinson (1970)	CBD	III	Multicomponent	Normal	P2
Mayur et al. (1970)	CBD	III[1]	Binary	Recycle	P1
Mayur and Jackson (1971)	CBD	III	Binary	Normal	P2
Kerkhof and Vissers (1978)	CBD	III[1]	Binary	Normal	P3
Murty et al. (1980)	CBD	III	Binary	Normal	P2
Hansen and Jorgensen (1986)	CBD	III[1]	Binary	Normal	P1
Christensen and Jorgensen	CBD	III[1]	Binary	Recycle	P1
Diwekar et al. (1987, 1989)	CBD	II	Multicomponent	Normal	P2, P3
Farhat et al. (1990)	CBD	III[1]	Multicomponent	Normal	P2
Mujtaba and co-workers (1988-2003)	CBD, IBD, MVC, BED, BREAD	III, IV, V	Binary & Multicomponent	Normal, Recycle	P1, P2, P3
Logsdon & Biegler (1993)	CBD	II	Binary	Normal	P2
Diwekar (1992, 1995)	CBD, MVC	II, II[2]	Binary & Multicomponent	Normal	P1, P2, P3
Sorensen & Skogestad (1996)	CBD, IBD	III	Binary	Normal	P1
Bonny et al. (1996)	CBD	III[1]	Multicomponent	Recycle	Maximum Productivity
Betlem et al. (1998)	CBD	III	Binary & Multicomponent	Normal	P3
Li et al. (1998b)	Semi-batch, BREAD	IV	Multicomponent	Normal	P1
Furlonge et al. (1999)	MultiBD	IV[3]	Multicomponent	Normal	Minimum Energy
Wajge and Reklaitis (1999)	CBD, BREAD	V	Multicomponent	Normal	Maximum Conversion

Key: 1 – without holdup   2 – with holdup   3 – with tray hydraulics

5.4. Summary of the Past Work on the Solution of Optimisation Problems

Depending on the numerical techniques available for solving optimal control or optimisation problems the model reformulation or development of simplified version of the original model was always the first step. In the Sixties and Seventies simplified models represented by a set of Ordinary Differential Equations (ODEs) were developed. The explicit Euler or Runge-Kutta methods (Huckaba and Danly, 1960; Domenech and Enjalbert, 1981) were used to integrate the model equations and the Pontryagin's *Maximum Principle* was used to obtain optimal operation policies (Coward, 1967; Robinson, 1969, 1970 etc.).

In the Eighties and Nineties ODE models were discretized and solved as a set of algebraic equations using orthogonal collocation techniques and the optimal operation problems were solved using the *Maximum Principle* (Hansen and Jorgensen, 1986; Christensen and Jorgensen, 1987). Diwekar and co-workers (1986, 1987, 1991a, 1991b, 1992) used the short-cut method in conjunction with the *Maximum Principle* in their optimisation studies.

Also in the Eighties and early Nineties Morison (1984), Cuthrell and Biegler (1987) and Renfro et al. (1987) developed dynamic optimisation algorithms and applied them to processes described by a set of Differential and Algebraic Equations (DAEs). Morison used a sequential model solution and optimisation strategy (feasible path approach) while others used collocation technique to discretize DAEs to a large set of algebraic equation and applied simultaneous model solution and optimisation strategy (infeasible path approach). Logsdon (1990) extended the approach of Renfro et al. to include path constraints and higher index DAE problems. Each of the methods has their own merits and demerits. In all cases the optimisation problem was formulated as a nonlinear programming problem (NLP), which were solved using efficient techniques (Chen, 1988).

Mujtaba and co-workers (1988, 1992, 1993, 1996, 1998, 1999) applied the technique of Morison (1984) while Jang (1993), Logsdon and co-workers (1990, 1993) and Farhat et al. (1990) used other techniques in their optimisation studies of batch distillation.

The *Maximum Principle* and *Nonlinear Programming (NLP)* based approaches will be discussed in the following.

5.5. Maximum Principle Based Dynamic Optimisation Technique

The *Maximum Principle* of Pontryagin (Pontryagin et al., 1964) may be stated briefly as follows. For a given set of ODEs:

$$\frac{dx_i}{dt} = f_i(x_1...x_n, u_1...u_n), \qquad i=1...n \qquad \text{(P.1)}$$

with initial conditions represented by:

$$x_i(t_0) = x_{i0} \qquad \text{(P.2)}$$

and terminal conditions:

$$x_i(t_1) = x_{i1} \qquad \text{(P.3)}$$

It is required to find the control scheme $u = u(t)$ which transfers the system from conditions x_{i0} to conditions x_{i1} in the shortest time.

To find the solution it is required to introduce adjoint variables $\psi_1, \psi_2 .... \psi_n$ where each variable is defined by:

$$\frac{d\psi_i}{dt} = -\sum_{j=1}^{n} \psi_j \frac{\partial f_j(x,u)}{\partial x_i} \qquad i=1....n \qquad \text{(P.4)}$$

It is further necessary to define the Hamiltonian function:

$$H(\psi, x, u) = \sum_{j=1}^{n} \psi_j f_j(x,u) \qquad \text{(P.5)}$$

From which it follows that:

$$\frac{dx_i}{dt} = \frac{\partial H}{\partial \psi_i} \qquad \text{(P.6)}$$

and

$$\frac{d\psi_i}{dt} = -\frac{\partial H}{\partial x_i} \qquad i=1....n \qquad \text{(P.7)}$$

The maximum principle now states that the function $H(\psi, x, u)$ attains its maximum value at the point $u = u(t)$ where u is the value of the control parameter along an optimal route. Furthermore it may be shown that this maximum value of the Hamiltonian is both positive and constant in the range $t_0 \leq t \leq t_1$. This means $\dfrac{dH}{du} = 0$ in the range $t_0 \leq t \leq t_1$.

5.5.1. Application to Batch Distillation: Minimum Time Problem

The application of Equations P.1-P.7 to batch distillation enables the solution of time optimal control problems.

5.5.1.1. Simple Model

Robinson (1969) considered the type III model but with zero-column holdup and without a condenser holdup tank for the purpose. The model equations in the form of ODEs are:

$$\frac{dB}{dt} = \frac{-V}{R+1} \tag{P.8}$$

$$\frac{dx_B}{dt} = \frac{-V}{R+1} \cdot \frac{(y - x_B)}{B} \tag{P.9}$$

where B and x_B are the amount and composition of the material left the reboiler at any time t, y is the composition of the vapour entering the condenser (same as instant distillate composition) which can be obtained by plate to plate calculations using vapour liquid equilibrium relationship (see Chapter 4), R is the external reflux ratio and V is the constant vapour boilup rate.

The *minimum time* problem (shown as a negative of a maximum time problem) can be written as:

$$\underset{R}{Maximise} \quad J = -\int_0^{t_F} dt$$

subject to:

$$D = \int_0^{t_F} \frac{V}{R+1} dt = \underline{D}$$

$$x_D = \frac{\displaystyle\int_0^{t_F} y \frac{V}{R+1} dt}{D} = \underline{x}_D$$

Note that the initial amount and composition (B_0, x_{B0}) in the reboiler at time $t = 0$ will provide the initial condition given in Equation P.2. The final desired amount of distillate and composition ($\underline{D}$, $\underline{x_D}$) can be used in the overall total mass and component balance to give the final (at time t_F) reboiler condition (B_F, x_{BF}) as required by the condition given in Equation P.3.

The Hamiltonian function according to Equation P.5, which should be maximised, is:

$$H = -\psi_1 \cdot \frac{V}{R+1} - \psi_2 \frac{V}{R+1} \frac{(y - x_B)}{B} \tag{P.10}$$

where the two adjoint variables ψ_1, ψ_2 may be defined according to Equation P.4 or P.7 as

$$\frac{d\psi_1}{dt} = -\psi_2 \frac{V}{R+1} \cdot \frac{(y - x_B)}{B^2} \tag{P.11}$$

$$\frac{d\psi_2}{dt} = \psi_2 \frac{V}{R+1} \frac{\left(\frac{\partial y}{\partial x_B} - 1\right)}{B} \tag{P.12}$$

The Equations P.8, P.9, P.11, P.12 must now be solved simultaneously subject to the conditions that at all times the Hamiltonian function defined in Equation P.10 is maximised. Diwekar (1992, 1995) shows that the optimality condition $\frac{dH}{du} = 0$ in the range $0 \leq t \leq t_F$ will give the time optimal reflux ratio R as:

$$R = \frac{B + (\psi_2 / \psi_1)(y - x_B)}{(\psi_2 / \psi_1) \frac{\partial y}{\partial R}} - 1 \tag{P.13}$$

5.5.1.2. Simple Model with Variable Boilup Rate

The boilup rate in actual columns is seldom kept constant but falls off as the distillation proceeds (as discussed in Chapter 3, also see Greaves et al., 2001; Greaves, 2003). Robinson (1969) assumed that the boilup rate is a linear function of the still composition (with boilup V_0 at time $t = 0$ and k is a constant)

$$V = V_0 + kx_B \tag{P.14}$$

This changes the definition of ψ_2 in Equation P.12 which is now given by:

$$\frac{d\psi_2}{dt} = \frac{\psi_1 k}{R+1} + \frac{V}{R+1} \left\{ \frac{V\left(\frac{\partial y}{\partial x_B} - 1\right)}{B} + \frac{k(y - x_B)}{B} \right\} \tag{P.15}$$

5.5.1.3. Simple Model for Multicomponent System

The model equations for multicomponent distillation under constant boilup rate (Robinson, 1969) are:

$$\frac{dB}{dt} = \frac{-V}{R+1} \tag{P.16}$$

$$\frac{dx_{Bi}}{dt} = \frac{-V}{R+1} \frac{(y - x_{Bi})}{B}, \qquad i = 1 .... (n_c - 1) \tag{P.17}$$

where, n_c is the number of components in the mixture.

These lead to a new definition of the Hamiltonian function.

$$H = -\psi_1 \frac{V}{R+1} - \sum_{i=1}^{n_c - 1} \psi_{2i} \frac{V}{R+1} \cdot \frac{(y_i - x_{Bi})}{B} \tag{P.18}$$

and

$$\frac{d\psi_1}{dt} = -\sum_{i-1}^{n_c - 1} \psi_{2i} \cdot \frac{V}{R+1} \cdot \frac{(y_i - x_{Bi})}{B^2} \tag{P.19}$$

$$\frac{d\psi_{2i}}{dt} = \sum_{j=1}^{n_c - 1} \psi_{2j} \frac{V}{(R+1)} \cdot \frac{\frac{\partial y_i}{\partial x_{Bi}}}{B} - \frac{\psi_{2i} V}{(R+1)B} \tag{P.20}$$

Equations P.16, P.17, P.19, P.20 must be solved simultaneously subject to the condition that Equation P.18 is again maximised at all times.

5.5.1.4. Simple Model with Condenser Holdup

For a system in which condenser holdup may not be neglected, the equations based on the assumption of complete mixing within the condenser are (Robinson, 1969):

$$\frac{dB}{dt} = \frac{-V}{R+1} \tag{P.21}$$

$$\frac{dx_D}{dt} = \frac{V}{H_c}(y_T - x_D) \tag{P.22}$$

$$\frac{dx_B}{dt} = \frac{RVx_D}{(R+1)B} - \frac{Vy_T}{B} + \frac{Vx_B}{(R+1)B} \tag{P.23}$$

These lead to the Hamiltonian definition:

$$H = -\psi_1 \frac{V}{R+1} + \psi_2 \frac{V}{H_c}(y_T - x_D) + \psi_3 \left\{ \frac{RVx_D}{(R+1)B} - \frac{Vy_T}{B} + \frac{Vx_B}{(R+1)B} \right\} \tag{P.24}$$

and

$$\frac{d\psi_1}{dt} = \psi_3 \left\{ \frac{RVx_D}{(R+1)B^2} - \frac{Vy_T}{B^2} + \frac{Vx_B}{(R+1)B^2} \right\} \tag{P.25}$$

$$\frac{d\psi_2}{dt} = -\psi_2 \frac{V}{H_c}\left(\frac{\partial y_T}{\partial x_D} - 1\right) - \psi_3 \left\{ \frac{RV}{(R+1)B} - \frac{V}{B}\frac{\partial y_T}{\partial x_D} \right\} \tag{P.26}$$

$$\frac{d\psi_3}{dt} = -\psi_2 \frac{V}{H_c}\left(\frac{\partial y_T}{\partial x_B}\right) + \psi_3 \left(\frac{V}{B}\cdot\frac{\partial y_T}{\partial x_B} - \frac{V}{(R+1)B} \right) \tag{P.27}$$

Again Equations P.21-P.23, P.25-P.27 must be solved simultaneously subject to condition of Equation P.24 being maximised at all times.

5.5.1.5. Example 1

Robinson (1969) considered the following example problem. A binary feed mixture with an initial amount of charge, $B_0 = 100$ kmol and composition $x_{B0} = <0.50, 0.50>$ molefraction, having constant relative volatility of 2.0 was to be processed in a batch distillation column with 8 theoretical stages. The aim was to produce 40 kmol of distillate product ($\underline{D}$) with composition ($\underline{x_D}$) of 0.5 molefraction for component 1 in minimum time (t_F) using optimal reflux ratio (R).

Table 5.2 summarises the results for two cases: (i) constant vapour boil-up rate, (ii) variable vapour boilup rate. The initial and final time optimal reflux ratio values are shown in Table 5.2 for both cases. The optimal reflux ratios between these two points follow according to Equation P.13 for each case. See details in the original reference (Robinson, 1969).

5.5.1.6. Example 2

Robinson (1970) considered an industrial 10-component batch distillation operation. The feed condition is shown in Table 5.3. The distillation column was currently producing the desired product using constant reflux ratio scheme. Table 5.4 summarises the results of the application of *minimum time* problem using simple model with and without column holdup.

The existing operation used constant reflux ratio ($R=7.0$) throughout. The optimal operation using simple model starts with an initial reflux ratio ($R=3.2$) which increases exponentially until the final reflux ratio is $R=12.2$. The optimal operation using holdup model starts with an initial reflux ratio ($R=3.24$ which increases exponentially until the final reflux ratio is $R=18.0$. See the original reference (Robinson, 1970) for the optimal reflux ratio profiles. It is interesting to note here that a zero holdup model may provide substantially inaccurate results. Robinson (1970) implemented the optimal reflux ratio profiles (as shown in Table 5.4) in the real plant and the results are presented in Table 5.5. It clearly shows that the optimal reflux profile (obtained using the zero holdup model) when implemented in the real plant could not achieve the required product purity. On the other hand implementation of the optimal policy obtained using the holdup model in the real plant achieves the product purity close to the desired value.

Table 5.2. Summary of *Minimum Time* Problem
Using Pontryagin's *Maximum Principle*

Case	Boilup Rate kmol/hr	Optimal Reflux Ratio at		Minimum Time, hr
		t_0	t_F	
Constant Boilup Rate	$V = 100.0$	2.75	6.5	1.90
Variable Boilup Rate	$V = 30+200x_3$	2.95	5.8	1.92

Table 5.3. Feed Condition for Example 2 (Robinson, 1970)

Component	Boiling Point (°C)	Relative Volatility (approx.)	Initial Composition (Wt. %)
Tricyclene	153.0	1.40	0.4
α – Pinene	156.0	1.32	52.2
Camphene	159.0	1.20	0.9
β – Pinene	164.0	1.00	6.5
Δ^4 Carene	170.0	0.85	1.0
Δ^3 Carene	170.0	0.85	33.0
M-Cymene	175.2	0.75	0.7
P-Cymene	177.1	0.72	1.0
Dipentene	177.5	0.70	2.4
Terpinolene	186.0	0.60	1.9

Table 5.4. Summary of the Results for Example 2

The total amount of feed, B_0	= 7800 gal.
The total column holdup	= 300 gal
Average vapour boilup rate, V	= 2500 gal/hr
Desired amount of product	= 3800 gal
Desired product purity	= 93% (by weight) α-pinene

Case	Reflux ratio at time		Batch Time, hr
	t_0	t_F	
Existing plant operation	7.0	7.0	12.35
Optimal operation (simple model)	3.2	12.2	9.30
Optimal operation (hold-up model)	3.4	18.0	10.66

Table 5.5. Implementation of Optimal Reflux Profiles in the Real Plant

Case	Batch time, hr	α-pinene composition in distillate (wt%)
Existing	12.35	93.1.
Optimal (simple model)	9.30	90.1
Optimal (holdup model)	10.66	93.5

5.5.2. Application to Batch Distillation: Maximum Distillate Problem

Referring to Simple Model (with zero column holdup) presented in section 5.5.1, a maximum distillate problem can be formulated as:

$$\underset{R}{Maximise} \quad J = \int_0^{t^*} D\,dt = \int_0^{t^*} \frac{V}{R+1}\,dt$$

subject to:

$$x_D = \frac{\displaystyle\int_0^{t^*} y\,\frac{V}{R+1}\,dt}{D}\Bigg|_{t=t^*} = \underline{x}_D$$

where t^* is the fixed batch time.

The Hamiltonian, the adjoint equations and the optimal reflux ratio correlation will be same as those in Equations P.10-P.13 (Diwekar, 1992). However, note that the final conditions (stopping criteria) for the *minimum time* and the *maximum distillate* problems are different. The stopping criterion for the *minimum time* problem is when $(\underline{D}, \underline{x_D})$ is achieved, while the stopping criterion for the *maximum distillate* problem is when $(t^*, \underline{x_D})$ is achieved. See Coward (1967) for an example problem.

5.5.3. Application to Batch Distillation: Maximum Profit Problem

Kerkhof and Vissers (1978) considered the following profit function:

$$P = \frac{DP_r - B_0 C_0}{t_F + t_s} - C_f \tag{5.2}$$

where, P = Average profit, $/hr
   D = Amount of distillate product, kmol
   P_r = sales value of the product, $/kmol
   B_0 = initial raw material charge, kmol
   C_0 = cost of raw material, $/kmol
   C_f = fixed operating cost, $/hr
   t_F = batch time, hr
   t_s = set up time, hr

Referring to the simple zero holdup model (described in 5.5.1.1), the *maximum profit* problem can now be written as:

$$\underset{R,\, t_F}{Maximise}\ \ J = \frac{DP_r - B_0 C_0}{t_F + t_s}$$

subject to:

$$x_D = \frac{\displaystyle\int_0^{t^*} y\, \frac{V}{R+1}\, dt}{D} = \underline{x}_D$$

Table 5.6. Summary of Maximum Profit Problem using *Maximum Principle*
(Kerkhof and Vissers, 1978)

Case	N	α	x_{B0}	$\underline{x}_D$	q_k	P_r, C_0	D_{opt}	T_{opt}	P_{max}
1	10	2.0	.70	.95	0.36	30, 12	69.39	3.50	45.91
2	9	1.75	.40	.90	4.60	90, 24	40.13	5.25	43.90
3	8	1.65	.70	.90	3.20	30, 12	70.55	4.20	26.20
4	7	1.80	.50	.90	7.25	45, 12	47.34	4.30	25.61
5	6	2.0	.50	.90	6.60	45, 12	48.03	3.60	59.00

5.5.3.1. Example

Table 5.6 presents the results of a set of 5 *maximum profit* problems considered by Kerkhof and Vissers (1978). Each problem has different column configuration (i.e. different number of plates, N), handles different feed mixture (characterised by different relative volatility, α), has different distillate product quality ($\underline{x}_D$), has different cost values (i.e. P_r, C_0) and different set up times (t_s). The initial charge, B_0 = 100 kmol, V = 60 kmol/hr, t_s= 1 hr, and C_f = 150 $/hr. All cases deal with binary mixtures of different initial composition, x_{B0}.

Kerkhof and Vissers (1978) defined a dimensionless number q_k indicating **the degree of difficulty** of a separation.

$$q_k = \frac{\underline{x}_D - x_{B0}}{x_{B0}(1 - \underline{x}_D)(\alpha^{N+1} - 1)} \tag{5.3}$$

For each case the value of q_k is also shown in Table 5.6. Note that although the maximum profit (shown in Table 5.6) has no direct correlation with q_k, Kerkhof and Vissers (1978) showed that $(D_{opt} / D_{max})100$ has a direct correlation with q_k. The value of $(D_{opt} / D_{max})100$ decreases with the increase of q_k. See original reference for further details. Note that D_{max} is defined as $D_{max} = (B_0 x_{B0}) / \underline{x}_D$.

5.5.4. Application to Batch Distillation: Short-cut Model

Diwekar (1992, 1995) has extensively used Pontryagin's *Maximum Principle* for solving all types of optimisation problems (section 5.2) using the short-cut model presented in Chapter 4. Refer to the original references for example problems.

5.6. Approaches to Nonlinear Dynamic Optimisation Technique

5.6.1. Feasible Path Approach

In this approach, the process variables are partitioned into dependent variables and independent variables (optimisation variables). For each choice of the optimisation variables (sometimes referred to as decision variables in the literature) the simulator (model solver) is used to converge the process model equations (described by a set of ODEs or DAEs). Therefore, the method includes two levels. The first level performs the simulation to converge all the equality constraints and to satisfy the inequality constraints and the second level performs the optimisation. The resulting optimisation problem is thus an unconstrained nonlinear optimisation problem or a constrained optimisation problem with simple bounds for the associated optimisation variables plus any interior or terminal point constraints (e.g. the amount and purity of the product at the end of a cut). Figure 5.2 describes the solution strategy using the feasible path approach.

The advantage of this approach is that the resulting optimisation problems are in general small in terms of the number of optimisation variables. Since each search point is feasible, if the process is terminated at the optimisation level before reaching a solution, the terminating point is still feasible and may be acceptable as a practical, although sub-optimal, solution of the problem. However, since the feasible path approach requires a complete solution of the ODEs or DAEs for each trial value of the optimisation variables this may be expensive. The simulation level may fail since a feasible region may not even exist for certain value of the optimisation variables. For example, for a choice of vapour boilup rate, reflux ratio and batch time, the reboiler content can take a negative value (not physically possible).

Morison (1984) and Vassialics (1993) developed sequential model solution and optimisation strategy which is commonly known as Feasible Path Approach.

5.6.2. Infeasible Path Approach

In this approach, the ODE or DAE process models are discretised into a set of algebraic equations (AEs) using collocation or other suitable methods and are solved simultaneously with the optimisation problem. Application of the collocation techniques to ODEs or DAEs results in a large system of algebraic equations which appear as constraints in the optimisation problem. This approach results in a large sparse optimisation problem.

Cuthrell and Biegler (1987) and Renfro et al. (1987) developed dynamic optimisation methods based on the infeasible path approach. The main advantage of this approach is that it avoids repetitive simulations during iteration of the optimisation problem and hence eliminates the possibility of premature termination due to a convergence failure of the simulation. Also, this approach is faster in terms of computation time compared to feasible path approach but at the expense of increasing the size of the optimisation problems in terms of variables.

However, in practice, a chemical process may not be physically well defined over a wide range of constraints. Therefore, some precautions must be taken to identify and to satisfy such constraints (e.g. non-negative flows) when an infeasible path method is used. Also, this approach is less familiar to design engineers since each point may not be feasible and acceptable, and a large number of variables may require to be initialised. In addition, it is difficult to identify the effect of optimisation variables on the objective function during the course of optimisation and thus providing diagnostic information becomes a more serious issue for the infeasible path approach.

Chen (1988) provided detailed accounts on feasible and infeasible path approaches in optimisation.

5.7. Nonlinear Programming (NLP) Based Dynamic Optimisation Problem-Feasible Path Approach

Here, a dynamic optimisation (also known as optimal control) problem formulation and solution proposed by Morison (1984) based on Sargent and Sullivan (1979) is presented. The process model can be described by a system of DAEs (model types III, IV and V presented in Chapter 4) as:

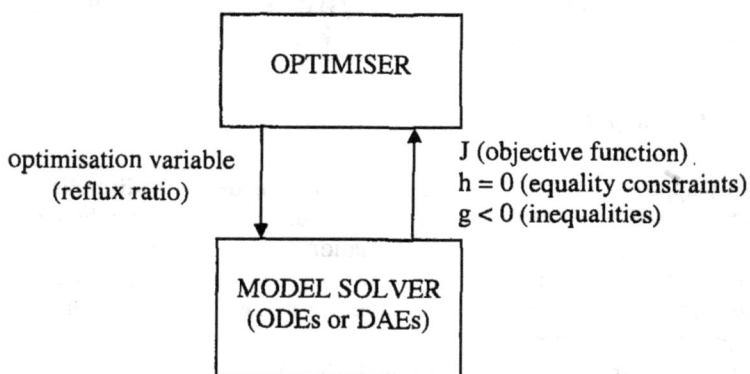

Figure 5.2. Feasible Path Optimisation Strategy

$$f(t, \dot{x}(t), x(t), u(t), v) = 0 ; \quad [t_0, t_F] \tag{5.4}$$

where t is the independent variable (time), $x(t) \ \varepsilon \ R^n$ is the set of all state variables, $\dot{x}(t)$ denotes the derivatives $x(t)$ with respect to time, $u(t) \ \varepsilon \ R^m$ is a vector of control variables, such as reflux ratio, and $v \ \varepsilon \ R^p$ is a vector of time invariant parameters (design variables), such as molar holdups on each plate. Suitable initial conditions $x(t)$ are defined at time $t = t_0$. The time interval of interest is $[t_0, t_F]$ and the function $f: Rx \ R^n x \ R^n x \ R^m x \ R^p \rightarrow R^n$ is assumed to be continuously differentiable with respect to all its arguments.

The system is subject to bounds on the controls $a^u(t) \leq u(t) \leq b^u(t)$, $t \ \varepsilon \ [t_0, t_F]$ where $a^u(t)$ and $b^u(t)$ are given continuous functions of time on $[t_0, t_F]$, and interior point or terminal constraints of the form:

$$a^F \leq F(t_p, \dot{x}(t_p), x(t_p), u(t_p), v) \leq b^F \tag{5.5}$$

where $F(t_p, \dot{x}, x, u, v) \ \varepsilon \ R^n$ and is continuously differentiable with respect to its arguments. At the terminal point $t_p = t_F$, the system performance is measured in terms of a scalar objective function to be minimised:

$$J = F(t_F, \dot{x}(t_F), x(t_F), u(t_F), v) \tag{5.6}$$

The optimal control problem is to choose an admissible set of controls $u(t)$, and final time, t_F, to minimize the objective function, J, subject to the bounds on the controls and constraints.

5.7.1. Control Vector Parameterisation (CVP)

To pose the optimal control problem as a nonlinear programming (NLP) problem the controls $u(t)$ are approximated by a finite dimensional representation. The time interval $[t_0, t_F]$ is divided into a finite number of subintervals (N_s), each with a set of basis functions involving a finite number of parameters $u(t) = \phi^j(t, z_j)$, $t\varepsilon[(t_{j-1}, t_j)$, $j = 1, 2, .. ,J]$, where $t_J = t_F$. The functions $\phi^j (t, z_j)$ are assumed to be continuously differentiable with respect to t and z_j, and the derivatives are uniformly bounded. The control is thus defined by the parameters z_j and the switching time $t_j, j = 1, 2, ...,$ J. The control constraints now become:

$$a^u(t) \leq \phi^j(t, z_j) \leq b^u(t); \qquad t \ \varepsilon \ [(t_{j-1}, t_j), j = 1, 2, .. J=2N_s] \tag{5.7}$$

The set of decision variables for the nonlinear programme can now be written as:

$$y = \{z_1, z_2, ..., z_J, t_1, t_2, ..., t_J\} \qquad (5.8)$$

5.7.2. NLP Optimisation Problem

The optimisation problem presented in the previous section can be written more formally as: Find the set of decision variables, y (given by Equation 5.7) which will

minimise $\qquad J(y)$

$\qquad\qquad\qquad\qquad\qquad\qquad\qquad\qquad\qquad\qquad (5.9)$

subject to   Equality constraints   (Eqn. 5.4)
                          Inequality constraints   (Eqn. 5.5 and 5.7)

    This constrained nonlinear optimisation problem can be solved using a Successive Quadratic Programming (SQP) algorithm. In the SQP, at each iteration of optimisation a quadratic program (QP) is formed by using a local quadratic approximation to the objective function and a linear approximation to the nonlinear constraints. The resulting QP problem is solved to determine the search direction and with this direction, the next step length of the decision variable is specified. See Chen (1988) for further details.

    Mujtaba and co-workers have extensively used the feasible path approach implemented by Morison (1984) in batch distillation. This is because of several reasons: the algorithm is general, can optimise design variables, initial conditions of the states in addition to control and control switching times. The optimisation of control switching times together with final time is most desirable to obtain time optimal operation (*minimum time* problem) in batch distillation. The optimisation of the initial conditions is also desirable in developing off-cut recycle policies (presented in Chapter 8). Also the algorithm allows other options to consider: fixed switching times and final time, variable switching times but fixed final time (which is desirable to obtain optimal operational policy by solving a *maximum distillate* problem), etc.

5.8. NLP Based Dynamic Optimisation Problem- Infeasible Path Approach

In this approach the set of DAEs presented in Equation 5.4 is discretised into a set of Algebraic Equations (AEs) by applying collocation method.

Cuthrell and Biegler (1989) considered the orthogonal collocation method which is described below. Two Lagrange polynomials one for the state variable (x) and one for the control variable (u) can be written as:

$$x_{k+1}(t) = \sum_{i=0}^{K} x_i \varphi_i(t) \qquad where, \varphi_i(t) = \prod_{k=0,i}^{K} \frac{(t-t_k)}{(t_i - t_k)} \qquad (5.10)$$

$$u_k(t) = \sum_{i=1}^{K} u_i \psi_i(t) \qquad where, \psi_i(t) = \prod_{k=1,i}^{K} \frac{(t-t_k)}{(t_i - t_k)} \qquad (5.11)$$

where the state vector approximation, x_{k+1}, is a $(K+1)^{th}$ order (degree $<K+1$) polynomial and the control vector approximation, u_k, is a Kth order (degree $<K$) polynomial. The notation $k = 0,i$ denotes k starting from zero and $k \neq i$. Also the Lagrange polynomial has the desirable property that $x_{k+1}(t_i) = x_i$. Discretisation of Equation 5.4 using Equations 5.10-5.11 will results in a set of algebraic equations of the form:

$$f_i(t_i, x_i, u_i, v) = 0; \qquad i = 1, \ldots, K \qquad (5.12)$$

In the NLP optimisation problem presented in section 5.7.2, the model equations (equality constraints) can now be replaced by a set of discrete AEs. Once the points $t_i, i = 1, \ldots, K$ are chosen, the NLP optimisation problem can then be solved using SQP based or other methods. Here, the location of these points corresponds to the shifted roots of an orthogonal Legendre polynomial of degree K (see Villadsen and Michelsen, 1978 for details). Cuthrell and Biegler assumed that all profiles are smooth (i.e. analytic functions in t). In the event when this assumption does not hold, Cuthrell and Biegler (1989) proposed to extend the NLP formulation to finite elements. Refer to Vasantharajan and Biegler (1990), Logsdon and Biegler (1993) for further details on finite element methods. Li et al. (1998a,b) used similar approaches in batch distillation and solved the NLP problems using SQP based methods. Hansen and Jorgensen (1986), Christensen and Jorgensen (1987), Logsdon and Biegler (1989), Logsdon et al. (1990) used collocation method to discretize ODEs resulting from the batch distillation model and/or the

Pontryagin's *Maximum Principle* to pose the dynamic optimisation problems as NLP problems. The problems were then solved by SQP based or other efficient methods.

5.9. Gradient Evaluation Methods in NLP Based Optimisation Techniques

Evaluation of gradients is one of the major tasks in NLP based optimisation techniques. Rosen and Luus (1991) reviewed a number of methods for the evaluation of gradients for dynamic optimisation (optimal control) problems which uses piecewise constant optimal control profile. Some of these methods are discussed here.

5.9.1. Gradient Evaluation for Infeasible Path Approach

When the ODE or DAE process models and the control policy are both discretized, as described in section 5.8, the gradients that are required for the NLP problems can be obtained directly from the resulting set of algebraic equations.

5.9.2. Gradient Evaluation for Feasible Path Approach

When only the control is discretized as described in section 5.7, integration of the model equations is required to evaluate the performance index and to obtain the gradients. The evaluation of such gradients will consume a significant part of the total computational time needed to solve the optimisation problem.

A number of methods can be found in the literature to evaluate the gradients that are required for the NLP problems:

- the finite difference method
- the adjoint system method (Sargent and Sullivan, 1979)
- the state trajectory sensitivity equations method (Li et al., 1990)

The adjoint system approach requires integration of the model equations forward in time before integration of the adjoint system equations backward.

On the other hand, the trajectory sensitivity equations method requires simultaneous integration of a greater number of equations than the adjoint system approach. However, it is more stable than the adjoint system approach due to the requirement of forward integration only. It is usually preferred in the area of parameter estimation and sensitivity (Kalogerakis and Luus, 1983; Caracotsios and

Stewart, 1985). Advent of high-speed computers has improved significantly the performance of the finite difference and the trajectory sensitivity approaches.

Here, three different gradient evaluation methods are discussed with reference to piecewise constant control discretization. If the final time is to be optimised then there are $J=2N_s$ variables in each control variable as shown in Equation 5.8. If r is the number of control variables in the system then the total number (σ) of decision variables in vector y (Equation 5.8) will be,

$$\sigma = 2rN_s \tag{5.13}$$

Note if the final time t_F is fixed then $\sigma = 2rN_s - 1$. If S is the number of end point constraints in Equation 5.5, each defined by:

$$a_i^F \le g_i(t_F, \dot{x}(t_F), x(t_F), u(t_F), v) \le b_i^F; \qquad i = 1, 2, ..., S \tag{5.14}$$

and if each parameter in vector y (Equation 5.8) is subject to linear bounds:

$$y_j^L \le y_j \le y_j^U; \qquad j = 1, 2, ..., \sigma \tag{5.15}$$

then there will be $S + \sigma$ number of constraints in the optimisation problem.

At the final time, the following function needs to be evaluated to obtain the objective function and the end point constraints values.

$$G_i(y) = g_i(t_F, \dot{x}(t_F), x(t_F), u(t_F), v); \qquad i = 1, 2, ..., S \tag{5.16}$$

Note, $G_0(y) = J(y)$ is the objective function shown in Equation 5.6.

Also note that if N_s is chosen sufficiently large, the piecewise constant optimal control policy will be sufficiently close to the continuous optimal control policy.

5.9.2.1. Finite Difference Method

If the analytical expressions for the derivatives of $f(t, \dot{x}(t), x(t), u(t), v) = 0$, are difficult to obtain, a forward difference approximation can be used so that:

$$\frac{\partial G_i}{\partial y_j} = \frac{G_i(y_1, y_2, ..., y_j + \Delta y_j, ..., y_\sigma) - G_i(y)}{\Delta y_j} \tag{5.17}$$

Use of central difference technique will give more accurate approximation but will require an extra function evaluation at the point ($y_j - \Delta y_j$).

The model Equation 5.4 must be integrated once with the nominal value of y to obtain the final state of the system to yield the functions $G_i(y), i = 0,1,....,S$. Similarly, for each change in the vector y, the model equations have to be integrated once to determine the perturbed final state. However, all these differential equations can be integrated simultaneously. Changes in (z_j) in $t \in [(t_{j-1}, t_j),$ do not affect the state trajectory for any t smaller than t_{j-1}, because (z_j) acts during the interval $t \in [(t_{j-1}, t_j)$. Therefore, to obtain the gradient $\partial G_i / \partial z_j$ it is required to integrate the respective model equations only from t_{j-1} to t_F using the final state of the model equations at t_{j-1} as the initial condition.

5.9.2.2. Adjoint System Method

Similar to those presented in section 5.5 the adjoint equations for the model represented by Equation 5.4 can be written as:

$$\frac{d\psi_i}{dt} = -\frac{\partial H}{\partial x} \qquad\qquad i = 0,1,2,......, S \qquad\qquad (5.18)$$

where, H is the Hamiltonian defined by:

$$H(\psi, t, \dot{x}, x, u, v) = \sum_{j=0}^{S} \psi_j f_j(t, \dot{x}, x, u, v) \qquad\qquad (5.19)$$

With the results of the forward integration of the model Equation 5.4 and then by integrating backward the adjoint equations (Equation 5.18) the gradients can be determined from:

$$\frac{\partial G_i}{\partial y_j} = \left(\frac{\partial g_i}{\partial t_F} + H(\psi, t, \dot{x}, x, z_N, v)t_F \right)\frac{\partial t_F}{\partial y_j} + \sum_{j=1}^{N} \int_{t_{j-1}}^{t_j} \left[\left(\frac{\partial H}{\partial u} \right)^T \frac{\partial z_j}{\partial y_j} \right] dt \quad (5.20)$$

See Sargent and Sullivan (1979), Morison (1984) and Rosen and Luus (1991) for further details. The adjoint approach has the advantage that, in addition to the adjoint equations (Equation 5.18), only one extra equation (Equation 5.20), has to be integrated for each of the NLP optimisation variables. It is especially useful for piecewise constant control scheme (all controls having the same time intervals) because then:

$$\frac{\partial G_i}{\partial z_j} = \int_{t_{j-1}}^{t_j} \left(\frac{\partial H_i}{\partial u} \right) dt, \qquad i = 0,1,....,S, \ \ j=1,2,...,N \tag{5.21}$$

and it is necessary to integrate only r equations (as opposed to rN_s) from 0 to t_F to obtain all these derivatives. However, for each constraint, it is necessary to integrate a set of adjoint equations. The total number of equations to be integrated will depend on the value of S.

5.9.2.3. State Trajectory Sensitivity

The changes in the state trajectories due to small changes in the NLP optimisation variables can be given by the vector equation:

$$\frac{ds_j(t)}{dt} = \left(\frac{\partial f^T}{\partial x} \right)^T s_j(t) + \left(\frac{\partial f^T}{\partial u} \right)^T \frac{\partial z_{k+1}}{\partial y_j} \tag{5.22}$$

$$t_k \le t < t_{k-1}, \qquad k = 0,1,....,N_s - 1, \qquad j = 1,2,....,\sigma,$$

where $s_j(t)$ is the trajectory sensitivity vector:

$$s_j(t) = \frac{\partial x(t)}{\partial y_j} \tag{5.23}$$

and has to satisfy, at the switching times, the vector equations:

$$s_j(t_0) = 0, \tag{5.24}$$

$$s_j(t_k^+) = s_j(t_k^-) + [f(t, \dot{x}, x, z_k, v) - f(t, \dot{x}, x, z_{k+1}, v)]_{t_k} \frac{\partial t_k}{\partial y_j} \tag{5.25}$$

$$k = 1,2,....,N_s - 1$$

The desired derivatives can be obtained by evaluating:

$$\frac{\partial G_i}{\partial y_j} = \left(\frac{\partial g_i}{\partial t_F}\right)\frac{\partial t_F}{\partial y_j} + \left(\frac{\partial g_i}{\partial x_F}\right)^T \left[s_j(t_F^-) + f(t_F, \dot{x}_F, x_F, z_N, v)\frac{\partial t_F}{\partial y_j}\right] \quad (5.26)$$

The model equation (Equation 5.4) is integrated only once simultaneously with the trajectory sensitivity equation (Equation 5.22) for each of the elements of y, to determine the final state sensitivity for all the gradients in equation (Equation 5.26), $i = 0, 1,....,S$. Changes in (z_j) in $t\varepsilon[(t_{j-1}, t_j)$, do not affect the state trajectory for any t smaller than t_{j-1}, because (z_j) acts during the interval $t\varepsilon[(t_{j-1}, t_j)$. Therefore, the gradient $\partial G_i / \partial z_j$ is obtained by integrating the respective trajectory sensitivity equations only from t_{j-1} to t_F using no sensitivity, at the initial condition at t_{j-1}.

See the work of Vassiliadis (1993) for gradient evaluation methods for linear and exponentially varying control profile. See Rosen and Luus (1991) for gradient evaluations with the time invariant parameters (v) optimised and for guidelines for selecting the appropriate gradient evaluation method.

5.10. Application of NLP Based Techniques in Batch Distillation

5.10.1. Example 1

Logsdon and Biegler (1993) considered a binary separation of cyclohexane-toluene mixture in a conventional batch distillation column. *Maximum distillate* problem was considered to maximise the amount of distillate with cyclohexane purity of 0.998 molefraction. The input data for the problem is given in Table 5.7.

Logsdon and Biegler (1993) considered shortcut no holdup model (Type II) and used infeasible path approach to solve the optimisation problem. The results reported by them are presented in Table 5.7 and Figure 5.3. For the sake of comparison, the same optimisation problem is solved here using the feasible path approach. A detailed model of Type IV (CMH) with piecewise constant (which are optimised) reflux ratio profile is used in this study. The results are shown in Table 5.7 and Figure 5.3. The differences in the amount of distillate and the optimum reflux ratio profiles using two different optimisation approaches are mainly due to the use of no holdup model and the use of constant relative volatility by Logsdon and Biegler.

Note using a holdup model, Logsdon and Biegler also reported an optimal reflux ratio profile for the same example but it was significantly different than that obtained by using no holdup model. Refer to the original reference for further details.

5.10.2. *Example 2*

Mujtaba (1989) considered the *minimum time* problem with a separation task (D^*= 1.16 kmol, x^*_D= 0.906). The task is same as those reported for ideal case in Table 4.8 (Chapter 4). The simulation used 4 reflux ratio levels including an initial total reflux operation. Mujtaba also used 4 reflux ratio levels to compare the simulation results. The lower and upper bounds on the reflux ratio are (0.3 and 1.0).

The simulation results reported in Table 4.8 used 2.54 hr of initial total reflux operation (also used in the experimental column by Nad and Spiegel, 1987) before any product was withdrawn from the column. Here, the aim was also to find out-

- whether an initial total reflux period was at all required or not for the given separation task
- and if it was, for how long the column was to be operated under the total reflux condition.

Table 5.7. Input Data and Results for Example 1

Input Data
N	= No. of Plates	= 10
B_0	= Initial Feed	= 100.0 mole
x_{B0}	= Initial Feed Composition	= <0.55, 0.45>
H_j	= Plate Holdup	= 1 mole
H_c	= Condenser Holdup	= 1 mole
V	= Vapour Boilup	= 120 mole/hr
t_F	= Batch Time (fixed)	= 1 hr
α	= Relative Volatility used by Shortcut Model	= 2.85

VLE by SRK Method in the Detailed Model

Maximum Amount of Distillate. mole

a. Using Shortcut Model and Infeasible Path Approach
   and as reported by Logsdon and Biegler   = 37.029

b. Using Detailed Model (Type IV CMH)
   and Feasible Path Approach   = 33.536

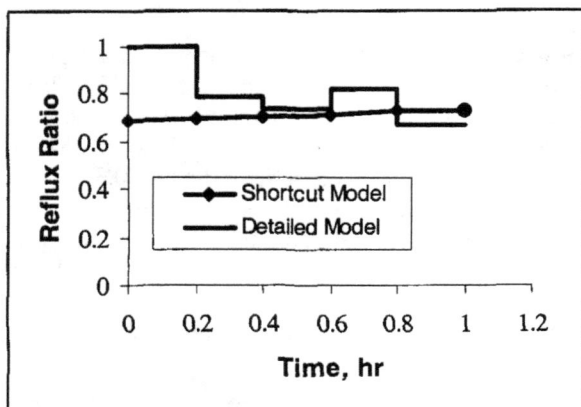

Figure 5.3. Reflux Ratio Profiles using Shortcut and Detailed Model

The optimisation results are summarized in Table 5.8. Figure 5.4 shows the reflux ratio profiles used in the simulation and those obtained by optimisation. Figure 5.5 shows the optimal accumulated and instant distillate composition profiles. The reflux ratio profile is increasing with time as expected. The results clearly show the benefit of optimising the reflux ratio. As can be seen from Table 5.8 the operation time is reduced by at least 50% compared to that in the simulation and experiment. The results also show that for the given separation task an initial total reflux operation is not at all required and the product can be collected from the very beginning of the process (see notes in section 3.3.1).

Table 5.8. Results of Example 2

	Objectives set (Separation Task)	Objectives attained
Amount of Distillate, D^*	1.16 kmol	1.16
Cyclohexane Composition, x^*_D	0.906 molefraction	0.906
Batch Time	5.16$^\$$ hr	2.61
% Operation Time Saved (compared to simulation)		= 50.58

$^\$$batch time as used in the simulation and in the experiment by Nad and Spiegel

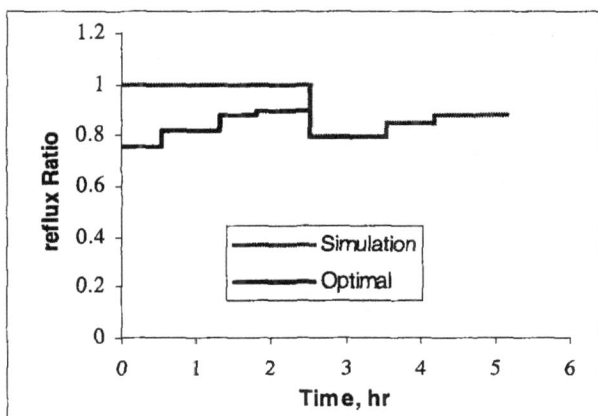

Figure 5.4. Reflux Ratio Profiles (Example 2)

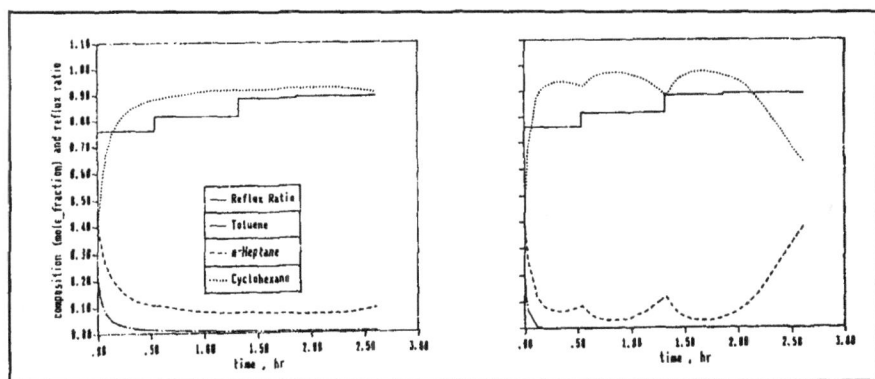

Figure 5.5. Accumulated (left) and Instant Distillate Composition Profiles (Example 2). [Mujtaba, 1989]

5.10.3. Example 3

Table 4.6 in Chapter 4 presents the simulation results for a quaternary batch distillation. The amount of product and the composition of key component of each cut were used by Mujtaba (1989) to formulate and solve a *minimum time* problem for each cut. Optimal reflux ratio in each operation step is obtained independently of other step with the final state of each step being the initial state of the next step.

Two cases are studied. For Case a, single time interval in each step is used, yielding an optimal constant reflux policy. For Case b, five time intervals are used in each step. Table 5.9 summarises the results using single constant (optimised) and variable constant (optimised) reflux ratio profiles. The instant and accumulated distillate profiles together with the reflux ratio profiles are shown in Figures 5.6 and 5.7.

It is clear from Table 5.9 that the results obtained are in very good agreement with the objectives set for each individual optimisation problem. Table 5.9 also cleary shows the advantages of optimal reflux policies over the conventional constant reflux operation. Table 5.9 shows that the time optimal control policy (variable reflux) saves about 63% of the operation time compared to that required in the simulation (Table 4.6). Even the time optimal constant reflux policy saves about 33% of the operation time compared to the original simulation

Table 5.9. Summary of Multi-CUT *Minimum Time* Optimisation Problem.
[Mujtaba, 1989]

		Objectives set (D^*, x^*_D)	Objectives attained	
			Case a	Case b
OPERATION	1. Distillate Amount, lbmol	8.139	8.117	8.117
STEP 1	2. Propane molefraction	0.981	0.981	0.981
	3. Batch time, hr	(4.07) @	4.01	2.82
STEP 2	1. Distillate Amount, lbmol	11.76	11.737	11.737
and	2. Propane molefraction	0.85	0.85	0.85
CUT 1	3. Batch time, hr	(1.81)	1.56	1.37
STEP 3	1. Distillate Amount, lbmol	36.548	36.566	36.566
and	2. Butane molefraction	0.988	0.988	0.988
CUT 2	3. Batch time	(18.27)	9.2	2.57
STEP 4	1. Distillate Amount, lbmol	8.619	8.62	8.62
	2. Pentane molefraction	0.940	0.941	0.941
	3. Batch time, hr	(4.31)	3.49	2.83
STEP 5	1. Bottom Amount, lbmol	59.38 (B^*)	59.44	59.44
and	2. Hexane molefraction	0.998 (x^*_B)	0.9972	0.9972
CUT 3	3. Batch time, hr	(1.78)	1.55	1.54

Note: Case a represents optimal constant reflux policy
Case b represents optimal variable reflux policy
@ Batch times within brackets are those used in the simulation (Table 4.6, Chapter 4)

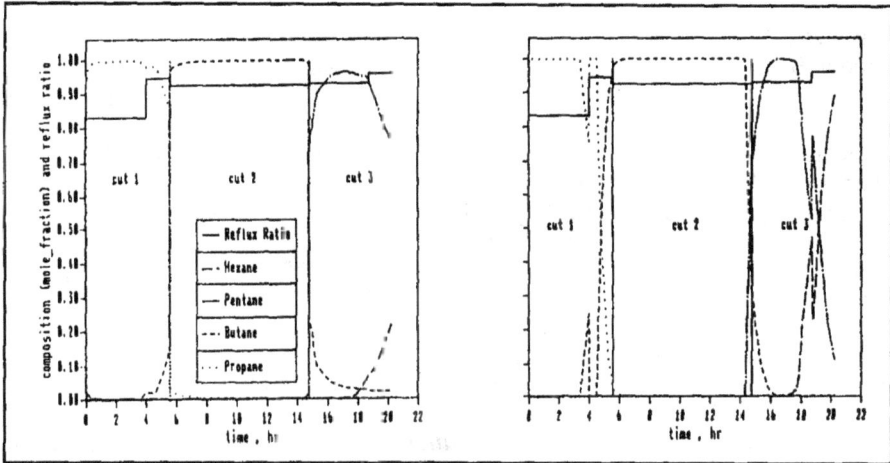

Figure 5.6. Accumulated (left) and Instant Distillate Composition and Reflux Ratio Profiles (Case a). [Mujtaba, 1989]

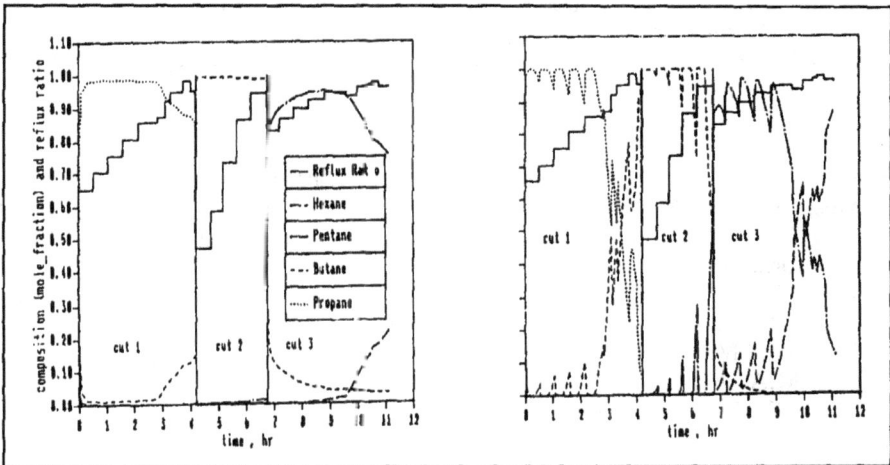

Figure 5.7. Accumulated (left) and Instant Distillate Composition and Reflux Ratio Profiles (Case b). [Mujtaba, 1989]

References

Bernot, C., Doherty, M.F. and Malone, M.F., *Chem. Eng. Sci.* **45** (1990), 1207.

Betlem, B.H.L., Krijnsen, H.C. and Huijnen, H., *Chem. Eng. J.* **71** (1998), 111.

Bonny, L., Domenech, S., Floquet, P. and Pibouleau, L., *Chem. Engng. Proc.* **35** (1996), 349.

Caracotsios, M. and Stewart, W. E., *Comput. chem. Engng.* **9** (1985), 359.

Chen, C.L., *A Class of Successive Quadratic Programming Methods for Flowsheet Optimisation*. PhD Thesis, (Imperial College, London, 1988).

Coward, I., *Chem. Eng. Sci.* **22** (1967), 503.

Converse, A.O. and Huber, C.I., *IEC Fund.* **4** (1965), 475.

Converse, A.O. and Gross, G.D., *IEC Fund.* **2** (1963), 217.

Christensen, F.M. and Jorgensen, S.B., *Chem. Eng. J.*, **34** (1987), 57.

Cuthrell, J.E. and Biegler, L.T., *AIChE J.* **33** (1987), 1257.

Domenech, S. and Enjalbert, M., *Comput. chem. Engng.* **5** (1981), 181.

Diwekar, U.M., *AIChE J.* **38** (1992), 1571.

Diwekar, U.M., *Batch Distillation: Simulation, Optimal Design and Control* (Taylor and Francis, Washington, D.C., 1995).

Diwekar, U.M., Malik, R.K. and Madhavan, K.P., *Comput. chem. Engng.* **11** (1987), 629.

Diwekar, U.M., Madhavan, K.P. and Swaney, R.E., *IEC. Res.* **28** (1989), 1011.

Diwekar, U.M. and Madhavan, K.P., *IEC. Res.* **30** (1991a), 713.

Diwekar, U.M. and Madhavan, K.P., *Comput. chem. Engng.* **15** (1991b), 833.

Edgar, T.F. and Himmelblau, D.M., *Optimization of Chemical Processes* (McGraw-Hill, 1983).

Farhat, S., Czernicki, M., Pibouleau, L. and Domenech, S., *AIChE J.* **36** (1990), 1349.

Furlonge, H.I., Pantelides, C.C. and Sorensen, E., *AIChE J.* **45** (1999), 781.

Greaves, M. A., *Hybrid Modelling, Simulation and Optimisation of Batch Distillation Using Neural Network Techniques*. Ph.D. Thesis, (University of Bradford, Bradford, UK, 2003).

Greaves, M. A., I. M. Mujtaba and M. A. Hussain (2001). in *Application of Neural Network and Other Learning Technologies in Process Engineering*, eds. Mujtaba, I.M. and Hussain, M.A. (Imperial College Press, London, 2001), 149.

Greaves, M.A., Mujtaba, I. M., Barolo, M., Trotta, A. and Hussain, M. A., *Trans. IChemE*, **81A** (2003), 393.

Hansen, T.T. and Jorgensen, S.B., *Chem. Eng. J.* **33** (1986), 151.

Huckaba, C.E. and Danly, D.E., *AIChE. J.* **6** (1960), 335.

Jang, S., *Chem. Eng. J.* **51** (1993), 83.

Kalogerakis, N. and Luus, R., *IEC Fund.* **22** (1983), 436.

Kerkhof, L.H. and Visseres, H.J.M., *Chem. Eng. Sci.* **33** (1978), 961.

Li, W.C., Biegler, L.T., Economou, C.G. and Morari, M., *Comput. chem. Engng.* **14** (1990), 451.

Li, P., Hoo, H.P. and Wozny, G., *Chem. Eng. Technol.* **21** (1998a), 853.

Li, P., Garcia, H.A., Wozny, G. and Reuter, E., *IEC Res.* **37** (1998b), 1341.

Logsdon, J.S., *Efficient determination of optimal control profiles for differential and algebraic systems*. PhD Thesis, (Carnegie Mellon University, USA, 1990).

Logsdon, J.S., and Biegler, L.T., *IEC Res.* **28** (1989), 1628.

Logsdon, J.S.and Biegler, L.T., *IEC Res.* **32** (1993), 700.

Logsdon, J.S., Diwekar, U.M. and Biegler, L.T., *Trans IChemE*, **68A** (1990), 434.

Mayur, D.N., May, R.A., and Jackson, R., *Chem. Eng. J.* **1** (1970), 15.

Mayur, D.N. and Jackson, R., *Chem. Eng. J.* **2** (1971),150.

Morison, K.R., *Optimal Control of Processes Described by Systems of Differential and Algebraic Equations*. PhD. Thesis, (Imperial College, University of London, 1984).

Mujtaba, I.M., *Optimal Operational Policies in Batch Distillation*. PhD Thesis, (Imperial College, University of London, 1989).

Mujtaba, I.M., *Trans. IChemE*, **75A** (1997), 609.

Mujtaba, I.M., *Trans. IChemE*, **77A** (1999), 588.

Mujtaba, I.M. and Macchietto, S. In *Recent Progres en Ginie de Procedes*, eds. Domenech, S. et al. (Lavoisier Techniche et Documentation, Paris, 1988) **2**, 191.

Mujtaba, I.M. and Macchietto, S., in *IMACS Annals on Computing and Applied Mathematics, 4: Computing and Computers for Control Systems*, eds. Borne, P. et al. (J.C. Baltzer AG, Scientific Publishing Co. Basel Switzerland, 1989), 55.

Mujtaba, I.M. and Macchietto, S., *Comput. chem. Engng.* **16S** (1992), S273.

Mujtaba, I.M. and Macchietto, S., *Comput. chem. Engng.* **17** (1993), 1191.

Mujtaba, I.M. and Macchietto, S., *J. Proc. Control.* **6** (1996), 27.

Mujtaba, I.M. and Macchietto, S., *IEC Res.* **36** (1997), 2287.

Mujtaba, I.M. and Macchietto, S., *Chem. Eng. Sci.* **53** (1998), 2519.

Murty, B.S.N., Gangiah, K and Hussain, A., *Chem. Eng. J.* **19** (1980), 201.

Pontryagin, L.S., Boltyanskii, V.G., Gamkrelidze, R.V. and Mishchenko, E.F., *The Mathematical Theory of Optimal Processes* (English Translation by D.E. Brown), (Macmillan, New York, 1964).

Reklaitis, G.V., Ravindran, A. and Ragsdell, K.M., *Engineering Optimization : Methods and Applications* (Wiley, New York, 1983).

Renfro, J.G., Morshedi, A. M. and Asbjornsen, O.A., *Comput. chem. Engng.* **11** (1987), 503.

Robinson, E.R., *Chem Eng Sci.* **24** (1969), 1661.

Robinson, E.R., *Chem Eng Sci.* **25** (1970), 921.

Rosen, O. and Luus, R., *Comput. chem. Engng.* **15** (1991), 273.

Sargent, R.W.H. and Sullivan, G.R., *IEC Proc. Des. Dev.* **18** (1979), 113.

Sorensen, E. and Skogestad, S., *Chem. Eng. Sci.* **51** (1996), 4949.

Vasantharajan, S. and Biegler, L.T., *Comput. chem. Engng.* **14** (1990), 1083.

Vassiliadis, V.S., *Computational Solution of Dynamic Optimization Problems with General Differential-Algebraic Constraints*. PhD thesis, (Imperial College, London, 1993).

Villadsen, J. and Michelsen, M.L., *Solution of Differential Equation Models by Polynomial Approximation* (Prentice-Hall, 1978).

Wajge, R.M. and Reklaitis, G.V., *Chem. Eng. J.* **75** (1999), 57.

Zamar, S.D., E. Salomone and O.A., *IEC Res.* **37** (1998), 4801.

CHAPTER 6

6. MULTIPERIOD OPERATION OPTIMISATION

6.1. Introduction

In batch distillation, as the overhead composition varies during operation, a number of main-cuts and off-cuts are made at the end of various distillation tasks or periods (see Chapter 3). Purities of the main-cuts are usually determined by the market or downstream process requirements but the amounts recovered must be selected based on the economic trade off between longer distillation times (hence productivity), reflux ratio levels (hence energy costs), product values, etc. Increasing the recovery of a particular species in a particular cut may have strong effects on the recovery of other species in subsequent cuts or, in fact, on the ability to achieve at all the required purity specifications in subsequent cuts. The profitable operation of such processes therefore requires consideration of the whole (multiperiod) operation.

Recall from Chapter 1 that, a single mixture (binary or multicomponent) can be separated into several products (*single separation duty*) and multiple mixtures (binary or multicomponent) can be processed, each producing a number of products (*multiple separation duties*) using only one CBD column thus leading to multiperiod operation in both cases.

For *single separation duty*, Diwekar et al. (1989) considered the multiperiod optimisation problem and for each individual mixture selected the column size (number of plates) and the optimal amounts of each fraction by maximising a profit function, with a predefined conventional reflux policy. For multicomponent mixtures, both single and multiple product options were considered. The authors used a simple model with the assumptions of equimolal overflow, constant relative volatility and negligible column holdup, then applied an extended shortcut method commonly used for continuous distillation and based on the assumption that the batch distillation column can be considered as a continuous column with changing feed (see Type II model in Chapter 4). In other words, the bottom product of one time step forms the feed of the next time step. The pseudo-continuous distillation model thus obtained was then solved using a modified Fenske-Underwood-Gilliland method (see Type II model in Chapter 4) with no plate-to-plate calculations. The short-cut method, also used by Diwekar et. al. (1989) and Diwekar and Madhavan (1991) requires the specification of the mole fraction of all components in each product cut in addition to that of the key component recovered in that cut. In practice it is very difficult to achieve a specification of this type with a multicomponent mixture and considerable differences in the results may be noticed compared with the case of variable non-key product composition (Diwekar and Madhavan, 1991). In fact, for a particular mixture and operating policy (reflux ratio, column pressure, etc.) the residue/distillate composition will follow well defined distillation maps (Bernot et al., 1990) and it is in general not possible to independently specify more than one distillate mole fraction. Furthermore, the shortcut method is limited to columns with large number of plates and with no holdup.

For *single separation duty*, Farhat et al. (1990) considered the operation of an existing column for a fixed batch time and aimed at maximising (or minimising) the amount of main-cuts (or off-cuts) while using predefined reflux policies such as constant, linear (with positive slope) and exponential reflux ratio profile. They also considered a simple model with negligible liquid holdup, constant molar overflow and simple thermodynamics, but included detailed plate to plate calculations (similar to Type III model).

For *single separation duty*, Al-Tuwaim and Luyben (1991) proposed a shortcut method to design and operate multicomponent batch distillation columns. Their method, however, required a great number of simulations, which must be computationally very expensive, before they could arrive at an optimum design and find an optimum reflux ratio. Further details are in Chapter 7.

For *single separation duty*, Bernot et al. (1991) presented a method to estimate batch sizes, operating times, utility loads, costs, etc. for multicomponent batch distillation. The approach is similar to that of Diwekar et al. (1989) in the sense that a simple short cut technique is used to avoid integration of a full column model. Their simple column model assumes negligible holdup and equimolal overflow. The authors design and, for a predefined reflux or reboil ratio, minimise the total annual cost to produce a number of product fractions of specified purity from a multicomponent mixture.

For *single separation duty*, Mujtaba and Macchietto (1993) proposed a method, based on extensions of the techniques of Mujtaba (1989) and Mujtaba and Macchietto (1988, 1989, 1991, 1992), to determine the optimal multiperiod operation policies for binary and general multicomponent batch distillation of a given feed mixture, with several main-cuts and off-cuts. A two level dynamic optimisation formulation was presented so as to maximise a general profit function for the multiperiod operation, subject to general constraints. The solution of this problem determines the optimal amount of each main and off cut, the optimal duration of each distillation task and the optimal reflux ratio profiles during each production period. The outer level optimisation maximises the profit function by manipulating carefully selected decision variables. These are chosen in such a manner that the need for specifying the mole fractions of all the components in the products, as required by previous methods is avoided. For values of the decision variables fixed by the outer loop, the multiperiod operation is decomposed into a sequence of independent optimal control problems, one for each distillation task. In the inner loop, a *minimum time* problem is then solved for each task to generate the optimal reflux ratio values, reflux switching times, and duration of the task. The procedure permits the use of very general distillation models (as those presented in Chapter 4) described by Differential and Algebraic Equations (DAEs), including rigorous thermodynamics if desired The model equations are integrated by using an efficient Gear's type method; the inner loop optimal control problems are solved using NLP based optimisation technique described in Chapter 5.

Mujtaba and Macchietto (1996) extended the work or Mujtaba and Macchietto (1993) to include *multiple separation duties*. Logsdon et al. (1990) also considered *multiple separation duties* using short-cut model. Some of these works are presented in Chapter 7.

In sections 6.2 to 6.4, the formulation and solution method proposed by Mujtaba and Macchietto (1993) will be presented in detail. Several example problems (involving binary and multicomponent mixtures) from Mujtaba and Macchietto are also presented to demonstrate the idea. Operational alternatives involving separations of binary and multicomponent mixtures are presented in detail in Chapter 3.

In section 6.5, the multiperiod optimisation problem formulation considered by Farhat et al. (1990) is presented with typical example problems.

6.2. Optimisation Problem Formulation- Mujtaba and Macchietto

The dynamic optimisation problem formulation is illustrated for representative multiperiod operations. The STNs in Figures 6.1 and 6.2 for binary and ternary mixtures undergoing *single separation duty* describe the multiperiod operations (see Chapter 3). For other networks, mixtures with larger number of components and other constraints the problem formulation requires only simple modifications of that presented in this section.

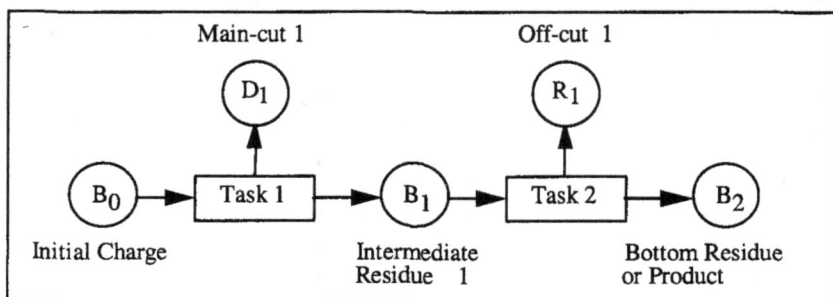

Figure 6.1. STN with One Main-Cut and One Off-cut (Binary Mixture). [Mujtaba and Macchietto, 1993][a]

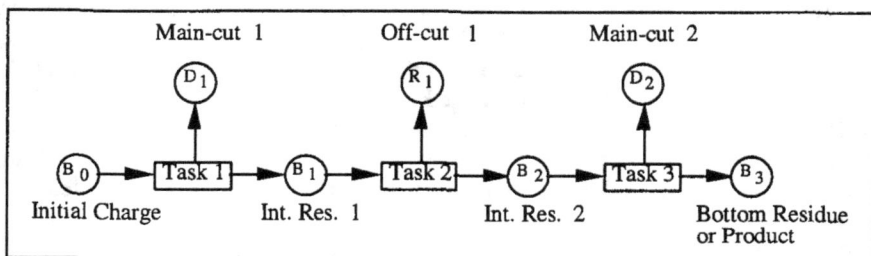

Figure 6.2. STN with Two Main-cuts and One Off-cut (Ternary Mixture). [Mujtaba and Macchietto, 1993][a]

6.2.1. *Binary Operation*

6.2.1.1. Degrees of Freedom Analysis

Refer to the STN shown in Figure 6.1. Given, B_0 and x_{B0} as the initial amount and composition of a mixture, we wish to obtain the main-cut 1 with specified purity in terms of the mole fraction of component 1 (x_{D1}^{1*}). The intermediate residue (B_1, x_{B1}) is further distilled off to obtain the off-cut 1 so as to satisfy the specification on the final bottom product composition for component 2 (x_{B2}^{2*}). Let D_1 denote the amount

---

[a] Reprinted from *Computers & Chemical Engineering*, **17**, Mujtaba, I.M. and Macchietto, S., Optimal operation of multicomponent batch distillation- multiperiod formulation and solution, 1191-1207, Copyright (1993), with permission from Elsevier Science .

of main-cut 1, (R_1, x_{R1}) denote the amount and composition of off-cut 1 and B_2 denote the amount of final bottom product (or residue).

The system can be considered with one input state (initial charge) and three output states (main-cut 1, off-cut 1, and bottom residue) defined by (B_0, x_{B0}), (D_1, x_{D1}), (R_1, x_{R1}) and (B_2, x_{B2}), respectively. For a mixture with n_C components each state is characterised by total amount of material and component mole fractions (n_C + 1 variables), with one summation equation ($\Sigma x^i = 1$), hence by n_C independent variables (two in this binary case). Since there is no accumulation of the intermediate residue an overall mass balance for this system gives n_C equations:

$$B_0 x_{B0}^i = D_1 x_{D1}^i + R_1 x_{R1}^i + B_2 x_{B2}^i; \qquad i = 1,2,...,n_c, \qquad n_c = 2 \quad (6.1)$$

For a given feed charge (n_C specifications), a degree of freedom analysis shows that there are $4n_C$ (state variables) - n_C (feed specifications) - n_C (mass balance) = $2n_C$ degrees of freedom.

Additional variables and defining equations may be introduced, if necessary, without changing the degree of freedom of this system. For example, a recovery of component 1 (Re_{D1}^1) in main-cut 1 over the distillation task 1 can be defined as:

$$D_1 x_{D1}^1 = Re_{D1}^1 B_0 x_{B0}^1 \qquad (6.2)$$

If one specifies the main-distillate and bottom residue purity (as x_{D1}^{1*} and x_{B2}^{2*}, respectively), two additional independent specifications are possible.

The choice of Re_{D1}^1 and $\underline{x}_{R1}^1$ as the two additional decision variables enables the solution of Equations 6.1-6.2 for the remaining unknowns as a function of these two variables, and in particular for the amounts of main-cut and off-cut to be produced. Let $\underline{D}_1$ and $\underline{R}_1$ denote the values obtained from specific values of Re_{D1}^1 and $\underline{x}_{R1}^1$. The whole multiperiod operation can then be decomposed into a sequence of 2 independent dynamic optimisation problems. In the first problem one can optimise the task 1 operation to obtain main-cut 1 of amount $\underline{D}_1$ and purity x_{D1}^{1*}. The final state of the first period gives the initial state for the second period. The values of the variables in the model at the final time of task 1 give the initial conditions of all variables for task 2. In the second problem, the task 2 operation is optimised to obtain off-cut 1 of amount $\underline{R}_1$ and purity $\underline{x}_{R1}^1$. Solution of these two problems permits the evaluation of performance measures and constraints for the overall operation (total time, productivity, etc.) as a function of the selected product purity specifications and decision variables.

6.2.1.2. Optimisation Problem Formulation

The optimisation problem formulation for the multiperiod operation given in Figure 6.1 can now be written as follows:

Problem P1

<u>Outer Loop Optimisation:</u>

P1-0   Max P
        Re_{D1}^1 and x_{R1}^1

         subject to $Re_{D1}^L \leq Re_{D1}^1 \leq Re_{D1}^U$

                             $x_{R1}^L \leq x_{R1}^1 \leq x_{R1}^U$

where P is a suitable overall profit function (Mujtaba and Macchietto, 1993) defined as:

$$P = \frac{C_{D1}D_1 + C_{B2}B_2 + C_{R1}R_1 - C_{B0}B_0}{t_1 + t_2} - C_{fc} \qquad (\$/hr) \qquad (6.3)$$

The recovery of component 1 (Re_{D1}^1) in main-cut 1 has constant lower and upper bounds Re_{D1}^L and Re_{D1}^U and the composition of component 1 in off-cut 1 (x_{R1}^1) has x_{R1}^L and x_{R1}^U as the constant lower and upper bounds. C_{D1}, C_{B2} and C_{R1} are the prices of the main-cut, bottom product and off-cut 1 product (\$/kmol) respectively. C_{B0} is the cost (\$/kmol) of the fresh feed mixture and C_{fc} is a fixed operating cost (\$/hr). The production times for task 1 and task 2 (t_1 and t_2, respectively) and the amounts of the product states (eg. D_1, R_1, etc.) are, of course, the functions of the main decision variables, and their values are given by the following inner loop optimisation problems solved in sequence.

Note, the profit function can be as general as desired (Logsdon et al., 1990), to reflect both annualised investment (e.g. column, reboiler, condenser, etc.) and operating cost (e.g. steam).

Inner Loop - Task 1 Optimisation:

With a given initial charge (B_0, x_{B0}) and for the current value of the outer loop decision variables and specifications:

i)   calculate $\underline{D}_1$ from Equation 6.2.

ii)   determine an optimal reflux ratio profile to obtain the required main-cut-1 $(\underline{D}_1, x_{D1}^{1*})$ in minimum time.

P1-1   Min t_1
            $r_1(t)$

        subject to $D_1(t_1) \geq \underline{D}_1$
                          $x_{D1}^1(t_1) \geq x_{D1}^{1*}$
                          $f(t, \dot{x}(t), x(t), u(t), v) = 0$

where $r_1(t)$ is the reflux ratio profile for the first period, and $f(t, \dot{x}(t), x(t), u(t), v) = 0$ is the set of DAEs representing the column model.

Inner Loop - Task 2 Optimisation:

With an intermediate residual (B_1, x_{B1}) being left over after the first inner loop problem and for the current value of the outer loop decision variables and specifications:

i)   calculate $\underline{R}_1$ from Equation 6.1

ii)   determine the optimal reflux profile $r_2(t)$ during task 2 to obtain the required off-cut 1 $(\underline{R}_1, x_{R1}^1)$ in minimum time.

P1-2   Min t_2
            $r_2(t)$

        subject to $R_1(t_2) \geq \underline{R}_1$
                          $x_{R1}^1(t_2) \geq x_{R1}^1$
                          $f(t, \dot{x}(t), x(t), u(t), v) = 0$

Note that if x_{R1}^1 is also specified, then the outer loop only optimises Re_{D1}^1 (the recovery of component 1 in main-cut 1) and vice versa.

Mujtaba and Macchietto used both Type III and IV-CMH models in the optimisation framework. This was to show that many models could be accommodated within the proposed optimisation framework, depending on the degree of "rigorousness" needed.

6.2.2. Multicomponent Operation

6.2.2.1. Degrees of Freedom Analysis

The degrees of freedom $(d.o.f)$ analysis is presented with reference to the STN given in Figure 6.2 which can then be generalised easily for other STNs.

Given, B_0 and x_{B0} as the initial amount and composition of a mixture, we wish to obtain 2 main-cuts of desired purity in terms of the key component mole fractions x_{D1}^{1*} (for main-cut 1) and x_{D1}^{2*} (for main-cut 2) with an off-cut with composition to be determined by the optimisation. As the bottom residue (B_3, x_{B3}) is not a desired product, hence no further separation task is considered after task 3.

To identify the maximum number of equality specifications that may be imposed on the problem, a steady state $d.o.f$ analysis could be carried out as for the binary case. For the linear networks considered here and assuming the feed charge is specified, the $d.o.f$ is (n_c x the number of distillation tasks). This analysis is, however, not very useful with multicomponent mixtures. One must take into consideration that additional constraints are imposed by the dynamics of the distillation process. For a particular mixture and operating policy (reflux ratio, column pressure, etc.), the residue and distillate composition will follow well-defined distillation maps (Bernot et al., 1990). Therefore, in general, it is not possible to independently specify more than one product (distillate or bottom) mole fraction and one intensive quantity (amount or recovery) per distillation task. The STN given in Figure 6.2 is a combination of three basic STNs, two main-cuts and one off-cut modules. If we define for each module two independent specifications-

- the mole fraction of the key component to be recovered (e.g. x_{D1}^{1*} in task 1, x_{R1}^{1} in task 2 and x_{D2}^{2*} in task 3)
- the recovery of the same key component with reference to the input state to the task (eg. Re_{D1}^{1} in task 1, Re_{R1}^{1} in task 2 and Re_{D2}^{2} in task 3, defined in Equations 6.4 and 6.5)

then, for a given feed charge, one can optimise each distillation task independently and sequentially, as for the binary case. The above decision variables may in turn be chosen in an outer optimisation loop, to optimise an overall objective function and to meet constraints involving variables from more than a single task. If some variables are fixed, (for example, the mole fraction of the key component in each main-cut) these can be eliminated from the decision variable list in the outer loop optimisation.

The recovery of component 1 in main-cut 1 (Re^1_{D1}) was defined in Equation 6.2. The recoveries for each subsequent operation task can be defined as:

Recovery of component 1 in off-cut 1 (Re^1_{R1}):

$$R_1 x^1_{R1} = Re^1_{R1}(B_0\, x^1_{B0} - D_1\, x^1_{D1})$$ (6.4)

Recovery of component 2 in main-cut 2 (Re^2_{D2}):

$$D_2 x^2_{D2} = Re^2_{D2}(B_0 x^2_{B0} - D_1 x^2_{D1} - R_1 x^2_{R1})$$ (6.5)

With a given values of Re^1_{D1} and x^{1*}_{D1} Equation 6.2 gives the amount, $\underline{D}_1$, of main-cut to be produced in task 1. Now one can optimise task 1 independently and the solution will give the mole fractions of all species in main-cut 1 and in the intermediate residue 1, the final value of all model variables at the end of task 1, and hence initial conditions for task 2. With a choice of Re^1_{R1} and x^1_{R1}, Equation 6.4 gives the amount of off-cut, $\underline{R}_1$, to be produced in task 2. The task 2 operation can now be independently optimised to obtain the compositions of off-cut 1 and intermediate residue 2, the final value of all model variables at the end of task 2, and hence initial conditions for task 3. Similarly with a choice of Re^2_{D2} and x^{2*}_{D2}, Equation 6.5 gives the amount, $\underline{D}_2$, of main-cut to be produced in task 3 and this will allow optimisation of task 3 independently. For the chosen outer loop decision variables it is thus possible to partition the whole multiperiod operation into a sequence of independent dynamic optimisation problems.

In general, if the amount and composition of the feed state are known then fixing two degrees of freedom (as mentioned above) for each distillation task (STN module) determines the thermodynamic distillation map which should be followed and results in a distillation profile which is well defined. The technique presented is general and can be easily extended to other networks involving larger number of operation tasks and components.

6.2.2.2. Optimisation Problem Formulation

The optimisation problem formulation for the operation shown in Figure 6.2 is presented in the following. Extension of the procedure illustrated to more components is a straightforward task.

Problem P2

Given: (B_0, x_{B0}), x_{D1}^{1*}, x_{D2}^{2*}

Outer Loop Optimisation:

P2-0   Max P
$\qquad$ Re_{D1}^1, Re_{R1}^1, x_{R1}^1, Re_{D2}^2

$\qquad\qquad$ subject to $\qquad$ $Re_{D1}^L \leq Re_{D1}^1 \leq Re_{D1}^U$

$\qquad\qquad\qquad\qquad\qquad$ $Re_{R1}^L \leq Re_{R1}^1 \leq Re_{R1}^U$

$\qquad\qquad\qquad\qquad\qquad$ $x_{R1}^L \leq x_{R1}^1 \leq x_{R1}^U$

$\qquad\qquad\qquad\qquad\qquad$ $Re_{D2}^L \leq Re_{D2}^2 \leq Re_{D2}^U$

where profit P is a suitable overall profit function, for example:

$$P = \frac{C_{D1}\,D_1 + C_{D2}\,D_2 + C_{B3}\,B_3 + C_{R1}\,R_1 - C_{B0}\,B_0}{t_1 + t_2 + t_3} - C_{fc} \quad (\$/hr) \quad (6.6)$$

All the decision variables are subject to constant lower and upper bounds. C_{D1}, C_{D2}, C_{B3} and C_{R1} are the prices of the main-cuts, bottom and off-cut 1 product ($/kmol). C_{B0} is the cost ($/kmol) of the fresh feed mixture and C_{fc} is a fixed operating cost ($/hr). The production times for task 1, task 2 and task 3 (t_1, t_2 and t_3, respectively) and the amounts of the product states (eg. D_1, R_1, D_2, etc.) are again functions of the main decision variables, and their values are given by the following optimisation problems solved in sequence.

Inner Loop - Task 1 Optimisation:

$\quad$ i) $\qquad$ calculate D_1 from Equation 6.2.
$\quad$ ii) $\qquad$ solve problem P2-1 (the same as P1-1)

Inner Loop - Task 2 Optimisation:

    i)   calculate $\underline{R}_l$ from Equation 6.4
    ii)   solve problem P2-2 (the same as P1-2)

Inner Loop - Task 3 Optimisation:

Given an intermediate residual $(B_2,\ x_{B2})$ (left after the second inner loop problem) and for the current value of the outer loop decision variables and specifications:

    i)   calculate $\underline{D}_2$ from Equation 6.5.
    ii)   determine the optimal reflux profile $r_3(t)$ in task 3 which yields main-cut-2 $(\underline{D}_2,\ x_{D2}^{2*})$ in minimum time.

P2-3   Min t_3
             $r_3(t)$

        subject to $D_2\,(t_3) \geq \underline{D}_2$
$$x_{D2}^2\,(t_3) \geq x_{D2}^{2*}$$
$$f(t, \dot{x}(t), x(t), u(t), v) = 0$$

Problem P2-3 is formally similar to P2-1 and P2-2. These are all solved using the same solution procedure. Any of the outer loop variables which have known or fixed values may be dropped from the outer loop optimisation.

Note that while recoveries and product purity specifications for the key components are selected as independent decision variables in the outer loop, these should not be assigned arbitrary values. It is possible that certain purities may not be achievable with the current column configuration (this can be checked against the minimum number of plates, N_{min} required for the given separation, Chapter 3), or due to the presence of azeotropes, etc. At the very least, the specifications must be consistent with the amounts of the various species present in the initial charge and in the feed state to each separation task (i.e. the overall mass balance must be satisfied). Thus care must be taken in selecting the outer loop specifications and bounds.

6.3. Solution Method – Mujtaba and Macchietto

6.3.1. Column Initialisation

The column initialisation is only required for the first inner loop optimisation problem (described in section 6.2). The liquid composition on the plates, condenser holdup tank and in the distillate accumulator (differential variables) at time $t = 0$ are set equal to the fresh charge composition (x_{B0}) to the reboiler. The DAE model equations are solved at time $t = 0$ to provide a consistent initialisation of all the remaining variables. The final values of all these variables at the end of the distillation task in each inner loop problem are stored and used for column initialisation for the subsequent inner loop optimisation problems. At the beginning of each task, the distillate accumulator holdup is set/reset to zero.

6.3.2. Inner Loop Optimisation Problems

Mujtaba and Macchietto (1993) used the solution method described by Mujtaba (1989) and Mujtaba and Macchietto (1991, 1998), based on the work by Morison (1984) to solve each inner loop optimisation problem presented in the previous section. The dynamic optimisation problem is posed as a nonlinear programming problem with the reflux ratio profile approximated by a finite dimensional representation (described in Chapter 5). The time interval of interest, say $[t_0, t_F]$ is divided into a finite number N_s of subintervals, each with a set of basis functions involving a finite number of parameters. Here, a piecewise constant representation of the reflux profile is used. Therefore, for N_s intervals, we have N_s constant reflux ratio levels and N_s reflux switching times, totalling $2N_s$ parameters to be optimised in each inner loop problem. Each *function evaluation* of the OPTIMISER (Figure 5.2) requires a full integration of the DAE system and this is achieved by using a Backward Differentiation Formula (BDF) method for integration, with special features to handle discontinuities. Solution of the inner loop optimisation problem requires calculating the gradients of the objective function and constraints with respect to all the optimisation variables. These gradients are evaluated in an efficient way using adjoint variables. Full details of the techniques are given in the original reference (Morison, 1894), in Mujtaba (1989) and in Chapter 5. The constrained nonlinear optimisation problem is solved using an efficient and robust Successive Quadratic Programming (SQP) technique (Chen, 1988).

6.3.3. Outer Loop Optimisation Problem

A complete solution of the sequence of inner loop problem is required for each *function evaluation* of the outer loop problem. The gradients of the objective function with respect to the decision variables are obtained by a finite difference technique but in an efficient way, i.e. by utilising the solution of each inner loop problem from the function evaluation stage. This is explained more clearly with reference to the STN shown in Figure 6.2. There are four decision variables (Re_{D1}^1, Re_{R1}^1, x_{R1}^1, Re_{D2}^2) to consider in the outer optimisation problem. At the *function evaluation* stage the solution (reflux profiles, column profile, duration of task) of each inner loop is stored as **A**, **B**, and **C**, respectively, as shown in Figure 6.3a.

For the gradient with respect to Re_{D1}^1, the variable is perturbed and the inner loop problems are solved as shown in Figure 6.3b. The new solutions **A1**, **B1**, and **C1** are used to calculate the gradient of the objective function with respect to Re_{D1}^1.

Re_{R1}^1 is then perturbed and only the inner loop problems P2-2 and P2-3 are solved (Figure 6.3c). The new solutions **B2**, **C2** and **A**, the base case solutions to problem P2-1 are used to calculate the gradient of the objective function with respect to the variable Re_{R1}^1. Note, calculations of the gradients with respect to variables x_{R1}^1 and Re_{D2}^2 only require solutions of problems P2-2 and P2-3 (Figure 6.3d) and P2-3 (Figure 6.3e), respectively.

The above-mentioned strategy requires the solution of just 8 inner loop problems for calculating the gradients with respect to the 4 decision variables. Note, additional efficiency can be achieved by using the corresponding optimal reflux ratio profiles from the previous pass as the initial estimate of the optimisation variables for each inner loop problem. This will significantly reduce the number of iterations required for each inner loop problem, and in particular for gradient evaluation.

The constraints in the outer loop optimisation problem are simple bounds on the decision variables and their gradients can be easily calculated. Mujtaba and Macchietto solved the outer loop optimisation problem using the efficient SQP technique due to Chen (1988). Mujtaba and Macchietto (1993) used 1.0E-3 and 1.0E-2 as the tolerances for the inner and outer loop optimisations respectively. These were quite tight considering the fact that all the optimisation variables and constraints were scaled within the range 0-10.

Decision Variables passed
to inner loop

Solution of inner loops
passed to outer loop

| Problem P2-0 |

| Problem P2-1 | **A** | Problem P2-2 | **B** | Problem P2-3 | **C** |

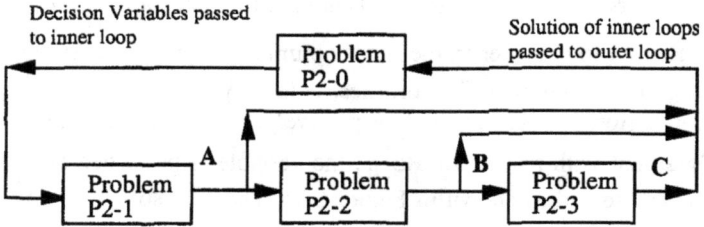

Figure 6.3a. Function evaluation. [Mujtaba and Macchietto, 1993][b]

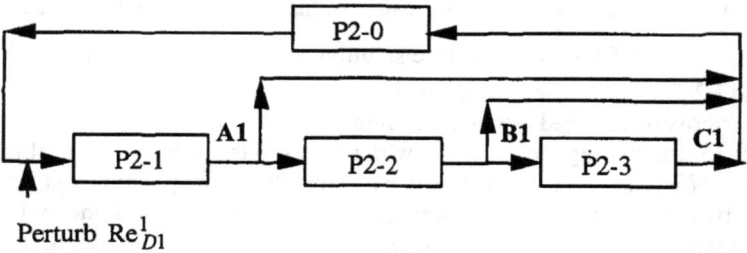

| P2-0 |

| P2-1 | **A1** | P2-2 | **B1** | P2-3 | **C1** |

Perturb Re_{D1}^1

Figure 6.3b. Gradient with respect to Re_{D1}^1. [Mujtaba and Macchietto, 1993][b]

---

[b] Reprinted from *Computers & Chemical Engineering*, **17**, Mujtaba, I.M. and Macchietto, S., Optimal operation of multicomponent batch distillation- multiperiod formulation and solution, 1191-1207, Copyright (1993), with permission from Elsevier Science .

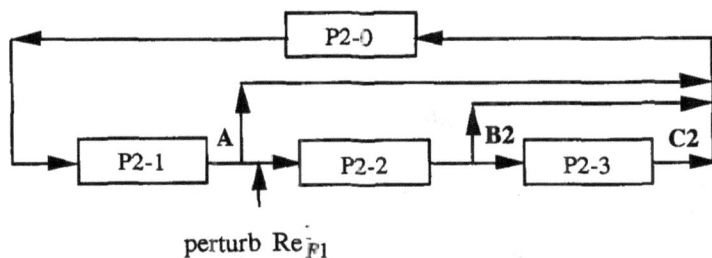

Figure 6.3c. Gradient with respect to Re_{R1}^1. [Mujtaba and Macchietto, 1993][c]

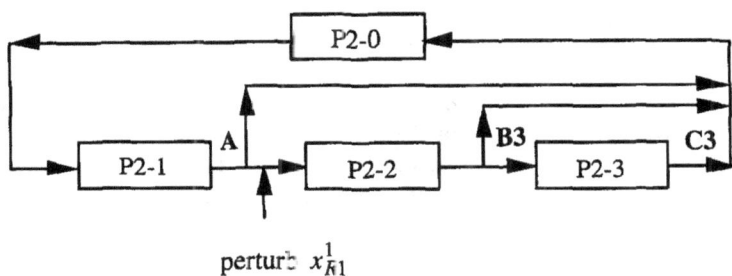

Figure 6.3d. Gradient with respect to x_{R1}^1. [Mujtaba and Macchietto, 1993][c]

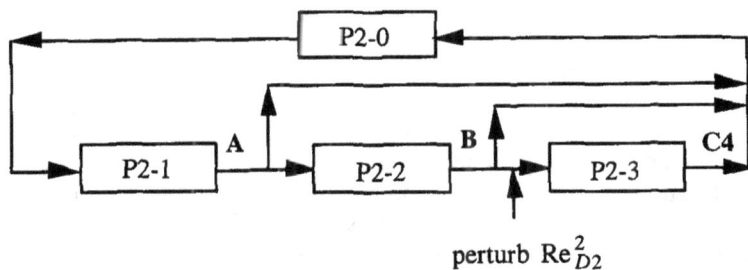

Figure 6.3e. Gradient with respect to Re_{D2}^2. [Mujtaba and Macchietto, 1993][c]

---

[c] Reprinted from *Computers & Chemical Engineering*, 17, Mujtaba, I.M. and Macchietto, S., Optimal operation of multicomponent batch distillation- multiperiod formulation and solution, 1191-1207, Copyright (1993), with permission from Elsevier Science .

6.4. Example Problems

6.4.1. Binary Distillation (Simple Model)

Mujtaba and Macchietto (1993) considered the operation shown in Figure 6.1 for a binary distillation with one main-cut and one off-cut. For simplicity the composition of the off-cut (x_{R1}^1) is fixed a priori which reduces the outer loop optimisation variables to one (Re_{D1}^1). It is also assumed that the value of the off-cut and the final bottom residue are negligible. However, the composition constraints have to be satisfied for the final bottom residue due to environmental restrictions. The input data for the problem are given in Table 6.1. A simple dynamic distillation model (Type III in Chapter 4) was used for this example.

The reflux ratio is discretised into two time intervals for task 1 and one time interval for task 2. Thus a total of 3 reflux ratio levels and 3 switching times are optimised for the whole multiperiod operation. Three cases are considered, corresponding to different values of the main-cut 1 product. For each case the optimal recovery (Re_{D1}^1) is obtained together with the optimal reflux profiles and the optimal duration of each task. These results are shown in Table 6.2 and Figure 6.4.

As expected, the results presented in Table 6.2 shows that increasing the value of the main product results in a higher recovery of component 1 in task 1, with correspondingly larger amount of main-cut 1 (of specified purity) collected. This reduces the amount of off-cut produced in task 2 since this product has no value. Longer operation at higher reflux ratios is also justified, in spite of the large hourly operation costs. Note that in all cases rising reflux ratio profiles were obtained in task 1 with lower reflux ratio obtained in task 2 (Figure 6.4). This trend is in agreement with practical experience (Rose, 1985).

Mujtaba and Macchietto (1993) reported that, in general, the outer loop requires 4-5 function evaluations and 3-4 gradient evaluations to converge. Each inner loop problem requires 6-8 function and 4-6 gradient evaluations. One complete function evaluation of the outer loop requires about 5-6 minutes (CPU) using a SPARC-1 Workstation. The outer loop gradient evaluation time is approximately 20-25% smaller than that of a function evaluation.

Table 6.1. Input Data for Binary Distillation. [Mujtaba and Macchietto, 1993][d]

No. of Ideal Separation Stages (including a reboiler and a total condenser)	= 6
Total Fresh Feed, B_0, kmol	= 10
Feed Composition, x_{B0}, mole fraction	= <0.3, 0.7>
Column Holdup, kmol:	
Condenser	= 0.1
Internal Plates	= 0.025
Relative Volatility, α	= <3.0, 1.0>
Vapour Flow rate, kmol/hr	= 3.0
Column Pressure, bar	= 1.013
Purity of Main-cut 1, x_{D1}^{1*}, mole fraction	= 0.95
Composition of Off-cut 1, x_{R1}^{1*}, mole fraction	= 0.40
Composition of Bottom Residual, x_{B2}^{2*}, mole fraction	= 0.95

Costs:
$C_{B0} = 1.0$ \$/kmol, $C_{R1} = C_{B2} = 0.0$, $C_{fc} = 5.0$ \$/kmol
$C_{D1} = 20, 30, 40$ \$/kmol

Initial Outer Loop Decision Variables: $\text{Re}_{D1}^1 = 0.75$
Initial Inner Loop Decision Variables:

Reflux Ratio Levels	0.80		0.90	0.65
Switching Times, hr	t = 0	2.5	5.0	6.0

Table 6.2. Results for Binary Distillation. [Mujtaba and Macchietto, 1993][d]

Case	C_{D1} $/kmol	Re_{D1}^1	D_1 kmol	t_1 hr	R_1 kmol	t_2 hr	P $/hr
1	20.0	0.63	1.99	3.43	2.03	1.53	1.00
2	30.0	0.66	2.09	3.74	1.78	1.41	5.20
3	40.0	0.71	2.25	4.34	1.36	1.19	9.46

[d] Reprinted from *Computers & Chemical Engineering*, 17, Mujtaba, I.M. and Macchietto, S., Optimal operation of multicomponent batch distillation- multiperiod formulation and solution, 1191-1207, Copyright (1993), with permission from Elsevier Science .

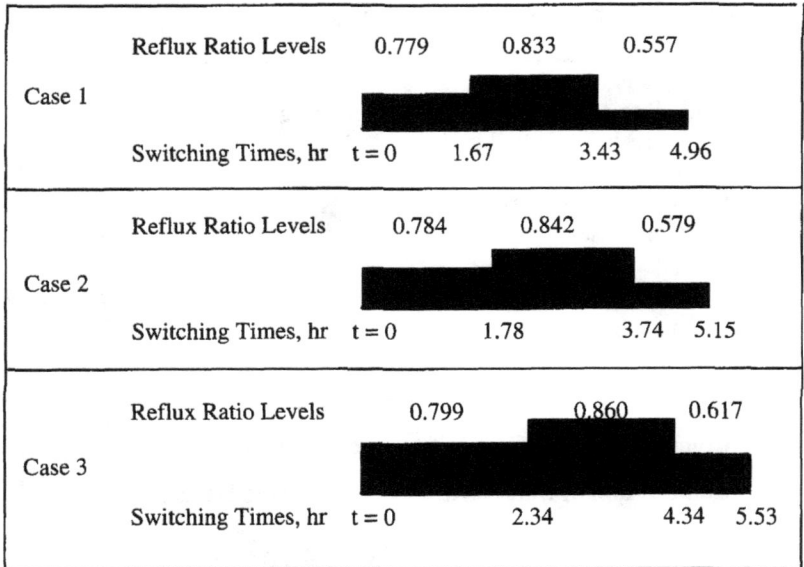

Figure 6.4. Optimal Reflux Ratio Profiles for Binary Distillation.
[Mujtaba and Macchietto, 1993][e]

6.4.2. *Ternary Distillation (Simple Model)*

Mujtaba and Macchietto (1993) considered the ternary operation described in Figure 6.2 where two main-cuts of specified purity and one off-cut are produced. In this problem, the recovery of component 1 in the off-cut 1 (Re_{R1}^1) is fixed at 0.95. The composition of the off-cut 1 (x_{R1}^1) is left as a decision variable to be optimised together with two the other decision variables, Re_{D1}^1 (the recovery of component 1 in main-cut 1) and Re_{D2}^2 (the recovery of component 2 in main-cut 2). Zero values for the off-cut and final bottom residue are assigned, with no constraints put on the compositions of either product. The input data for this problem are given in Table 6.3. The simple dynamic column model (Type III) was used for this example.

---

[e] Reprinted from *Computers & Chemical Engineering*, **17**, Mujtaba, I.M. and Macchietto, S., Optimal operation of multicomponent batch distillation- multiperiod formulation and solution, 1191-1207, Copyright (1993), with permission from Elsevier Science .

Table 6.3. Input Data for Ternary Distillation (Simple Model)[f]

No. of Ideal Separation Stages (including a reboiler and a total condenser)	= 10
Total Fresh Feed, B_0, kmol = 10	
Feed Composition, x_{B0}, mole fraction	= <0.3, 0.2, 0.5>
Column Holdup, kmol:	
Condenser	= 0.1
Internal Plates	= 0.0125
Relative Volatility, α	= <8.0, 4.0, 1.0>
Vapour Flow rate, kmol/hr	= 3.0
Column Pressure, bar	= 1.013
Purity of Main-cut 1, x_{D1}^{1*}, mole fraction	= 0.95
Purity of Main-cut 2, x_{D2}^{2*}, mole fraction	= 0.95
Recovery of component 1 in Off-cut 1, Re_{R1}^1	= 0.95

Costs:
$C_{B3} = C_{RI} = 0.0$, $C_{B0} = 1.0$ \$/kmol, $C_{fc} = 5.0$ \$/hr

Initial Outer Loop Decision Variables:
$Re_{D1}^1 = 0.85$, $x_{R1}^1 = 0.50$, $Re_{D2}^2 = 0.85$

Initial Inner Loop Decision Variables:

Reflux Ratio Levels	0.70	0.80	0.80	0.95		0.70	0.80

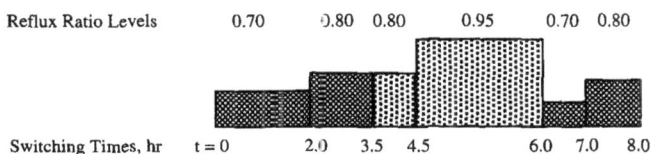

Switching Times, hr	t = 0		2.0	3.5	4.5		6.0	7.0	8.0

The reflux ratio is discretised into two control intervals for each operation task. Three cases were considered with different sales values for the two main-cuts. For all cases, the optimal recovery of component 1 in task 1 and that of component 2 in task 2, the optimal amounts of main-cut 1, off-cut 1 and main-cut 2, the optimal

[f] Reprinted from *Computers & Chemical Engineering*, **17**, Mujtaba, I.M. and Macchietto, S., Optimal operation of multicomponent batch distillation- multiperiod formulation and solution, 1191-1207, Copyright (1993), with permission from Elsevier Science .

duration of the distillation tasks and the optimal profit are shown in Table 6.4. For case 1, x_{R1}^1 was fixed a priori (= 0.6 molefraction) and Re_{D1}^1 and Re_{D2}^2 are optimised in the outer loop optimisation problem. The optimal reflux ratio policy for each case is presented in Figure 6.5, and accumulated and instant distillate composition profiles for case 2 are presented in Figure 6.6.

Case 2, where both main distillate products have the same unit value, is taken as the base case and used as a reference. Case 2 is similar to case 1 except that the composition of the off-cut was also optimised. The results are similar to case 1 in terms of recoveries and profit although the reflux ratio profiles are slightly different. Case 2 operation is slightly shorter and more profitable. The losses of component 2 in the off-cut for the two cases are close.

Table 6.4. Results For Ternary Distillation (Simple Model)[g]

case	C_{D1}, C_{D2} $/kmol	Re_{D1}^1	D_1 (kmol), t_1 (hr)	x_{R1}^1	R_1 (kmol), t_2 (hr)	Re_{D2}^2	D_2 (kmol), t_3 (hr)	P $/hr
1	20, 20	.83	2.61 3.56	.60*	.83 2.65	.77	1.24 1.46	3.74
2	20, 20	.82	2.58 3.55	.546	.95 2.53	.78	1.18 1.34	3.78
3	20, 60	.75	2.36 3.01	.70	1.03 2.90	.87	1.45 1.86	11.0
4	40, 20	.82	2.57 3.53	.502	1.05 2.40	.65	.93 .96	11.2

* Fixed

[g] Reprinted from *Computers & Chemical Engineering*, **17**, Mujtaba, I.M. and Macchietto, S., Optimal operation of multicomponent batch distillation- multiperiod formulation and solution, 1191-1207, Copyright (1993), with permission from Elsevier Science .

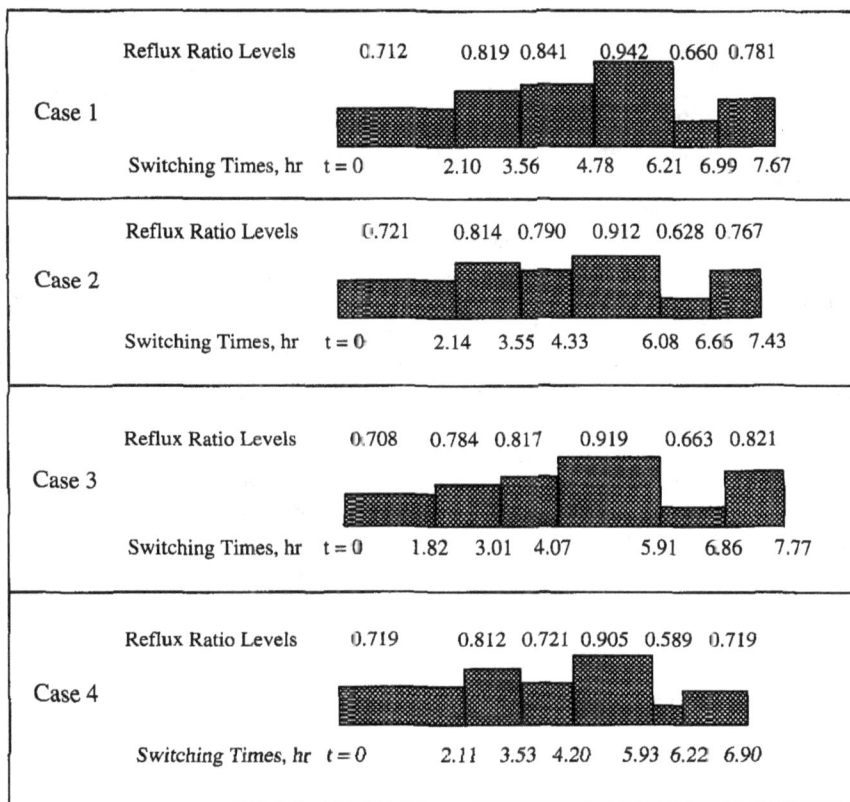

Figure 6.5. Reflux Ratio Profiles for Ternary Distillation (Simple Model)[h]

In case 3, the sales value of the second main-cut is much higher compared to that of the first main-cut. Therefore, a larger recovery of component 2 is now obtained to maximise the profit. Relative to the base case, less of main-cut 1 is produced in a shorter time but with very little loss of valuable component 2 and with low reflux ratios (Figure 6.5). In off-cut 1, a higher composition of $x_{R1}^1 = 0.70$ ensures a small loss of component 2. In case 4, although the sales value of main-cut 1 is increased relative to the base case, the recovery of component 1 in main-cut 1 was similar to that of case 2. Any higher recovery would require longer processing

---

[h] Reprinted from *Computers & Chemical Engineering*, **17**, Mujtaba, I.M. and Macchietto, S., Optimal operation of multicomponent batch distillation- multiperiod formulation and solution, 1191-1207, Copyright (1993), with permission from Elsevier Science .

time for task 1 and therefore would produce lower profit. The operation time for task 3 is much shorter than that in the base case, with lower recovery at lower reflux ratios. Also, a lower off-cut composition in this case significantly reduced the time required for task 2. In most cases high reflux ratios are required in task 2 (off-cut production) to ensure that the very high purity requirements for component 2 in the subsequent cut (task 3 operation) can be achieved.

Mujtaba and Macchietto (1993) reported that the outer loop solution typically required 4-6 function and 3-4 gradient evaluations with 7-10 function and 5-6 gradient evaluations in each inner loop problem. One complete function and gradient evaluation of the outer loop required about 8-10 minutes CPU and 15-18 minutes, respectively, on a SPARC-1 Workstation.

For the sake of comparison the case 2 problem of Table 6.4 was rerun with the same product specifications on the main-cuts but without any off-cut production (the STN for this operation is shown in Figure 6.7).

The results are summarised in Table 6.5. The accumulated and instant distillate composition profiles are shown in Figure 6.8. This operation requires a longer batch time and higher reflux ratios and gives more than 20% less profit compared to the results for case 2 (Table 6.4). This clearly shows the benefit of producing an intermediate off-cut for this specific separation problem.

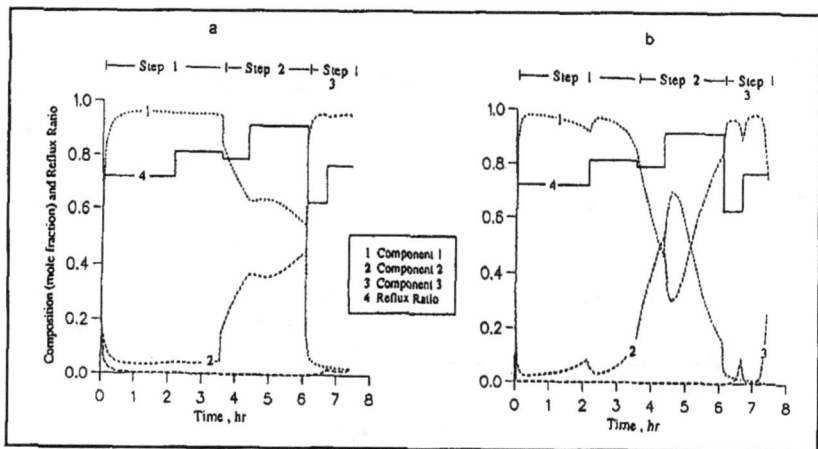

Figure 6.6. Ternary Distillation (Simple Model) with Off-cut Production (Case 2)
a) Accumulated Distillate Composition and Reflux Ratio Profile
b) Instant Distillate Composition and Reflux Ratio Profile[i].

---

[i] Reprinted from *Computers & Chemical Engineering*, **17**, Mujtaba, I.M. and Macchietto, S., Optimal operation of multicomponent batch distillation- multiperiod formulation and solution, 1191-1207, Copyright (1993), with permission from Elsevier Science .

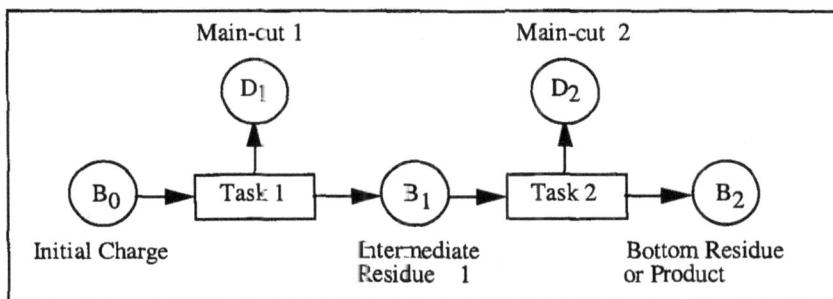

Figure 6.7. STN for Ternary Batch Distillation with Two Main Distillate Cuts[j]

Table 6.5. Results for Ternary Distillation (Simple Model without Off-cut Production)[j].

Optimal Recovery of Component 1 in Main-cut 1, Re_{D1}^{1}	$= 0.973$
Optimal D_1, kmol	$= 3.07$
Optimal Duration of Task 1, t_1, hr	$= 7.15$
Optimal Recovery of Component 2 in Main-cut 2, Re_{D2}^{2}	$= 0.952$
Optimal D_2, kmol	$= 1.85$
Optimal Duration of Task 2, t_2, hr	$= 3.94$
Optimal Profit, P, \$/hr	$= 2.97$

Reflux Ratio Profile:

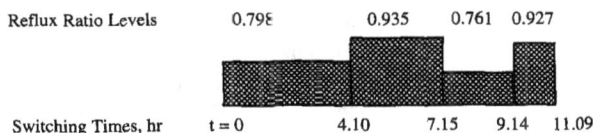

Costs: $C_{D1} = C_{D2} = 20.0$ \$/kmol, $C_{B2} = 0.0$, $C_{B0} = 1.0$ \$/kmol, $C_{fc} = 5.0$ \$/hr

---

[j] Reprinted from *Computers & Chemical Engineering*, **17**, Mujtaba, I.M. and Macchietto, S., Optimal operation of multicomponent batch distillation- multiperiod formulation and solution, 1191-1207, Copyright (1993), with permission from Elsevier Science .

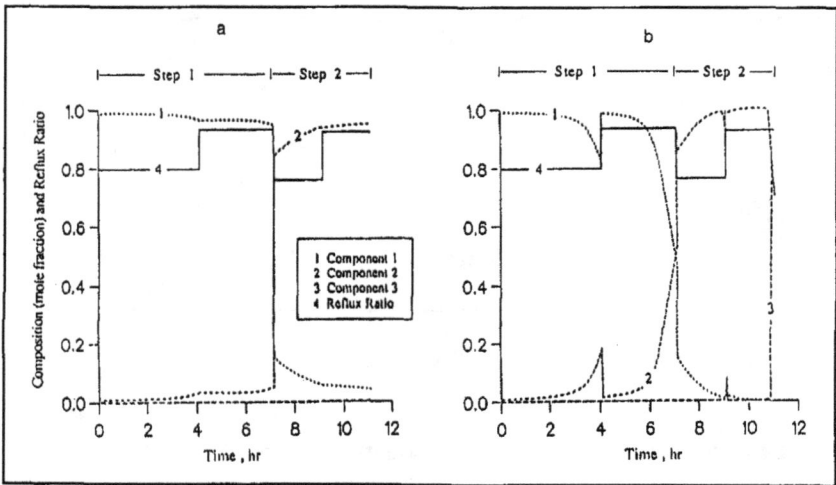

Figure 6.8. Ternary Distillation without Off-cut Production (Case 2).
a) Accumulated Distillate Composition and Reflux Ratio Profile
b) Instant Distillate Composition and Reflux Ratio Profile[k]

6.4.3. Ternary Distillation (Detailed Model)

This example is taken from Mujtaba and Macchietto (1993) which involves separation of a ternary mixture <Cyclohexane, n-Heptane, Toluene>. The initial amount of fresh feed is same as that used by Nad and Spiegel (1987) in an experimental column. However, for simplicity only a 10 stage (including reboiler and a total condenser) column is used instead of a 20 stage column used by the original authors. The multiperiod operation shown in Figure 6.2 is optimised, according to problem P2 (section 6.2.2.2). The problem specifications include the purity of the two main distillate products and the recovery of component 1 in the off-cut product. A detailed dynamic model (Type IV) was used here with rigorous non-ideal thermodynamics described by the Soave-Redlich-Kwong (SRK) equation of state. As before, 2 time intervals were used for the reflux ratio in task 1 and task 3 and one time interval for the off-cut production. The input data for this example are given in Table 6.6. The cost coefficients, also given in Table 6.6 were arbitrarily defined.

---

[k] Reprinted from *Computers & Chemical Engineering*, **17**, Mujtaba, I.M. and Macchietto, S., Optimal operation of multicomponent batch distillation- multiperiod formulation and solution, 1191-1207, Copyright (1993), with permission from Elsevier Science .

Table 6.6. Input Data for Ternary Distillation (Detailed Model)[1]

No. of Ideal Separation Stages	
(including a reboiler and a total condenser)	= 10
Total Fresh Feed, B_0, kmol	= 2.93
Feed Composition, x_{B0}, mole fraction	= <0.407, 0.394, 0.199>
Column Holdup, kmol:	
Condenser	= 0.02
Internal Plates	= 0.01
Condenser Vapour Load, kmol/hr	= 2.75
Column Pressure, bar	= 1.013
Purity of Main-cut 1, x_{D1}^{1*}, mole fraction	= 0.90
Purity of Main-cut 2, x_{D2}^{2*}, mole fraction	= 0.80
Recovery of Component 1 in Off-cut 1, Re_{R1}^1	= 0.95

Costs:
$$C_{D1} = C_{D2} = 20.0 \ \$/kmol$$
$$C_{B3} = C_{R1} = 0.0, \ C_{B0} = 1.0 \ \$/kmol, \ C_{fc} = 3.0 \ \$/hr$$

Initial Outer Loop Decision Variables:
$$Re_{D1}^1 = 0.80, \ x_{R1}^1 = 0.50, \ Re_{D2}^2 = 0.70$$

Initial Inner Loop Decision Variables:

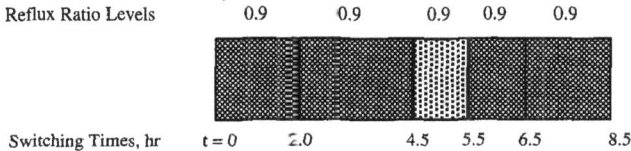

Reflux Ratio Levels	0.9	0.9	0.9	0.9	0.9

Switching Times, hr	$t = 0$	2.0		4.5	5.5	6.5		8.5

The optimal recoveries of component 1 and 2 in the main-cuts, off-cut composition, amount of each cut, duration of each task and reflux ratio profiles for each task are given in Table 6.7. They show that the desired product purities can be achieved to yield a maximum productivity for the operation of $2.0 per hr. The accumulated and instant distillate composition profiles for the optimal operation are

---

[1] Reprinted from *Computers & Chemical Engineering*, **17**, Mujtaba, I.M. and Macchietto, S., Optimal operation of multicomponent batch distillation- multiperiod formulation and solution, 1191-1207, Copyright (1993), with permission from Elsevier Science .

shown in Figure 6.9. This example demonstrates that the proposed method can indeed tackle realistic problems.

The computational statistics for the problem are:

- 5 function and 3 gradient evaluations outer loop problem.
- 10-15 function and 7-12 gradient evaluations for the first and third inner loop problems.
- 5-7 function and 4-5 gradient evaluations for the second inner loop problem.
- Approximately 30 minutes CPU for one complete function evaluation of the outer loop using a SPARC-1 Workstation.

This increased computation load is as expected for such a problem due to the use of a detailed model with rigorous thermodynamics.

Table 6.7. Results for Ternary Distillation (Detailed Dynamic Model)[m].

Optimal Recovery of Component 1 in Main-cut 1, Re_{D1}^1	= 0.779
Optimal D_1 , kmol	= 1.03
Optimal Duration of Task 1, t_1, hr	= 3.40
Optimal Off-cut 1 Composition, x_{R1}^1 , mole fraction	= 0.492
Optimal R_1 , kmol	= 0.508
Optimal Duration of Task 2, t_2, hr	= 2.77
Optimal Recovery of Component 2 in Main-cut 2, Re_{D2}^2	= 0.762
Optimal D_2 , kmol	= 0.784
Optimal Duration of Task 3, t_3, hr	= 2.18
Optimal Profit, P, \$/hr	= 2.00

Reflux Ratio Profile:

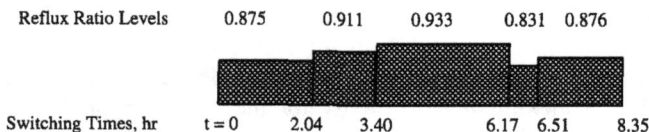

| Reflux Ratio Levels | 0.875 | 0.911 | 0.933 | 0.831 | 0.876 |

| Switching Times, hr | t = 0 | 2.04 | 3.40 | 6.17 | 6.51 | 8.35 |

---

[m] Reprinted from *Computers & Chemical Engineering*, **17**, Mujtaba, I.M. and Macchietto, S., Optimal operation of multicomponent batch distillation- multiperiod formulation and solution, 1191-1207, Copyright (1993), with permission from Elsevier Science .

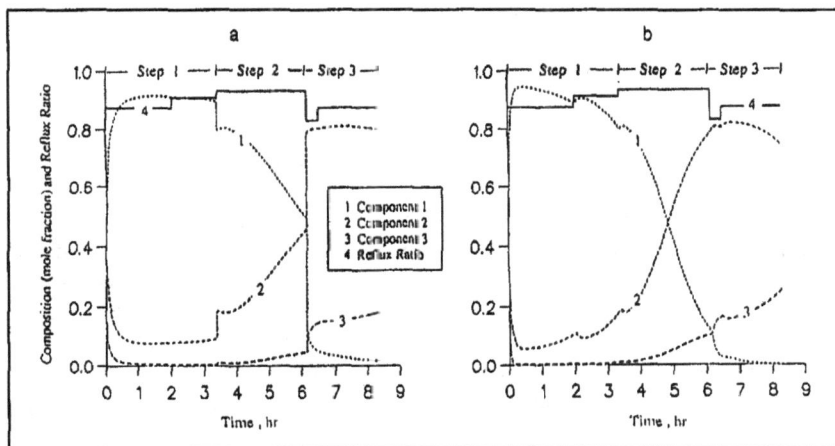

Figure 6.9. Ternary Distillation (Detail Model) with Off-cut Production
a) Accumulated Distillate Composition and Reflux Ratio Profile
b) Instant Distillate Composition and Reflux Ratio Profile[n]

6.4.4. Multiperiod Campaign Operation Optimisation - Industrial Case Study

Mujtaba and Macchietto (1994) presented an industrial case study in which dynamic optimisation method of Mujtaba and Macchietto (1993) is utilised for the development of the optimal operation of an entire batch distillation campaign where 100 batches of fresh charge have to be processed with secondary reprocessing of intermediate off-cuts. The process involved a complex separation of a five-component mixture of industrial interest, described using non-ideal thermodynamic models. In addition, the operation of the whole production campaign was subject to a number of resource constraints, for example -

(a) limited column capacity,
(b) limited number and size of intermediate off-cut storage vessels,

and operational constraints, such as,

(c) very high purity demands of individual products,

---

[n] Reprinted from *Computers & Chemical Engineering*, **17**, Mujtaba, I.M. and Macchietto, S., Optimal operation of multicomponent batch distillation- multiperiod formulation and solution, 1191-1207, Copyright (1993), with permission from Elsevier Science .

(d) maximum allowable reboiler temperature to avoid any thermal degradation of the products,

(e) fixed total operation time, etc.

The problem was to find the best operation of each batch so as to maximise the profit over the available plant time subject to these resource and operational constraints.

First, a dynamic model (Type IV) was validated against a more rigorous industrially supplied model which also accounted for hydraulics, pressure drop etc. showing that a good match was achieved. Then the optimal operation of individual batches was studied utilising the optimisation methods presented earlier. The reflux ratio was optimised as the operating variable. Finally, various policies for intermediate off-cut recovery, intermediate off-cut reprocessing policies and batch sequences were examined, resulting in the development of an effective campaign for the entire plant operation period.

This problem is considerably more complex than normally considered in the batch distillation literature. Significant problems arise due to the constraints mentioned previously and the significant interactions between product recoveries and batch time. A particular challenge results from the high product purity specifications since in many cases these cannot be achieved in a single pass through the still. An alternative policy was therefore considered where intermediate low purity cuts were produced and collected which were subsequently reprocessed in a second pass through the column. On the other hand, the number and capacity of intermediate storage vessels, which can be used, for this purpose is also limited, posing severe constraints on the operation.

The column description, product specifications, total plant operation time, number and size of available storage vessels, etc. are given in Table 6.8. For proprietary reasons the names of the components in the feed mixture are disguised and thermodynamic data ore omitted. Relative product values and feed costs are given in Table 6.10 (operating costs have also been scaled).

The following approach was used:

(a) A strategy for the processing of each batch (structural decisions) was selected consisting of a sequence of main-cut and off-cut tasks. STN in Figure 6.10 represents this.

(b) Additional specifications (mainly recoveries and/or purities) for each step were set, based on the required product purities.

(c) For given structural decision variables (as in (a)) and additional continuous decision variables (as in (b)), the remaining operating variables (reflux ratio profiles and times) are optimised for each task in sequence. In each step the reflux ratio profile was discretized into a finite number of intervals, each interval having a constant reflux ratio.

Reflux ratio levels and the length of the intervals were optimised to minimise the batch time for the task. Three reflux intervals were used for each main-cut and 2 for each off-cut operation. The final conditions of a preceding distillation task were used as the initial conditions for the next task.

(d) The overall constraints and economy of the entire campaign was assessed.

For a 5-component mixture a large number of alternative operation sequence is possible. However typical results of one such sequence, shown in Figures 6.10 and 6.11, and specifications, given in Table 6.9, are presented here.

Table 6.8. Column Description, Resource and Operational Constraints. [Adopted from Mujtaba and Macchietto, 1994]

Number of ideal stages	= 10
Total condenser	= 1
Partial reboiler	= 1
Stage holdup	= 0.35 cu ft
Condenser holdup	= 0.8 cu ft
Maximum reboiler duty	= 1 mmBtu/hr
Total charge (100 batches)	= 180,000 gal
Total plant operation time	= 800 hrs/yr
Feed composition	= <0.159, 0.256, 0.077, 0.013, 0.495> molefraction
Reboiler capacity	= 1800 gal.
Number of off-cut storage tanks	= 4
Capacity of each storage tank	= 3000 gal.

Product purities = 70 (mol %) for component 1 in product 1 (P1)
= 99 (mol %) for key components in other products
= 99.9% purity of component 5 in final bottom product (P5)
Maximum reboiler temperature: 360 F

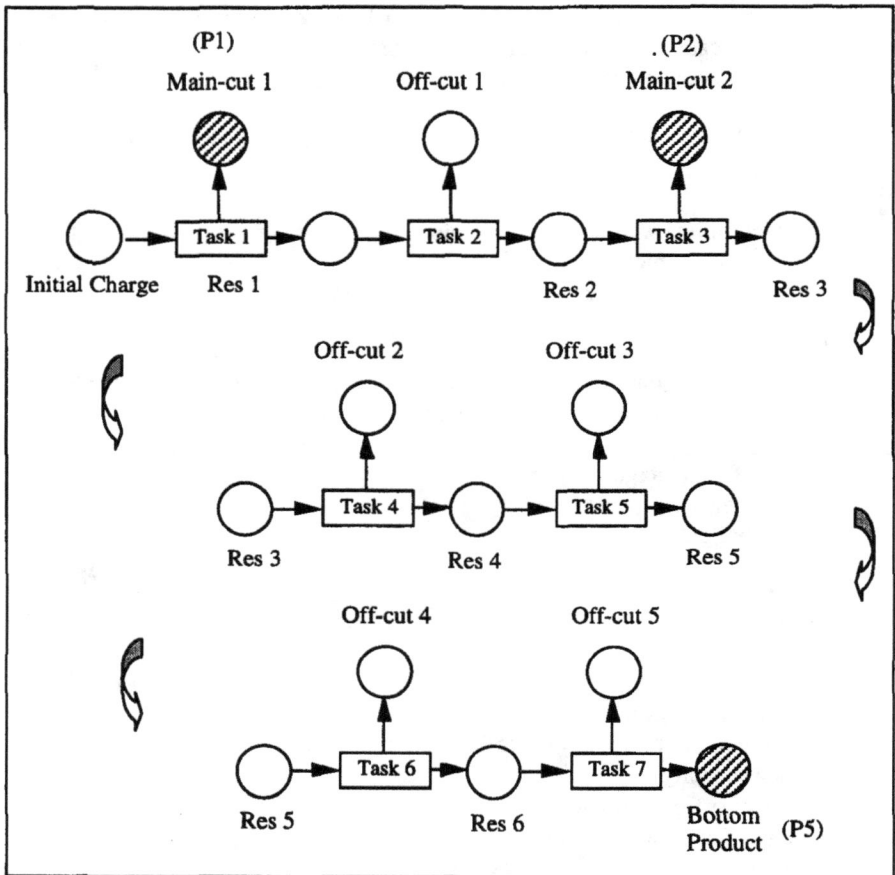

Figure 6.10. Operational Sequence for Fresh Feed Processing
(Product cuts in parenthesis)

The operation sequence for a single batch, shown in Figure 6.10, produces 2 main-cuts (P1 and P2), 5 off-cuts and a final bottom product (P5). Specifications for products P3 and P4 could not be achieved directly because of the high purity requirement, low amount of component 3 and 4 in the feed mixture and the proximity of their boiling points with other components. Therefore, low purity off-cuts 2-5 (essentially binary or a ternary mixtures) are produced, collected in separate storage vessels and reprocessed when the amount of in each vessel reaches the full capacity of the batch column (1800 gals). The sequences of operations for reprocessing off-cuts 2-5 are shown in Figures 6.11. Due to the availability of only 4 intermediate storage vessels and the low unit price for component 1 and 2, Off-cut 1 from each batch of fresh feed is discarded. Off-cuts 2/1, 3/1 and 4/1 are similarly discarded.

Optimum reflux ratio profiles for the STN shown in Figure 6.10 (one batch of fresh feed charge) are shown in Figure 6.12. The total campaign is summarised in Figure 6.13 and Table 6.10. For the selected decision and task recovery variables, the overall recovery of components 2 and 3 is not so good and the entire campaign requires approximately 10% more than the available time. Meeting these constraints and improving the overall economy of the operation can be done by optimising the structure of the operation, the continuous decision variables of type (b), and by refining the reflux ratio profiles within each task.

Table 6.9. Recoveries and Product Specifications in Each Distillation Task in Figures 6.10 and 6.11. [Adopted from Mujtaba and Macchietto, 1994] Purities in bold face indicate the desired product specifications. Other purity and recovery specifications (decision variables of type (b)) are set at typical values.

Task	Recovery	Purity	Spec
Task 1	: 97 % recovery of comp. 1 in the charge at	**70%** purity in Main-cut 1	(P1)
Task 2	: 70 % recovery of the remaining comp. 1 at	40% purity in Off-cut 1	
Task 3	: 90 % recovery of the remaining comp. 2 at	**99%** purity in Main-cut 2	(P2)
Task 4	: 75 % recovery of the remaining comp. 2 at	50% purity in Off-cut 2	
Task 5	: 80 % recovery of the remaining comp. 2 at	10% purity in Off-cut 3	
Task 6	: 90 % recovery of the remaining comp. 3 at	95% purity in Off-cut 4	
Task 7	: 90 % recovery of the remaining comp. 4 at	60% purity in Off-cut 5	(P5, bottoms)
Task 8	: 85 % recovery of the comp. 2 in the charge at	**99%** purity in Main-cut 2/1	(P2)
Task 9	: 95 % recovery of the remaining comp. 2 at	70% purity in Off-cut 2/1	(P3, bottoms)
Task 10	: 95 % recovery of the comp. 2 in the charge at	20% purity in Off-cut 3/1	(P3, bottoms)
Task 11	: 85 % recovery of the comp. 2 in the charge at	15% purity in Off-cut 4/1	
Task 12	: 98 % recovery of the remaining comp. 3 at	**99%** purity in Main-cut 4/1	(P3)
Task 13	: 96 % recovery of the comp. 3 in the charge at	**99%** purity in Main-cut 5/1	(P3)
Task 14	: 98 % recovery of the remaining comp. 4 at	**99%** purity in Main-cut 5/2	(P4)

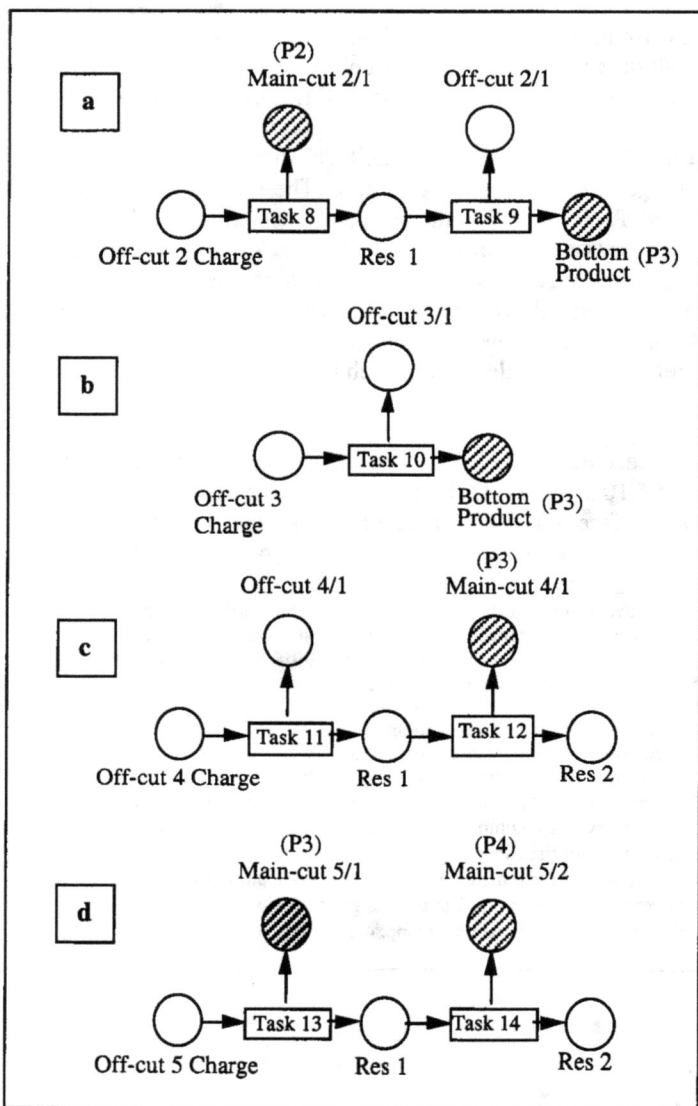

Figure 6.11. Off-cut Reprocessing (Product cuts in parenthesis)
a. Off-cut 2 Reprocessing   b. Off-cut 3 Reprocessing
c. Off-cut 4 Reprocessing   d. Off-cut 5 Reprocessing

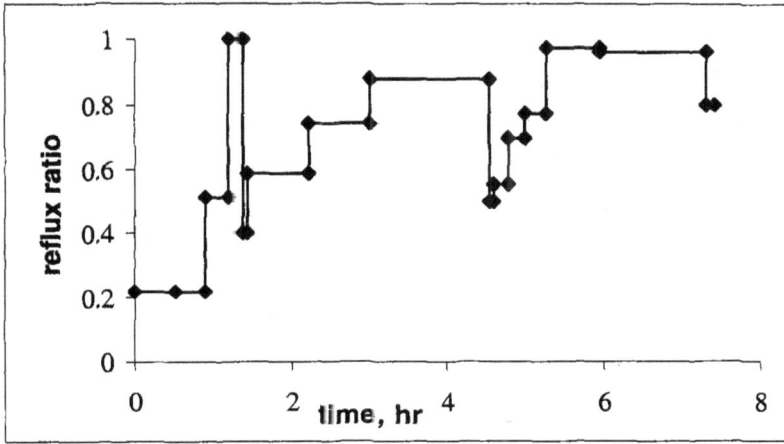

Figure 6.12. Optimal Reflux Ratio Profiles in Different Tasks of Figure 6.10

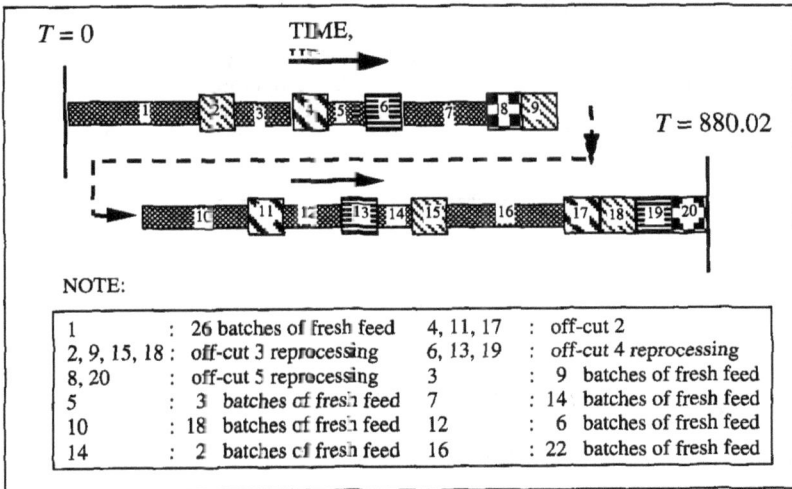

NOTE:

1	:	26 batches of fresh feed	4, 11, 17	:	off-cut 2
2, 9, 15, 18	:	off-cut 3 reprocessing	6, 13, 19	:	off-cut 4 reprocessing
8, 20	:	off-cut 5 reprocessing	3	:	9 batches of fresh feed
5	:	3 batches of fresh feed	7	:	14 batches of fresh feed
10	:	18 batches of fresh feed	12	:	6 batches of fresh feed
14	:	2 batches of fresh feed	16	:	22 batches of fresh feed

Figure 6.13. Overall Production Campaign.
[Adopted from Mujtaba and Macchietto, 1994]

Table 6.10. Summary of the Campaign. [Adopted from Mujtaba and Macchietto, 1994]. Total Fresh Feed Processed = 7800 lbmole

Total Products from the Campaign	lbmol	$/lbmol
P1 (70% comp. 1) = 1,720.0	1.0
P2 (99% comp. 2) = 1,425.0	0.94
P3 (99% comp. 3) = 428.0	13.2
P4 (99% comp. 4) = 84.2	0.0
P5 (99.9% of comp. 5) = 3,812.0	10.0

Component	Product	Recovery lbmole	Feed (in 100 fresh batches) lbmol	% Recovery
1	P1	1,204.0	1,240.0	97.08
2	P2	1,411.0	1,997.0	70.66
3	P3	424.0	600.6	70.59
4	P4	83.4	84.2	82.25
5	P5	3,812.0	3,861.0	98.73

Total Campaign Time, hr

Total Time for fresh feed processing (100 batches)	= 743.0
Total Time for off_cut 2 reprocessing (2.86 batches) ^	= 54.83
Total Time for off_cut 3 reprocessing (3.85 batches)	= 36.0
Total Time for off_cut 4 reprocessing (2.63 batches)	= 33.33
Total Time for off_cut 5 reprocessing (1.92 batches)	= 13.06
Total	= 880.02

Total Steam Requirement = 245.26 MMBtu, Cost = $7.8/MMBtu
Total Raw Material (Fresh Feed) = 100x78 = 7800 lbmol, Cost = $2.98/lbmol

Total Product Value	= $46,909.1
Total Raw Material cost	= $23,244.0
Total Operating Cost (O.C.)	= $ 1,913.02

Profit = (ΣDistillate products x value - raw material cost) - O.C (steam)
        = $21,752.08 / yr

^   2.86 batches means 2 batches of full capacity + 1 batch at 86% capacity

6.5. Optimisation Problem Formulation- Farhat et al.

For *single separation duty* Farhat et al. (1990) presented multiple criteria decision-making (MCDM) NLP based problem formulations for multiperiod optimisation. This involves either maximisation (Problem 1) of specified products (main-cuts) or minimisation (Problem 2) of unspecified products (off-cuts) subject to interior point constraints. These two optimisation problems are described below.

6.5.1. Problem 1 - Maximisation of Main-cuts

For a linear reflux ratio policy in each of the cuts, the problem *OP1* can be written as:

$$
\begin{aligned}
OP1: \quad & \underset{A,B,T}{Max} \left\{ \beta_{2n-1} P_r + \sum_j \beta_j P_k \right\} \\
& j \in [1,3,5,...,2n-3]; \qquad k = (j+1)/2
\end{aligned}
\tag{6.7}
$$

$$
\text{subject to} \qquad g_k = XSPEC_k \text{ (main-cut purity)}
$$

where, β is a binary variable $(0,1)$ and determines the existence of a main-cut; P_k is the amount of main-cut with k specifying the key-component in the cut; j relates the time interval for the main-cut; n is the number of components.

The amount of main-cut in $t \, \varepsilon \, [t_{j-i}, \, t_j]$ can be given by:

$$
\begin{aligned}
P_k \quad &= \int_{t_{j-1}}^{t_j} \frac{Vdt}{R_j + 1} \\
&= \int_{t_{j-1}}^{t_j} \frac{Vdt}{a_j t + b_j + 1}
\end{aligned}
\tag{6.8}
$$

where, V is the vapour boilup rate; R_j is the reflux ratio; t is the time; a and b are constant parameters; A, B and T are vectors defined as:

$$A = [a_1, a_2, \ldots, a_{2n-2}]$$
$$B = [b_1, b_2, \ldots, b_{2n-2}] \qquad (6.8)$$
$$T = [t_1, t_2, \ldots, t_{2n-2}]$$

The last production term is isolated from the general summation in Equation 6.7, because it can be expressed from the overall mass balance:

$$P_n = B_0 - \sum_{i=1}^{n-1} (P_i + S_i) \qquad (6.9)$$

where S is the amount of off-cut.

6.5.2. Problem 2 - Minimisation of Off-cuts

For a linear reflux ratio policy in each of the cuts, the problem $OP2$ can be written as:

$$OP2: \qquad \underset{A,B,T}{Min} \left\{ \sum_j \beta_j S_k \right\} \qquad (6.10)$$

$$j \in [2, 4, 6 \ldots, 2n - 2]; \qquad k = j/2$$

The amount of off-cut in $t \, \varepsilon \, [t_{j-1}, t_j]$ can be given by:

$$S_k \quad = \int_{t_{j-1}}^{t_j} \frac{V dt}{R_j + 1}$$

$$= \int_{t_{j-1}}^{t_j} \frac{V dt}{a_j t + b_j + 1} \qquad (6.11)$$

Farhat et al. (1990) used augmented Lagrangian method to solve the optimisation problem presented above. See the original reference for further details.

6.5.3. Example

Farhat et al. (1990) considered the separation of a ternary mixture with an objective to minimise two off-cuts. The operation sequence considered is shown as an STN in Figure 6.14. There are two main-cuts $(P_1 \simeq D_1; P_2 \simeq D_2)$, two off-cuts $(S_1 \simeq R_1; S_2 \simeq R_2)$ and the final specified bottom product $(P_3 \simeq B_4)$. Farhat et al. solved the optimisation problem $OP2$ which is equivalent to solving $OP1$. The input data for the problem is given in Table 6.11.

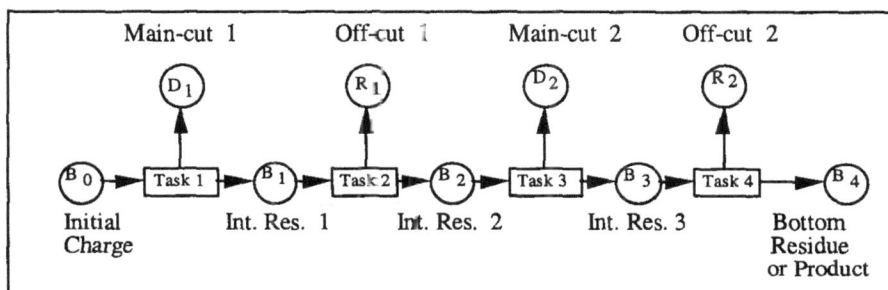

Figure 6.14. Operation Sequence Considered by Farhat et al. (1990)

Table 6.11. Input Data for Farhat et al. (1990) Problem

No. of Plates (including a reboiler and a total condenser)	= 7
Total Fresh Feed, B_0, mol	= 100
Feed Composition, x_{B0}, molefraction	= <0.33, 0.33, 0.34>
Column Holdup, mol	= negligible
Vapour Boilup, mol/hr	= 110
Column Pressure, torr	= 761.654
Purity of Main-cut 1, $XSPEC_1$ $(\sim x_{D1}^{1*})$, molefraction	= 0.95
Purity of Main-cut 2, $XSPEC_2$ $(\sim x_{D2}^{2*})$, molefraction	= 0.925
Purity of Bottom Product, $XSPEC_3$ $(\sim x_{B4}^{3*})$, molefraction	= 0.95
Total Batch Time, hr (fixed)	= 2.5

Farhat et al. considered both optimal constant and optimal linear reflux ratio for this problem (Figure 6.15). Final time was fixed and 4 time intervals were considered. The length of each time interval was also optimised. Table 6.12 presents the summary of the optimisation results using both options of reflux ratio profiles. A significant gain of 10.7% in specified products can be observed between the optimal linear reflux policy and the optimal constant reflux policy.

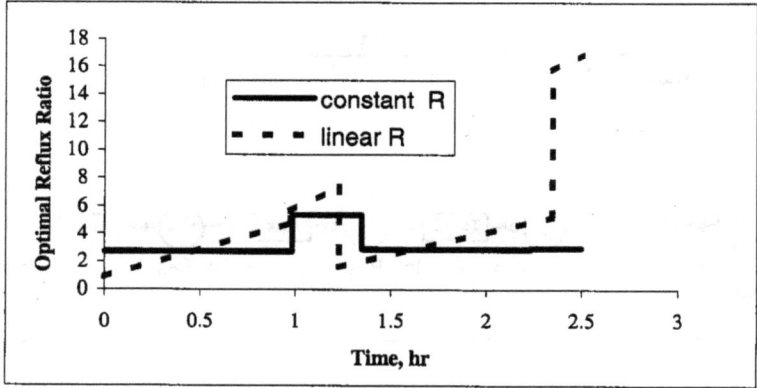

Figure 6.15. Optimal Reflux Ratio Profiles.

Table 6.12. Summary of Optimisation Results

Cuts	Amount of Cuts (mol) Collected Using	
	Constant R	Linear R
Main-cut 1 (P_1)	29.48	31.23
Off-cut 1 (S_1)	6.31	3.68
Main-cut 1 (P_2)	25.31	29.62
Off-cut 1 (S_2)	7.55	0.98
Bottom product (P_3)	31.35	95.34
Total products = $P_1 + P_2 + P_3$	86.14	95.34
Total off-cuts = $S_1 + S_2$	13.86	4.66

References

Al-Tuwaim, M.S., and Luyben, W.L., *IEC Res.* **30** (1991), 507.

Bernot, C., Doherty, M.F. and Malone, M.F., *Chem. Eng. Sci.* **45** (1990), 1207.

Bernot, C., Doherty, M.F. and Malone, M.F., *Design and operating policies for multicomponent batch distillation. AIChE Annual Meeting*, Los Angeles, USA (1991).

Chen, C.L., *A Class of Successive Quadratic Programming Methods for Flowsheet Optimisation*. PhD Thesis, (Imperial College, London, 1988).

Diwekar, U.M., Madhavan, K.P. and Swaney, R.E., *IEC. Res.* **28** (1989), 1011.

Diwekar, U.M. and Madhavan, K.P , *IEC. Res.* **30** (1991), 713.

Farhat, S., Czernicki, M., Pibouleau, L. and Domenech, S., *AIChE J.* **36** (1990), 1349.

Logsdon, J.S., Diwekar, U.M. and Biegler, L.T., *Trans. IChemE*, **68A** (1990), 434.

Morison, K.R., *Optimal Control of Processes Described by Systems of Differential and Algebraic Equations*. PhD. Thesis, (Imperial College, University of London, 1984).

Mujtaba, I.M., *Optimal Operational Policies in Batch Distillation*. PhD Thesis, (Imperial College, University of London, 1989).

Mujtaba, I.M. and Macchietto, S. In *Recent Progres en Ginie de Procedes*, eds. Domenech, S. et al. (Lavoisier Techniche et Documentation, Paris, 1988) **2**, 191.

Mujtaba, I.M. and Macchietto, S., in *IMACS Annals on Computing and Applied Mathematics, 4: Computing and Computers for Control Systems*, eds. Borne, P. et al. (J.C. Baltzer AG, Scientific Publishing Co. Basel Switzerland, 1989), 55.

Mujtaba, I.M. and Macchietto, S., in *Proceedings PSE'91 - 4th International Symposium on Process Systems Engineering*, **1**: Design, (Quebec, Canada, 1991), 19.1

Mujtaba, I.M. and Macchietto, S., *Comput. chem. Engng.* **16S** (1992), S273.

Mujtaba, I.M. and Macchietto, S., *Comput. chem. Engng.* **17** (1993), 1191.

Mujtaba, I.M. and Macchietto, S., *In Proceedings of the 1994 IChemE Research Event*, IChemE, UK, **2** (1994) 911.

Mujtaba, I.M. and Macchietto, S., *Chem. Eng. Sci.* **53** (1998), 2519.

Nad, M. and Spiegel, L., *In Proceedings CEF 87* , Sicily, Italy, April (1987), 737.

Rose, L.M., *Distillation Design in Practice* (Elsevier, New York, 1995).

CHAPTER 7

7. DESIGN AND OPERATION OPTIMISATION

7.1. Introduction

There are many papers in the literature dealing with the optimisation of the operation of batch distillation for a given column design and for a *single separation duty* (Farhat et al. 1990; Bernot et al., 1991, 1993; Logsdon and Biegler, 1993; Mujtaba and Macchietto, 1993). However, the design of batch columns has attracted much less attention. The main design parameters (Al-Tuwaim and Luyben, 1991) are the:

- Number of stages (plates) in the column
- Vapour boilup rate
- Diameter
- Still capacity (batch size)
- Reboiler and condenser size (heat transfer areas).

Previous literatures on batch distillation operation shows that it is possible to obtain a desired separation using a column with a large number of equilibrium stages and short batch times at low reflux ratio, or alternatively using a column with fewer stages but at higher reflux ratios and for longer batch time. Al-Tuwaim and Luyben (1991) reported on this issue in detail using binary and ternary separations.

Clearly, the design and operations issues are interdependent and must be considered simultaneously.

Diwekar et al. (1989), Diwekar and Madhavan (1991) considered the design of batch columns using a predefined operation policy (reflux ratio profiles) for a *single separation duty*, simple operations (no intermediate cuts) and very simplistic short-cut models. When treating multicomponent mixtures, their design algorithm is based either on (i) independent specification of all the components in the distillate product and thus unrealistic or (ii) independent specification of one of the components in the distillate product and the rest calculated using Hengestebeck-Geddes' equation (presented in Chapter 4) which is more realistic. Logsdon et al. (1990) addressed the

192

simultaneous design and operation for *single and multiple separation duties* for binary mixtures producing only main-cuts (no off-cuts). They used short cut models and their profit function does not reflect the different importance of each separation duty.

For *single separation duties*, Al-Tuwaim and Luyben (1991) provided a short-cut method for simultaneous optimisation of design and operation for binary and ternary separations. Using repetitive simulation strategy they have explained in detail the interaction between design and operation with an objective to maximise a capacity factor (total amount of specified products over unit time).

Sharif et al. (1998) considered simultaneous design and operation of multicomponent batch distillation for *single separation duties*. They considered a detailed model (similar to type IV-CMH) and treated the number of plates as an integer decision variable (design variable) in the optimisation framework that resulted in a Mixed Integer Nonlinear Programming (MINLP) problem. The optimisation problem was solved via decomposition using an outer approximation and augmented penalty (OA/AP) solution algorithm based on Viswanathan and Grossmann (1990). See the original references and Sharif (1990) for further details.

Mujtaba and Macchietto (1996) presented a more general formulation for optimal design and operation, dealing in particular with *multiple separation duties*, multicomponent mixtures, more complex operations (involving off-cuts) and more general objective functions. The method utilises a dynamic model (Type IV, Chapter 4) of the column in the form of a generic system of DAEs. Models of various "rigor" (type III and V, etc. of Chapter 4) can therefore be used.

In this chapter first, the optimisation method of Al-Tuwaim and Luyben (1991) for *single separation duty* is presented. Then the optimisation problem formulation and solution considered by Mujtaba and Macchietto (1996) is explained. Finally, the optimisation problem formulations considered by Logsdon et al. (1990) and Bonny et al. (1996) are presented.

7.2. Design and Operation Optimisation for Single Separation Duty by Repetitive Simulation

Single Separation Duty refers to the situation, where a single mixture (binary or multicomponent) is separated into several products using only one batch distillation column. Figures 7.1 and 7.2 show the operation sequences in STN form considered by Al-Tuwaim and Luyben (1991) for binary and ternary mixtures.

The following optimisation problem was considered:

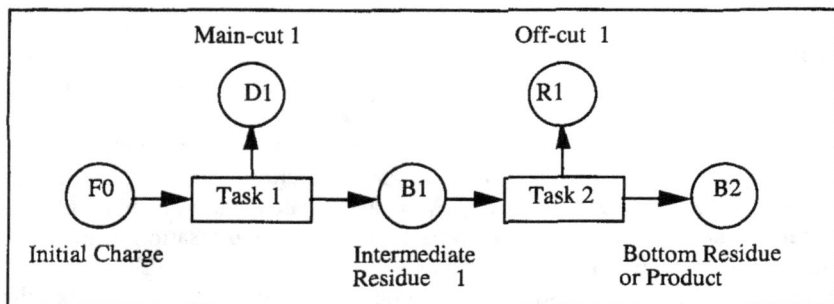

Figure 7.1. State Task Network for Binary Batch Distillation with One Main-Cut and One Off-cut. [Mujtaba and Macchietto, 1993][a]

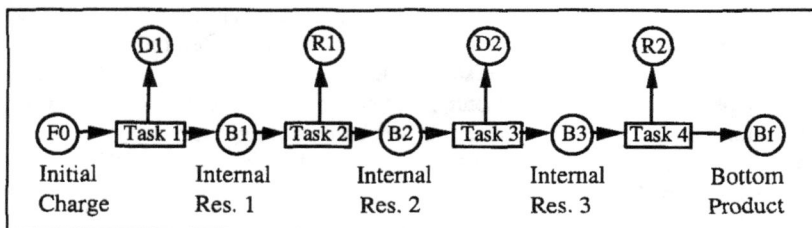

Figure 7.2. STN for Ternary Batch Distillation with Two Main-cuts and Two Off-cuts. [Mujtaba and Macchietto, 1996][b]

---

[a] Reprinted from *Computers & Chemical Engineering*, **17**, Mujtaba, I.M. and Macchietto, S., Optimal operation of multicomponent batch distillation- multiperiod formulation and solution, 1191-1207, Copyright (1993), with permission from Elsevier Science .

b Reprinted from *Journal of Process Control*, **6**, Mujtaba, I.M. and Macchietto, S., Simultaneous optimisation of design and operation of multicomponent batch distillation column-single and multiple separation duties, 27-36 , Copyright (1996), with permission from Elsevier Science .

given: the feed mixture and thermophysical properties, the amount and composition of the feed, the operation sequence, the distillate flow rate, the vapour boilup, the product specifications, the materials of construction of the column

determine: the optimum number of stages, column diameter, reboiler and condenser heat transfer areas and the optimum reflux ratio of operation which will maximise the capacity factor.

so as to minimise: the total investment and operating cost

subject to: lower and upper bounds on the number of stages

The capacity factor is defined as:

$$CAP = \frac{\sum_{j=1}^{n_c} Pr_j}{t_F + 0.5}$$ (7.1)

where Pr_j is the total moles of the jth product (on specification) collected during the batch, t_F is the total batch time, n_c is the number of components in the mixture. The batch set up time is 0.5 hr. For a binary separation the capacity factor becomes:

$$CAP = \frac{D1 + B2}{t_F + 0.5}$$ (7.2)

and for the ternary mixture it becomes:

$$CAP = \frac{D1 + D2 + Bf}{t_F + 0.5}$$ (7.3)

Al-Tuwaim and Luyben (1991) adopted the following strategy for solving the optimisation problem:

1. Start with the lower bound of the number of stages

2. Determine the optimum reflux ratio for the whole operation that will maximise the CAP

3. Calculate the column vapour flow rate using Equation 7.4

4. Calculate the column diameter using Equation 7.5
5. Calculate the energy consumption using Equation 7.7
6. Calculate the steam flow rate using Equation 7.8
7. Calculate the energy cost using Equation 7.9
8. Calculate the reboiler heat transfer area using Equation 7.10
9. Calculate the condenser heat transfer area using Equation 7.11
10. Calculate the column cost using Equation 7.12
11. Calculate the reboiler and condenser cost using Equation 7.13
12. Calculate the total design and operation cost (using Equations 7.11-7.13)

13. Increment the number of stages and repeat from step 2

14. Choose the design which will give the lowest design and operation cost.

Al-Tuwaim and Luyben (1991) used the following equations to calculate:

a. The column vapour flowrate (V', lb mole/hr)

$$V' = \frac{VF}{CAP} \tag{7.4}$$

where, V = vapour boilup rate, lb mole/hr, F = Distillate rate, lb mole/hr

b. The column diameter (DIA, ft)

$$DIA = 2\left(\frac{M_w V'}{\rho_v \pi V_m}\right)^{1/2} \tag{7.5}$$

where M_w is the vapour molecular weight, ρ_v is the vapour density (lb/ft^3), V_m is the maximum allowable superficial velocity (ft/s) given by

$$V_m = K_v\left(\frac{\rho_L - \rho_v}{\rho_v}\right)^{1/2} \tag{7.6}$$

where K_v = empirical constant = 0.3 ft/s for 24-in spacing between plates; ρ_L is the liquid density (lb/ft3).

c. The energy consumption (Q', Btu/hr)

$$Q' = V' H_v'$$ (7.7)

where, H_v' is the molal heat of vaporisation of the feed (Btu/lb mole)

d. The steam flowrate (W lb/hr)

$$W = Q'/H_v$$ (7.8)

where, H_v is the heat of vaporisation of steam = 915.5 Btu/lb

The Energy Cost (C_E, \$/hr) can be calculated using

$$C_E = C_{steam} W$$ (7.9)

where C_{steam} is the price of steam.

e. The reboiler heat transfer are (A_R, ft^2)

$$A_R = Q'/U\Delta T$$ (7.10)

where, U is the overall heat transfer coefficient, Btu/(hr ft^2 F), ΔT is the temperature gradient (F).

f. The condenser heat transfer are (A_C, ft^2)

$$A_C = Q'/U\Delta T$$ (7.11)

where, U is the overall heat transfer coefficient, Btu/(hr ft^2 F), ΔT is the temperature gradient (F).

g. The column cost (C_t, \$/hr)

$$C_t = C_b F_M + N_T C_{bt} F_{TM} F_{TT} F_{NT} + C_{pl}$$ (7.12)

where,
1. F_M is the material of construction factor (shell);

2. N_T is the number of trays;

3. C_{bt} is the cost of trays as a function diameter,
     $C_{bt} = 278.38 \exp(0.1739(DIA))$

4. C_b is the cost of the shell as a function of shell weight, height and diameter; C_{pl} is the platform and ladder cost as a function of diameter.

for $N_T < 40$:

$$C_b = \exp[6.329 + 0.18255(\ln W_s) + 0.0229(\ln W_s)^2]$$

$$C_{pl} = 182.50(DIA)^{0.73960} L^{0.70684}$$

where $W_s \equiv$ shell weight (lbs); $L \equiv$ tangent to tangent length (ft).

for $N_T \geq 40$

$$C_b = \exp[6.823 + 0.1.178(\ln W_s) + 0.02468(\ln W_s)^2]$$

$$C_{pl} = 151.81(DIA)^{0.63316} L^{0.80161}$$

5. F_{TM} is the materials of construction factor as a function of diameter (trays);

6. F_{TT} is the cost factor for tray type (valve, grid, bubble cap, sieve);
     $F_{TT} = 1.0$ for valve trays.
     For other trays see Al-Tuwaim and Luyben (1991).

7. F_{NT} is the cost factor for number of trays;
     for $N_T < 20$,

$$F_{NT} = 2.25 / (1.0414)^{N_T}$$

Otherwise, $F_{NT} = 1.0$

h. Condenser and Reboiler Capital Costs ($)

$$\text{Equipment cost} = 10^3 (0.73 + 0.064 A^{0.65}) F_{EX} \qquad (7.13)$$

where A is condenser or reboiler heat transfer area (Equation 7.10 or 7.11) F_{EX} is the condenser or reboiler material of construction factor. All the capital costs mentioned in Al-Tuwaim and Luyben (1991) were updated with cost indices for the year 1987.

7.2.1. Example - Single Separation Duty

Al-Tuwaim and Luyben (1991) considered the following optimisation problem.

given: an initial amount of binary feed, $F0 = 200$ lb mole; composition $x_{F0} = $ <0.5, 0.5> molefraction, vapour boilup rate, $V = 100$ lbmole/hr, relative volatility $= $ <3.0, 1.0>; distillate product purity, $x_{DI} = 0.95$ molefraction; stainless steel as the materials of column construction; energy cost $= \$5$ per 1000 lb of steam; payback period $= 5$ yrs; bounds on the number of stages $= $ <10, 100> and total operating hours $= 8500$ hrs per year.

determine: Optimum number of stages and optimum reflux ratio (Opt. RR) which will minimise the total design and operating cost.

Table 7.1 summarises the results of repetitive simulation following the strategy described in section 7.2. Table 7.1 shows that a column with 70 plates will give the best design and operation for the binary mixture being processed in the column in a campaign mode for 8500 hrs a year. However, it was not clear in Al-Tuwaim and Luyben (1991) whether the actual optimum value of N lies in between 60-70 or 70-80 plates.

See the original reference for more examples with binary and ternary mixtures. Also the sensitivities of the steam cost, materials of constructions and pay back period on the design and operation are presented in detail in Al-Tuwaim and Luyben (1991).

7.3. Design and Operation Optimisation for Single and Multiple Separation Duties: Problem Formulation and Solution

Mujtaba and Macchietto (1996) presented a general design and operation optimisation problem formulation and solution for *single and multiple separation duties*.

Table 7.1. Summary of Design and Operation Optimisation of Binary Mixture Using *Single Duty Separation*. [Adopted from Al-Tuwaim and Luyben, 1991]

N	Opt. RR	CAP	DIA	AR, AC	Capital Cost	Steam Cost	Total Cost
10	1.70	51.7	2.4	1059, 1324	2.233	1.230	1.677
20	1.40	59.6	2.3	919, 1149	2.295	1.067	1.526
30	1.10	63.5	2.2	862, 1078	2.454	1.002	1.492
40	0.98	68.1	2.1	804, 1005	2.591	0.934	1.452
50	0.82	72.4	2.1	756, 946	2.724	0.878	1.423
60	0.68	76.5	2.0	716, 895	2.854	0.831	1.402
70 (opt)	**0.58**	**80.2**	**2.0**	**683, 854**	**2.984**	**0.793**	**1.390**
80	0.54	81.9	1.9	669, 836	3.138	0.777	1.404
90	0.54	82.2	1.9	666, 833	3.309	0.774	1.435
100	0.54	82.2	1.9	666, 833	3.481	0.774	1.470

Note: All costs are in $100 000.

7.3.1. Representation of Design, Operations and Separations Duties

7.3.1.1. Design and Operation Decision Variables

It is aimed to optimally design a batch column capable of processing M number of feed mixtures, each mixture constituted of NC_m number of components (***Multiple Separation Duties***, Figure 7.3). The most important variables characterising a batch column design are the still capacity (B_0), the number of theoretical stages (N) and the vapour boilup rate (V). The latter can be used to obtain both the column diameter and reboiler and condenser duties (as described in section 7.2), hence it represents a convenient design variable. Here storage issues are ignored and therefore the initial set of design decision variables considered is $D^D = \{N, B_0, V\}$.

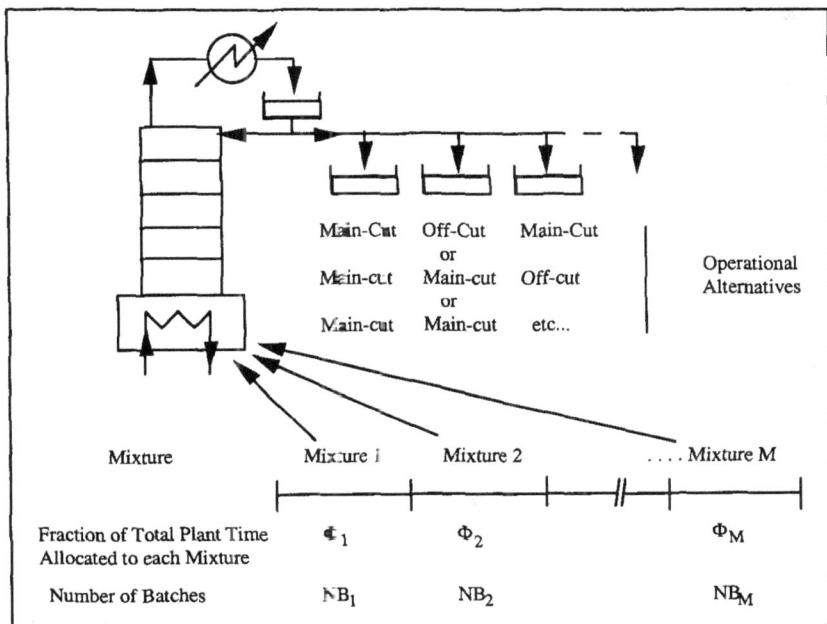

Figure 7.3. Batch distillation Column with *Multiple Separation Duty* and Multiple Operational Alternatives. [Mujtaba and Macchietto, 1996][c]

The procedure for processing a given batch charge of mixture m (*operation m*), can be viewed as a sequence of NT_m distillation *tasks* to produce one or more main-cuts, possibly some intermediate off-cuts and a final bottom residue or product (Figure 7.1). For a ternary mixture this can be represented in the form of a STN shown in Figure 7.2. Each *state* s is characterised by a name (e.g. $D1$), an amount S_S (e.g. S_{D1}) and a composition vector x_S (e.g. x_{D1}). The molar fraction of an individual component j in state s is denoted by x_s^j. The sets of external feed states, main-cuts and off-cuts states in *operation m* are defined as EF_m, MP_m, and OP_m, respectively. For example, Figure 7.2 shows operation 1 for a ternary mixture distillation with NT_1=4 tasks, EF_1={$F0$}, MP_1={$D1$, $D2$, Bf} and OP_1={$R1$, $R2$}. Several feed states could occur, for example in the preparation of a mixed charge or

[c] Reprinted from *Journal of Process Control*, **6**, Mujtaba, I.M. and Macchietto, S., Simultaneous optimisation of design and operation of multicomponent batch distillation column-single and multiple separation duties, 27-36 , Copyright (1996), with permission from Elsevier Science .

in the subsequent addition of material to the still. These mixing tasks can be considered but are ignored in the following. Overall, a batch charge B_{Om} of feed mixture m produces a quantity Θ_s, $\forall s \in \{MP_m, OP_m\}$, of each product. A material in state s is assigned a unitary value C_s. Off-cut materials may have a positive residual value, zero value or a negative value accounting for disposal cost.

A name and a set of input and output states (SI_i and SO_i, respectively) characterise each distillation *task* i. Each task is modelled by an associated set of DAEs of the form $f(t, \dot{x}_i(t), x_i(t), u_i(t), v) = 0$, where $u_i(t)$ is a vector of time varying control variables (e.g. reflux ratio, actual vapour boilup rate). With each task an *initialisation procedure* and an *integration procedure* are associated. The *initialisation procedure* permits the calculation of a consistent initial set of all model variables ($x_i(t)$ and $\dot{x}_i(t)$), given the amounts and compositions of all input states fed to the task. From these initial conditions and given a profile of all control variables, u_i, the *integration procedure* permits calculation of all trajectories up to some termination condition (e.g. final time), and hence all final column states, product amounts, utility requirements, etc. and output states for the task. The time for processing each batch (batch time), t_{bm}, is assumed to consist of the processing time t_{pm} (the sum of all task times) and a set up time t_{su}. A fixed set up cost C_{su} per batch is assumed.

Mujtaba and Macchietto (1996) assumed that the *structure* of each operation (a specific sequence of tasks and states and choice of main and off-cut states) is given a priori but the operating control variables, u_i (in particular reflux ratio profile) and termination conditions for each task are to be optimised. It is usually best to operate at the maximum available energy input rate, unless hydraulic problems are encountered. For this reason, the vapour load is assumed to be always at its design value V during operation (Diwekar et al., 1989, 1991) and is dropped from the optimisation variables vector.

Chapter 6 shows that-

- **given** the feed state (amount and composition) and V, and
- **specification** of a set of two additional operations decision variables for each task i, denoted by the vector d^o_i (typically the mole fraction of a key component in a distillate or bottom product and a key component recovery, two product purities for the same component, one purity and a product amount, etc.)

the optimisation problem permits calculation of the amount of distillate, intermediate or bottom product and their composition, recoveries, energy requirements, etc. *for the task*, as the solution of a well defined dynamic optimisation problem (Mujtaba and Macchietto, 1993). This is discussed in detail in Chapter 6.

Thus with given BO_m and V, specification of the set of operations decision variables $D^O_m = \{d^O_i, i = 1, .., NT_m\}$ (a total of $2* NT_m$ decision variables) and control variables $U_m = \{u_i, i = 1, .., NT_m\}$ for all distillation tasks in *operation m*, it will be possible to calculate the overall performance measures for the batch (total distillation time, overall separation, products and intermediates amounts, recoveries, energy, etc.). The same decision variables may be *optimised* to achieve some overall objective for the operation, (e.g. overall profit) subject to overall constraints (e.g. overall time, energy, etc.).

7.3.1.2. Separation Duty

It is necessary to specify the relative importance of each duty in order to fully characterise the design requirements. The total time horizon for completing all separation tasks is denoted by H and the available plant operation time over a year by H_{yr}. For each mixture m, a total amount F_m is processed in $NB_{m,H}$ number of batches (all equal) of size B_{0m} according to *operation m*. This requires a total operation time $T_m = \Phi_m *H$ (Φ_m is the fraction of production time allocated to mixture m, Figure 7.3). Since it is always optimal to use the column at its maximum capacity (full charges), the amount processed in each batch (feed charge or batch size) for all batches are equal to the still capacity. A capacity utilisation factor may be easily introduced to account for any density differences between mixtures. Hence the mixture index can be dropped and B_0 is used instead of B_{0m} to denote both the pot capacity and the batch size for each mixture. The following constraints given by Equations 7.14-7.18 therefore hold (for $m = 1, .., M$).

$$T_m = \Phi_m *H \tag{7.14}$$

$$H = \Sigma_m T_m \tag{7.15}$$

$$NB_{m,H} = F_m / B_0 = T_m / t_{bm} \tag{7.16}$$

$$B_0 < F_m \tag{7.17}$$

$$t_{bm} < T_m \tag{7.18}$$

The *multiple separation duty* specifications can be made in several ways. Two of which are:

Strategy I

H and F_m given - Larger B_0 will always be favoured (for finite set up times), because the idle time for the plant is reduced and because the capital costs typically increase with capacity by an exponent less than 1 (economy of scale). The optimum value of B_0 will depend on constraints 7.14-7.18.

Strategy II

H and Φ_m given - In this case more material per batch can always be processed just by increasing B_0 indefinitely, up to the rather useless bound of constraint 7.17. Therefore a specification of the pot size is required a priori.

7.3.2. Objective Function

Mujtaba and Macchietto (1996) considered a general profit function P (the annual profit) which includes net revenues (products, raw materials), operating costs and annualised capital costs. The functions given by Logsdon et al. (1990) are used here for the latter two terms, so as to enable a direct comparison. Other cost models could easily be utilised. Operating costs include utilities (mainly steam) and capital costs include a column with N intermediate plates, a reboiler (with pot) and a condenser. The capital and operating costs of storage for intermediate products are ignored.

$$P \ (\$/yr) = \sum_m (\text{Re}\,v_m - OC_m)NB_m - ACC \qquad (7.19)$$

where, net revenue ($/batch) is:

$$\text{Re}\,v_m = (\sum_s \Theta_s C_s) - B_0 C_{s'} \qquad s \in \{MP_m, OP_m\}, s' \in EF_m \qquad (7.20)$$

no. of batches / yr is:

$$NB_m = NB_{m,H}*(H_{yr}/H) = (\Phi_m*H/t_{bm})*(H_{yr}/H) \qquad (7.21)$$

operating costs ($/batch) is:

$$OC_m = (K_3*V/A \ \$/hr)*(t_{pm} \ hr/batch) + C_{su} \ (\$/batch) \qquad (7.22)$$

annualised capital cost ($/yr) is:

$$ACC = K_1 V^{0.5} N^{0.8} + K_2 V^{0.65} \qquad (7.23)$$

with V in kmol/hr, $K_1 = 1500$, $K_2 = 9500$, $K_3 = 180$ and $A = 8000$. $C_{s'}$ is the price for the feed ($/kmol).

7.3.3. Optimisation Problem Formulation and Solution

The optimisation problem is expressed as follows:

given: M feed mixtures, each with NC_m components to be separated according to a predefined operation structure (STN),
a set of product specifications and constraints for each separation duty (purities of key components in main products, amounts, etc.),
product specifications and constraints for the multiple duty operation (production horizon and fraction allocated to each mixture)

determine: the optimum set of design decision variables $D^D = \{B_0, N, V\}$, operating decisions $D^O = \{D^O{}_1, ..., D^O{}_M\}$, (total number of operating decision variables, $NV = 2 * \Sigma_m NT_m$), and operating control variables $U = \{U_1, .., U_M\}$ (reflux ratios and times) for all tasks in all operations

so as to maximise: the objective function P
subject to: any constraints

Mujtaba and Macchietto (1996) formulated a two-level optimisation problem with D^D and D^O optimised in an outer loop and U optimised in the inner loop, using an extension of the method of Mujtaba and Macchietto (1993). The outer loop problem can be written as:

P1-0   Max P
             D^D, D^O

     s. t. $h(D^D, D^O) = 0$ (overall equality constraints)
                     $g(D^D, D^O) \leq 0$ (overall inequality constraints)
                     $D^{Dl} \leq D^D \leq D^{Du}$ (bounds on design variables)
                     $D^{Ol} \leq D^O \leq D^{Ou}$ (bounds on operating variables)

Equality constraints $h(D^D, D^O) = 0$ may include, for example, a ratio between the amounts of two products, etc. Inequality constraints $g(D^D, D^O) \leq 0$ for the overall operation include Equations 7.14-7.18 (the first two of which are easily eliminated when Φ_m and H are specified) and possibly bounds on total batch time for individual mixtures, energy utilisation, etc. Any variables of D^D and D^O which are fixed are simply dropped from the decision variable list. Here, *Strategy II* was adopted for the multiple duty specification, requiring B_0 to be fixed a priori. Similar considerations hold for V, the vapour boilup rate. The batch time is inversely proportional to V for a specified amount of distillate. Also alternatively, for a given batch time, the amount of product is directly proportional to V. This can be further explained through Equations 7.24-7.26):

Reflux ratio (r) definition:

$$r = \text{liquid refluxed/vapour rate} = L_R/V \qquad (7.24)$$

Rate of distillate, L_D (kmol hr):

$$L_D = V - L_R = V(1-r) \qquad (7.25)$$

Batch time, t (hr) for a given amount of distillate (S_D^* , kmol):

$$t = S_D^* /L_D \qquad (7.26)$$

An increase of V will increase L_D (Equation 7.25) and vice versa for a given reflux ratio profile r. An increase in L_D will decrease batch time t and vice versa (Equation 7.26) for a given amount of distillate to be produced (S_D^*).

As the operating and capital costs grow with V by an economic factor less than 1, a column with large V will always be profitable (Diwekar et al., 1989) and the problem becomes unbounded (also proven in Logsdon et al., 1990). Hence V is also fixed a priori. This leaves just N (the number of internal ideal separation stages or plates) as the only design variable to be optimised ($D^D = \{N\}$). Out of all possible operation decision variables, it is common to specify the mole fraction of key components in main-cuts and sometimes some recoveries or amounts for off-cuts. Assuming NSP such specifications are made, there are $(NV + 1 - NSP)$ outer optimisation problem decision variables.

Apart from N (an integer) the optimisation problem is a standard Nonlinear Programming (NLP) problem, with the inner optimisation problem providing the values of the outer objective function and constraints. In the outer problem Mujtaba and Macchietto (1996) proposed to use N as a continuous variable (within integer bounds). When N is passed to the inner loop problems its value is rounded off to the nearest integer (e.g. $N=10.66$ is rounded to 11, but 10.45 is rounded to 10), as required for plate-to-plate calculations. The gradient with respect to N is calculated by finite difference using function and constraint values at the rounded off value of N and at its next higher integer value. At the final iteration the immediately lower and upper integer values are checked and the best solution is taken as the optimum. Eliceche and Sargent (1981) also considered a similar strategy.

Having design parameters fixed in the outer problem and with a specific choice of D^O (discussed in section 7.2) the inner loop optimisation can be partitioned into M independent sequences (one for each mixture) of NT_m dynamic optimisation problems. This will result to a total of $ND = \Sigma_m NT_m$ problems. In each (one for each task) problem the control vector u_i for each task is optimised. This can be clearly explained with reference to Figure 7.3 which shows separation of M (=2) mixtures (mixture 1 = ternary and mixture 2 = binary) and number of tasks involved in each separation duty (3 tasks for mixture 1 and 2 tasks for mixture 2). Therefore, there are 5 (= ND) independent inner loop optimal control problems. In each task a parameterisation of the time varying control vector into a number of control intervals (typically 1-4) is used, so that a finite number of parameters is obtained to represent the control functions. Mujtaba and Macchietto (1996) used a piecewise constant approximation to the reflux ratio profile, yielding two optimisation parameters (a control level and interval length) for each control interval. For any task i in operation m the inner loop optimisation problem (problem Pl-i) can be stated as:

given:	an initial charge from *task* i-1
determine:	an optimal reflux ratio profile $r(t)$ to obtain the required product (S_{Di}^* , x_{Di}^*)
so as to minimise:	the task time
subject to:	any constraints

Mathematically it can be written as:

Pl-i
$$\min_{r(t_i)} \quad J = t_i$$

$$\text{subject to} \quad S_{Di} \geq S_{Di}^*$$

$$x_{Di} \geq x_{Di}^*$$

DAE model equations
etc.

where S_{Di}, x_{Di} are the amount of distillate and its composition in *task* i at the end

of *task* time t_i, S_{Di}^*, x_{Di}^* are the specified amount of distillate product and its purity.

The solution of the inner loop problems is achieved using rigorous dynamic optimisation algorithms detailed in Chapter 5 and 6 and also by Mujtaba and Macchietto (1993) (except for minor nomenclature changes). However, the solution of the outer loop optimisation problem especially calculating gradients with respect to the decision variables are slightly different and therefore the method will be explained here with reference to Figure 7.4.

A complete solution of the sequence of inner loop problems (Figure 7.5a) is required for each "function evaluation" of the outer loop problem. Mujtaba and Macchietto (1996) obtained the gradients of the objective function with respect to the decision variables by finite difference technique but in an efficient way. This is achieved by utilising the solution of each inner loop problem from the function evaluation stage.

During the function evaluation step, the solution (reflux profiles, column profile, duration of task) of each inner loop is stored as A, B, C, D and E respectively, as shown in Figure 7.5a. For the gradient with respect to design decision variable N, the variable is perturbed and the inner loop problems are solved as shown in Figure 7.5b. The new solutions A1, B1, C1, D1 and E1 are used to calculate the gradient of the objective function with respect to the design variable (N).

It is very important to note that the operating decision variables are selected in such a way that the variables associated with one separation duty do not affect the operations involved in other separation duties. Also operating decision variables associated with any tasks within any separation duty do not affect preceding tasks but only affect the following distillation tasks. Therefore, within any separation duty, the evaluation of gradients with respect to operating decision variables of a particular task requires the solution of the inner loop problem concerning that task and the tasks ahead. For example, the evaluation of the gradient with respect to d^O_2

in Task 2 of separation duty 1, solutions of inner loop problems P1-2 and P1-3 concerning Task 2 and 3 respectively are required (B2, C2, Figure 7.5c). The new solutions B2, C2 together with A, D and E from function evaluation step are used to calculate the gradient with respect to d^O_2. Similarly solutions D3, E3 together with A, B and C are used to calculate the gradient with respect to d^O_1 of separation duty 2 (Figure 7.5d).

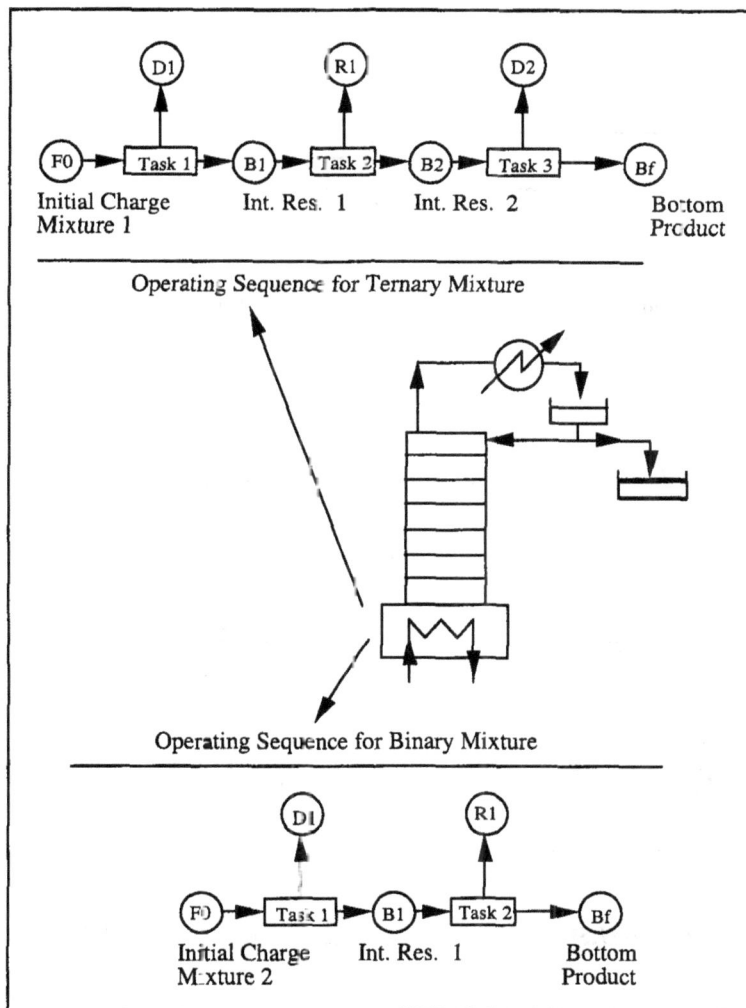

Figure 7.4. Operating Sequences of Batch Distillation Column with Two Separation Duties [Mujtaba and Macchietto, 1996][d]

---

[d] Reprinted from *Journal of Process Control*, **6**, Mujtaba, I.M. and Macchietto, S., Simultaneous optimisation of design and operation of multicomponent batch distillation column-single and multiple separation duties, 27-36 , Copyright (1996), with permission from Elsevier Science .

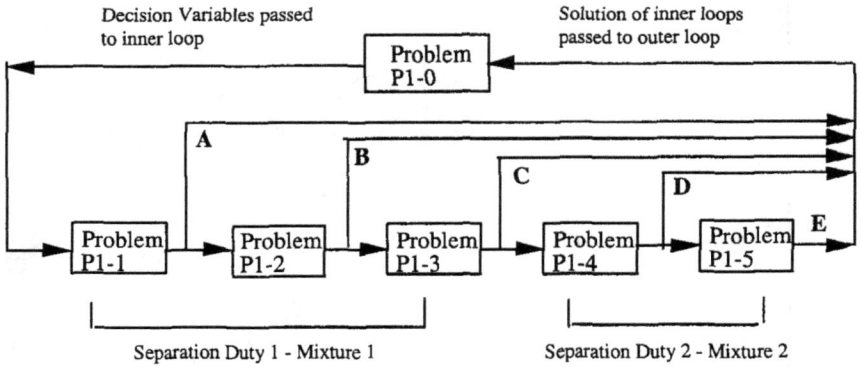

Figure 7.5a. Function Evaluation. [Mujtaba and Macchietto, 1996][e]

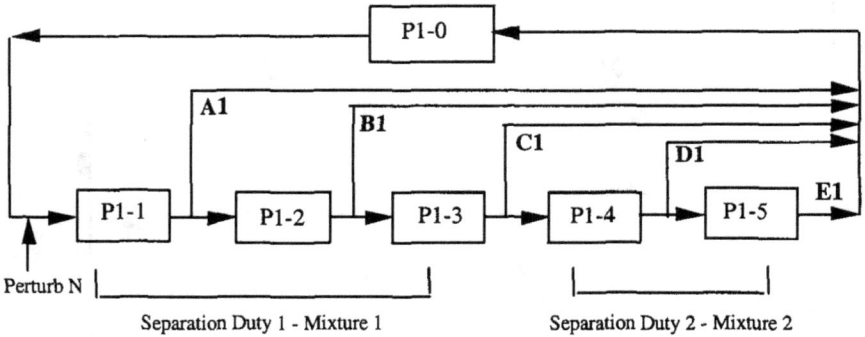

Figure 7.5b. Gradient with respect to design variable N^e

---

[e] Reprinted from *Journal of Process Control*, **6**, Mujtaba, I.M. and Macchietto, S., Simultaneous optimisation of design and operation of multicomponent batch distillation column-single and multiple separation duties, 27-36 , Copyright (1996), with permission from Elsevier Science .

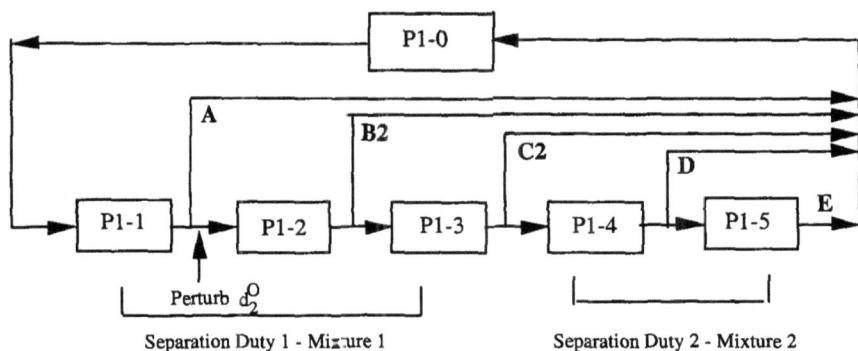

Figure 7.5c. Gradient with respect to operating variable (separation duty 1). [Mujtaba and Macchietto, 1996]^f

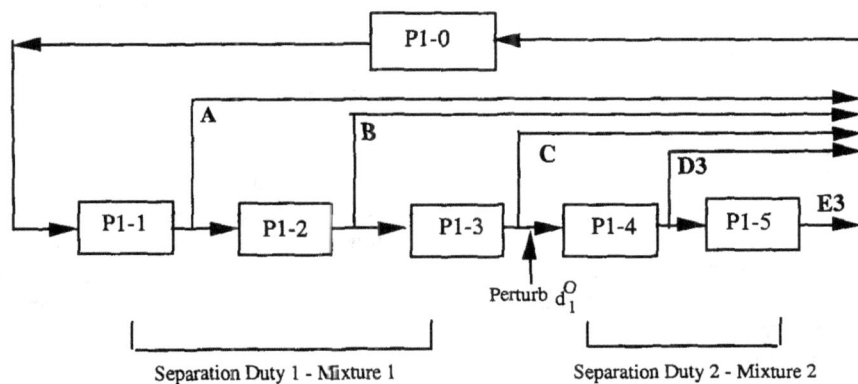

Figure 7.5d. Gradient with respect to operating variable (separation duty 2). [Mujtaba and Macchietto, 1996]^f

---

^{f} Reprinted from *Journal of Process Control*, **6**, Mujtaba, I.M. and Macchietto, S., Simultaneous optimisation of design and operation of multicomponent batch distillation column-single and multiple separation duties, 27-36 , Copyright (1996), with permission from Elsevier Science .

The solution procedure described above is generally applicable for *single and multiple separation duties*. The M sequences of inner loop problems for different mixtures may be solved in parallel if desired. The *single separation duty* optimal design and operation problem is simply obtained by setting $M=1$. The *single separation duty* optimal operation problem presented in Chapter 6 can simply be obtained by setting $M=1$ and fixing variables in D^D.

7.3.4. Examples

7.3.4.1. Single Separation Duty, $M=1$

The example is taken from Mujtaba and Macchietto (1996). As described in Chapter 4 (section 4.2.4.2.3), Nad and Spiegel (1987) considered a *single duty* ternary separation in an 18-plate (excluding the condenser and reboiler) batch distillation column using the operation sequence shown in Figure 7.2. The total operation time was 8.86 hr. Mujtaba and Macchietto developed the optimal design and operating policy for this ternary separation using the operation sequence and product specifications used by Nad and Spiegel. The results of Nad and Spiegel are used here as the base case for comparison.

In this separation, there are 4 distillation tasks ($NT=4$), producing 3 main product states $MP=\{D1, D2, Bf\}$ and 2 off-cut states $OP=\{R1, R2\}$ from a feed mixture $EF=\{F0\}$. There are a total of 9 possible outer decision variables. Of these, the key component purities of the main-cuts and of the final bottom product are set to the values given by Nad and Spiegel (1987). Additional specification of the recovery of component 1 in Task 2 results in a total of 5 decision variables to be optimised in the outer level optimisation problem. The detailed dynamic model (Type IV-CMH) of Mujtaba and Macchietto (1993) was used here with non-ideal thermodynamics described by the Soave-Redlich-Kwong (SRK) equation of state. Two time intervals for the reflux ratio in Tasks 1 and 3 and 1 interval for Tasks 2 and 4 are used. This gives a total of 12 (6 reflux levels and 6 switching times) inner loop optimisation variables to be optimised. The input data, problem specifications and cost coefficients are given in Table 7.1.

The optimum number of plates, the optimum values of the decision variables for both outer and inner loop optimisation problems, and optimal amounts and composition of all products are shown in Table 7.2. Typical composition profiles in the product accumulator tank are shown in Figure 7.6. Bold faced mole fractions in Table 2 are the specifications (all satisfied) and underlined mole fractions are decision variables which were optimised. Although the optimum number of plates is almost close to that of the base case, the optimal total operation time is 14% lower than the base case. The profit with the optimal design and operation is 35% higher than that for the base case (calculated using the same cost model). This is obtained with only 6 reflux intervals while the base case operation required about 20 reflux changes. There is no need for initial total reflux in the optimal operation as required in the base case (for 2.54 hrs). It is reasonable to assume that the use of large total reflux operation period (2.54 hrs) affects the overall profit for the base case.

For each outer loop "function" and "gradient" evaluation 4 and 14 inner loop problems were solved respectively (a total of 124 inner loop problems). For the inner loop problems 12-14 iterations for Tasks 1 and 3 and 5-7 iterations for Tasks 2 and 4 were usually required. For this problem size and detail of dynamic and physical properties models the computation time of slightly over 5 hrs (using SPARC-1 Workstation) is acceptable. It is to note that the optimum number of plates and optimum recovery for Task 1 (Table 7.2) are very close to initial number of plates and recovery (Table 7.1). This is merely a coincidence. However, during function evaluation step the optimisation algorithm hit lower and upper bounds of the variables (shown in Table 7.1) a number of times. Note that the choices of variable bounds were done through physical reasoning as explained in detail in Chapter 6 and Mujtaba and Macchietto (1993).

7.3.4.2. Multiple Separation Duties, Two Mixtures ($M=2$): Effect of Different Allocation Time

This example is taken from Mujtaba and Macchietto (1996). The problem is to design a column for 2 binary separation duties. One of the separations is very easy compared to the other one. The fraction of production time for each duty is specified together with the still capacity (B_0) and the vapour load (V). Each binary mixture produces only one main distillate product and a bottom residue (states $MP_1=\{D1, Bf1\}$ and $MP_2=\{D2, Bf2\}$) from feed states $EF_1=\{F1\}$ and $EF_2=\{F2\}$, respectively, with only one distillation task in each separation duty. Desired purities are specified for the two main-cuts (x^l_{D1} and x^l_{D2}). Also obtain the optimal operating policies in terms of reflux ratio for the separations.

In this problem, there are 3 outer loop decision variables, N and the recovery of component 1 from each mixture ($Re^l_{D1,B0}$, $Re^l_{D2,B0}$). Two time intervals for reflux ratio were used for each distillation task giving 4 optimisation variables in each inner loop optimisation making a total of 8 inner loop optimisation variables. A series of problems was solved using different allocation time to each mixture, to show that the optimal design and operation are indeed affected by such allocation. A simple dynamic model (Type III) was used based on constant relative volatilities but incorporating detailed plate-to-plate calculations (Mujtaba and Macchietto, 1993; Mujtaba, 1997). The input data are given in Table 7.3.

Table 7.1. Input Data, Product Specifications and Decision Variables for Ternary Distillation (Detailed Dynamic Model, Single Duty)[g].

Recoveries $Re^j_{a,b}$ are defined as: amount of component j in state a / amount of component j in state b. For $M=1$, $\Phi_1 = 1$. The mixture index is omitted for clarity.

Column:	Still Capacity = Batch size, $F0$, kmol	= 2.93
	Condenser Vapor Load, V, kmol/hr	= 2.75
	Column Holdup, kmol:	
	Condenser = 1.2% of $F0$ Internal Plates (total) = 1.9% of $F0$	

Duties: Total Plant Operation Time, H_{yr}, hr/yr   = 8000.0
Total time horizon, H, hr   = 8000.0
No. of mixtures and fraction of horizon for each: $M=1$, $\Phi_1 = 1$

Mixture: Components: Cyclohexane, n-Heptane, Toluene
Feed Composition, x_{F0}, mole fraction   = <0.407, 0.394, 0.199>

Specifications:
Cyclohexane mole fraction in $D1$, x^1_{D1} = 0.895
n-Heptane mole fraction in $D2$, x^1_{D2} = 0.863
Toluene mole fraction in Bf, x^3_{Bf} = 0.990
Cyclohexane recovery in $R1$, $Re^1_{R1,B1}$ = 0.95

Outer Loop Decision Variables, initial value <lower and upper bounds>:

Number of Plates, N	= 16	<12	22>
Cyclohexane recovery in $D1$, $Re^1_{D1,F0}$	= 0.85	<0.70	0.92>
Cyclohexane mole fraction in $R1$, x^1_{R1}	= 0.40	<0.30	0.45>
Toluene recovery in $D2$, $Re^2_{D2,B2}$	= 0.85	<0.70	0.92>
Toluene mole fraction in $R2$, x^2_{R2}	= 0.30	<0.30	0.45>

Costs:

C_{D1} = 30.0 \$/kmol, C_{D2} = 26.0 \$/kmol, C_{R1} = C_{R2} = - 1.0 \$/kmol
C_{Bf} = 24.0 \$/kmol, C_{F0} = 2.0 \$/kmol, C_{su} = 0.0
OC and ACC as described in section 7.3

---

[g] Reprinted from *Journal of Process Control*, **6**, Mujtaba, I.M. and Macchietto, S., Simultaneous optimisation of design and operation of multicomponent batch distillation column-single and multiple separation duties, 27-36 , Copyright (1996), with permission from Elsevier Science

Table 7.2. Summary of Results - Example 1^h

Optimal Outer Loop Decision Variables:
$N = 17$ $Re^1_{D1,F0} = 0.842$ $x^1_{R1} = 0.373$ $Re^2_{D2,B2} = 0.892$ $x^2_{R2} = 0.362$

Optimal Inner Loop Decision Variables:

Task	1		2	3		4
Interval	1	2	1	1	2	1
Reflux Ratio	0.799	0.879	0.922	0.853	0.889	0.875
End Time, hr	1.24	2.56	4.79	5.82	7.01	7.64
Task Time, tp hr		2.56	2.27		2.22	0.63

Optimal Products Amount (per batch) and Composition:

Product	Amount, kmol	Composition
D1	1.12	<**0.895**, 0.100, 0.005>
R1	0.48	<<u>0.373</u>, 0.601, 0.026>
D2	0.78	<0.012, **0.863**, 0.125>
R2	0.22	<0.000, <u>0.362</u>, 0.638>
Bf	0.33	<0.000, 0.010, **0.990**>

Optimal No. of Batches: $NB = 1047.1$

Profit:

Optimal Design and Operation	= 15,168.0 \$/yr
Base Case Operation (Nad and Spiegel, 1987)	= 11,165.2 \$/yr

Solution Statistics:

No. of Outer Loop Function Evaluations, NF	= 10
No. of Outer Loop Gradient Evaluations, NG	= 6
Total CPU time, hr on SPARC-10	= 5.23

---

[h] Reprinted from *Journal of Process Control*, **6**, Mujtaba, I.M. and Macchietto, S., Simultaneous optimisation of design and operation of multicomponent batch distillation column-single and multiple separation duties, 27-36 , Copyright (1996), with permission from Elsevier Science

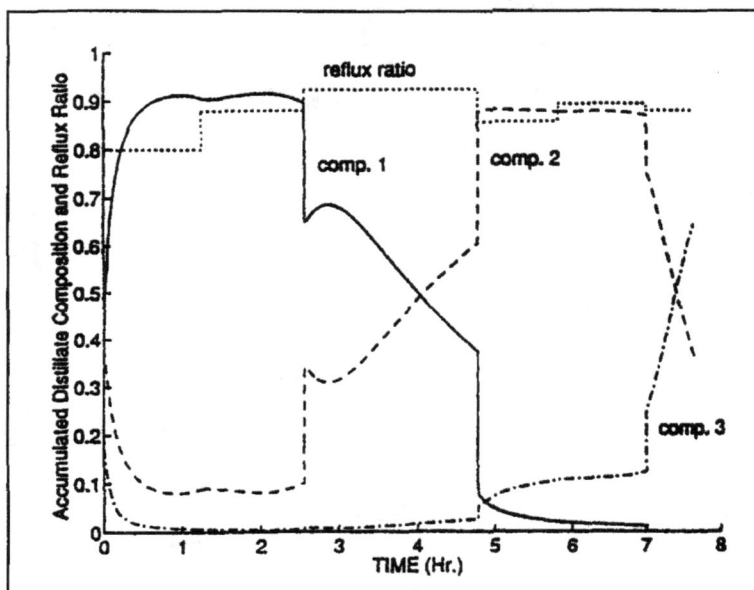

Fig. 7.6. Composition and Reflux Ratio Profiles (Single Separation Duty)[i]

The optimum results for different times allocated to each separation duty are summarised in Tables 7.4 and 7.5. They clearly show that the time allocation plays an important role in determining the size of the column as well as the operation policy. The more difficult the separation (Separation 1) is the larger is the number of plates required and the lower is the yearly profit achieved. The situation reverses as the separation becomes easy (Separation 2). Table 7.4 also presents two extreme cases, where a column designed for separation 2 is allocated 100 % of the time to Separation 1 (1a) and one designed for separation 1 is allocated 100% to Separation 2 (5a). Case 1a shows a significant drop (66 %) in profit compared to case 1, however for case 5a this is only 3 % when compared to case 5. This clearly shows that an easy separation task can easily be accommodated in a column which was designed for a more difficult separation without losing much profit, but the reverse situation might have significant impact on the profit, and in many cases might not even give a feasible separation.

---

[i] Reprinted from *Journal of Process Control*, **6**, Mujtaba, I.M. and Macchietto, S., Simultaneous optimisation of design and operation of multicomponent batch distillation column-single and multiple separation duties, 27-36 , Copyright (1996), with permission from Elsevier Science

Table 7.3. Input Data and Product Specifications using Binary Mixtures (Simple Dynamic Model, Multiple Duties). [Mujtaba and Macchietto, 1996][j]

Column:	Still Capacity = Batch size, BO, kmol	= 10.0
	Condenser Vapour Load, kmol/hr	= 10.0
	Column Holdup, kmol:	
	Condenser	= 1.0 % of BO
	Internal Plates (total)	= 3.0 % of BO

Duties:	Total Plant Operation Time, H_{yr}, hr/yr	= 8000.0
	Total time horizon, H, hr	= 8000.0

No. of mixtures and fraction of time horizon for each mixture:
$M=2$; $\Phi_1, \Phi_2 = $ variable
Set up time for each batch, $t_{su1} = t_{su2}$,hr   = 0.4

Mixtures:

Feed Composition for Mixture 1 and 2, (mole fraction)
$x_{F1} = x_{F2} = <0.5, 0.5>$
Relative Volatility: Mixture 1 = $<1.5, 1.0>$, Mixture 2 = $<2.5, 1.0>$

Specifications:

Purity of distillate products (mole fraction) $x^1_{D1} = x^1_{D2} = 0.95$

Costs:

$C_{D1} = C_{D2} = 20.0$ \$/kmol, $C_{Bf1} = C_{Bf2} = C_{su1} = C_{su2} = 0.0$
$C_{F1} = C_{F2} = 2.0$ \$/kmol, OC_1, OC_2 and ACC as in section 7.3

---

[j] Reprinted from *Journal of Process Control*, **6**, Mujtaba, I.M. and Macchietto, S., Simultaneous optimisation of design and operation of multicomponent batch distillation column-single and multiple separation duties, 27-36 , Copyright (1996), with permission from Elsevier Science

Table 7.4. Summary of the Results for Multiple Separation Duties
[Mujtaba and Macchietto, 1996][k]

Case	Φ_1 / Φ_2	N	$Re^l{}_{D1,F1} / Re^l{}_{D2,F2}$	NB_1 / NB_2	P, \$/yr
1	100 / 0	20	0.692 /	2,861.4 /	55,306.0
1a	100 / 0	9 (fixed)	0.536 /	1,448.7 /	18,799.8
2	75 / 25	18	0.776 / 0.829	1,759.5 / 1,219.5	98,423.9
3	50 / 50	16	0.752 / 0.857	1,152.7 / 2,314.8	143,012.2
4	25 / 75	11	0.502 / 0.866	638.6 / 3,311	181,934.4
5	0 / 100	9	/ 0.853	/ 4,410.0	236,562.8
5a	0 / 100	20 (fixed)	/ 0.785	/ 5,235.6	232,144.0

Table 7.5. Optimal Product Amounts per Batch and Reflux Ratio Profiles -
Multiple Separation Duties Example. [Mujtaba and Macchietto, 1996][k]

Case	Optimal reflux ratio profile				Optimal amount (per batch) (kmol)
	Separation 1		Separation 2		
	Time (hr)	Reflux ratio	Time (hr)	Reflux ratio	D1/D2
1	0.00-1.31	0.833			3.644 /
	1.31-2.40	0.865			
1a	0.00-4.70	0.945			2.821 /
	4.70-5.12	0.942			
2	0.00-1.70	0.845	0.00-0.58	0.578	4.083 / 4.363
	1.70-3.01	0.889	0.58-1.24	0.709	
3	0.00-1.80	0.856	0.00-0.63	0.591	3.960 / 4.512
	1.80-3.07	0.901	0.63-1.33	0.723	
4	0.00-2.04	0.904	0.00-0.77	0.615	2.642/ 4.559
	2.04-2.73	0.901	0.77-1.40	0.747	
5			0.00-0.79	0.632	/ 4.490
			0.79-1.41	0.748	
5a			0.00-0.52	0.576	/ 4.133
			0.52-1.13	0.682	

---

[k] Reprinted from *Journal of Process Control*, **6**, Mujtaba, I.M. and Macchietto, S., Simultaneous optimisation of design and operation of multicomponent batch distillation column-single and multiple separation duties, 27-36 , Copyright (1996), with permission from Elsevier Science

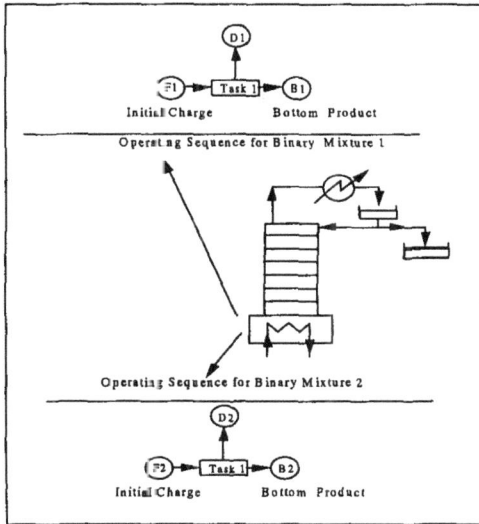

Figure 7.7. Multiperiod Operating Sequences used by Logsdon et al.

7.4. Multiperiod Design and Operation Optimisation by Logsdon et al.

Logsdon et al. (1990) formulated multiperiod *multiple separation duties* design and operation optimisation problem for two binary mixtures. Only one distillate product per mixture was considered according to the operation shown in Figure 7.7.

Therefore only two time periods are involved in the optimisation problem formulation, one for each mixture. Instantaneous switching occurs between batches and therefore set up time between the batches is ignored. The profit function does not include the allocation of time to each separation. For each mixture (m) individual profits (P_m) were maximised to maximise the overall profit.

$$P_m = \frac{D_m C_m - F_m CF_m}{t_m} - \frac{ACC}{H_{yr}} \qquad (7.27)$$

where, (referring to Figure 7.7) for mixture 1, $D_1 = D1$ and $F_1 = F1$; for mixture 2, $D_2 = D2$ and $F_2 = F2$. C_m is the sales value of D_m, CF_m is the raw material cost of F_m. t_m is the batch time for processing mixture m. ACC and H_{yr} are as defined earlier. Note that there is no value or cost attached to the final bottom product from each mixture. The short-cut model of Diwekar (1988) was used in the optimisation problem formulation. The maximum profit problem is posed as a nonlinear programming problem using the *Maximum Principle* and ODE discretization technique based on orthogonal collocation on finite elements (described in Chapter 5).

7.4.1. Example 1

Two binary mixtures are being processed in a batch distillation column with 15 plates and vapour boilup rate of 250 moles/hr following the operation sequence given in Figure 7.7. The amount of distillate, batch time and profit of the operation are shown in Table 7.6 (base case). The optimal reflux ratio profiles are shown in Figure 7.8. It is desired to simultaneously optimise the design (number of plates) and operation (reflux ratio and batch time) for this multiple separation duties. The column operates with the same boil up rate as the base case and the sales values of different products are given in Table 7.6.

Using the problem formulation and solution given in Logsdon et al. (1990) the optimal design, operation and profit are shown in Table 7.6. The optimal reflux ratio profiles are shown in Figure 7.8. Period 1 refers to Task 1 and Period 2 refers to Task 2. The results in Table 7.6 and Figure 7.8 clearly show the benefit of simultaneous design and operation optimisation for multiple separation duties. The benefit has been obtained due to reduction in batch time.

7.4.2. Example 2

This example is taken from Mujtaba and Macchietto (1996). Here, the profit function (Equation 7.27) used by Logsdon et al. (1990) is considered for the *multiple separation duties* presented in section 7.3.4.2. Using the input data presented in Table 7.3 with ($\Phi_1 = \Phi_2 = 0$ and $t_{su1} = t_{su2} = 0$ hr) the optimal design and operation policies are obtained and the results are presented in Table 7.7. The profit per batch for each separation is calculated by multiplying the profit per hr and the batch time.

It is important to note that the use of Logsdon et al.'s objective function will not change the number of plates and operating policies if the mixtures were to share the column by a certain fraction of time (the profit function of Logsdon et al. does not include the allocation time). However, for different allocation time to each mixture, the total number of batches for each separation, individual and total yearly profit, etc. can be evaluated by using the results presented in Table 7.7. These are summarised in Table 7.8. Comparison of the results in Table 7.4 and 7.8 clearly shows the effects of set-up time and Logsdon et al.'s objective function on the number of batches and profit.

Table 7.6. Multiperiod Optimisation of Design and Operation
[Adopted from Logsdon et al., 1990]

Base Case: Input Data

The number of plates, N	$= 15$
The vapour boilup rate, V moles/hr	$= 250$
Column holdup	$= 0$

Binary mixture 1 & 2:	
Initial feed, $F1$ & $F2$, moles	$= 116$
Initial composition, molefraction, mixture 1	$= <0.75, 0.25>$
Relative volatility, mixture 1	$= <1.5, 1.0>$
Initial composition, molefraction, mixture 2	$= <0.50, 0.50>$
Relative volatility, mixture 2	$= <2.0, 1.0>$
Distillate composition in Task 1 and Task 2,	
molefraction of component 1	$= 0.95$

Sales values and raw material costs:
$C_1 = C_2 = 115$ ($/moles), $\quad CF_1 = 75.8$ ($/moles) $\quad CF_2 = 55.1$ ($/moles)

Base Case: Results (only operation is optimised)

Distillate amounts:	$D1$	$= 89.186$ moles	$D2$	$= 60.241$ moles
Batch time:	Task 1	$= 2.053$ hr	Task 2	$= 1.565$ hr
Profit:	$\Sigma P = 932.35$ ($/hr)			

Optimal Design and Operation:

Optimal number of plates, $N = 20$

Distillate amounts:	$D1$	$= 89.572$ moles	$D2$	$= 60.314$ moles
Batch time:	Task 1	$= 1.829$ hr	Task 2	$= 1.453$ hr
Profit:	ΣP	$= 1061.91$ ($/hr)		

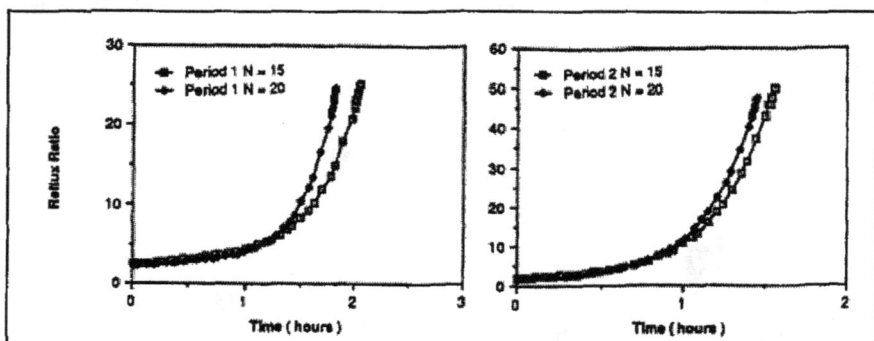

Figure 7.8. Optimal Reflux Ratio Policies for the Base Case (N=15) and Optimal Design Case (N=20). [Logsdon et al., 1990][1]

7.4.3. Combination of Allocation Time with Zero Set up Time

Here, the same example problem (as in section 7.4.2) is run using the objective function of Mujtaba and Macchietto (1996) with the allocation times $\Phi_1 = 0.75$, $\Phi_2 = 0.25$. However, the set up times are $t_{su1} = t_{su2} = 0$ as in Logsdon et al. The results are presented in Table 7.9.

Comparison of the results in Table 7.9 with those in Table 7.7 (and Table 7.8) and Table 7.4 show significant differences in the design, operating policies, optimal recoveries of products, number of batches to be processed for each duty and total yearly profit. This clearly shows the importance of including allocation time and set up time between batches in the objective function. It is to be noted that in all cases a simple model but detailed plate-to-plate calculations (Type III) with reasonable column holdup is used (unlike a short-cut model ignoring column holdup as in Logsdon et al.).

---

[1] Reprinted with permission from IChemE, UK. Full reference is at the end of the chapter.

Table 7.7. Optimal Design and Operation Policies using Logsdon et al.'s Objective Function. [Mujtaba and Macchietto, 1996][m]

Optimal Outer Loop Decision Variables: $N = 20$, $Re^1_{D1,F1} = 0.761$, $Re^1_{D2,F2} = 0.822$

Optimal Product Amount:

Product	Amount, kmol/batch
D1	4.00
D2	4.33

Optimal Reflux Ratio Profiles:

Cut	Separation 1		Separation 2	
Interval	1	2	1	2
Reflux Ratio	0.835	0.879	0.571	0.703
End Time	1.49	2.78	0.536	1.22

Profit:

$/hr:	Separation 1	= 9.63	Separation 2	= 42.57
$/batch:	Separation 1	= 26 72	Separation 2	= 51.85

Table 7.8. Effect of Allocation Time and Logsdon et al.'s Objective Function on the Number of Batches and Profit

Case	Φ_1 / Φ_2	NB_1	NB_2	P, $/yr
1	100/0	2878		76892.1
2	75/25	2162	1642	142928.8
3	50/50	1439	3279	208446.0
4	25/75	720	4918	274223.0
5	0/100		6557	340000.0

[m] Reprinted from *Journal of Process Control*, 6, Mujtaba, I.M. and Macchietto, S., Simultaneous optimisation of design and operation of multicomponent batch distillation column-single and multiple separation duties, 27-36, Copyright (1996) with permission from Elsevier Science.

Table 7.9. Optimal Design and Operation Policies for using Mujtaba and
Macchietto's Objective Function[n]
(Input Data same as in Table 7.3) $t_{su1} = t_{su2} = 0$, $\Phi_1 = 0.75$, $\Phi_2 = 0.25$.

Optimal Outer Loop Decision Variables: $N = 21$, $Re^l_{D1,F1} = 0.663$, $Re^l_{D2,F2} = 0.736$

Optimal Product Amount :	Product	Amount, kmol/batch
	$D1$	3.49
	$D2$	3.87

Optimal Reflux Ratio Profiles:

Cut	Separation 1		Separation 2	
Interval	1	2	1	2
Reflux Ratio	0.829	0.856	0.583	0.666
End Time	1.16	2.21	0.55	1.02

Optimal Number of Batches: $NB_1 = 2707$ $NB_2 = 1953$

Profit:
$/batch: Separation 1   = 22.49   Separation 2 = 44.91
$/yr (Total)   = 148,798.7

7.5. Multiperiod Operation Optimisation by Bonny et al.

For a fixed column design, Bonny et al. (1996) considered multiperiod operation
optimisation but with **multiple separation duties**. That is why this is presented in
this chapter rather than in Chapter 6. The optimisation problem can be stated as:

given: a set of multicomponent mixtures, column configuration,
number of batches, product specifications

---

[n] Reprinted from *Journal of Process Control*, **6**, Mujtaba, I.M. and Macchietto, S., Simultaneous
optimisation of design and operation of multicomponent batch distillation column-single and multiple
separation duties, 27-36, Copyright (1996), with permission from Elsevier Science.

determine:	the optimal fractional amount of each mixture that should be processed in a particular batch, the optimal reflux ratio for each cut in each batch
so as to maximise:	the productivity of the campaign or maximise the total amount of products over the campaign in a given campaign time or minimise the processing time for a given amount of products
subject to:	any constraints.

Mathematically the *maximum productivity* optimisation problem can be presented as:

$$\underset{f_{i,j},R_j}{Max} \quad J = \frac{\sum_{j}\sum_{k}\beta_k P_{jk}}{t}$$

subject to   product specification, $x^*_{j,k}$
model equations
linear bounds on decision variables, etc.

where, P_{jk} is the desired product k (k also refers the key component in the product) in batch j with product specification $x^*_{j,k}$; β_k is the economical factor; t is the total campaign time; R_j is the reflux ratio profile in batch j; $f_{i,j}$ is the fraction of mixture i in batch j (explained below). Type III model was used but without column holdup. For other types of optimisation problem formulation and solution methods see the original reference.

Figure 7.9 shows the policy of mixing from a set of M multicomponent mixtures to make the initial charges to a set of NB batches in campaign mode operation. A_i ($i = 1, M$) denotes the amount of each mixture, $f_{i,j}$ ($j = 1, NB$) denotes the fraction of A_i charged in batch j.

For a given reboiler size B_0, the following constraints apply:

$$\sum_{i=1}^{M} A_i \le B_0 NB \qquad (7.28)$$

Bonny et al. (1996) considered that the whole quantity of each mixture must be processed, therefore,

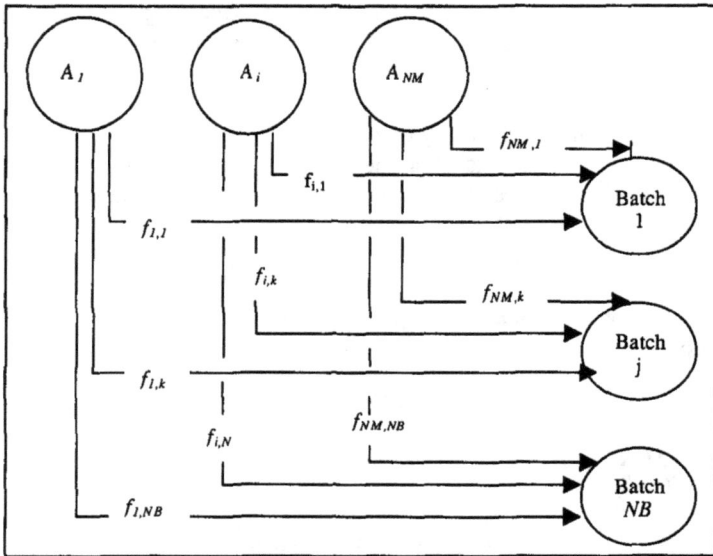

Figure 7.9. Initial Charging Policy by Bonny et al. (1996)

$$\sum_{j=1}^{NB} f_{i,j} = 1; \qquad \text{for all } i = 1, M \qquad (7.29)$$

For a particular mixture i, the limit on $f_{i,j}$ is given by,

$$0 \le f_{i,j} \le Min\left(\frac{B_0}{A_i}, 1\right) \qquad (7.30)$$

The constraints given by Equations 7.29 and 7.30 guarantee that in each batch the reboiler will not be overcharged.

It is not clear from Bonny et al. (1996) the exact reason for such mixing policy. However, this type of mixing before the separation may be suitable when the composition of a key component in one particular mixture may not be suitable for efficient separation of that component in a particular product cut. Further investigation of this argument is certainly necessary but is beyond the scope of this book.

7.5.1. Example

Bonny et al. (1996) considered separation of 3 multicomponent mixtures composed of same components (Cyclohexane, n-Heptane, Toluene) but with different compositions in 3 batches. The input data for the problem is given in Table 7.10. A single constant reflux ratio is considered in each batch which is optimised together with the batch time. The objective of the operation is to maximise the productivity. The results are summarised in Table 7.11 and Figure 7.10.

See the original reference for details and other example problems.

Table 7.10. Input Data

	A_i mol	Composition, molefraction
Mixture 1	150	<0.40, 0.40, 0.20>
Mixture 2	100	<0.50, 0.30, 0.20>
Mixture 3	50	<0.10, 0.50, 0.40>

$B_0 = 100$ mol	V, Vapour Boilup $= 91.77$ mol/hr
$NB = 3$	$N =$ No of plates $= 10$
Pressure $= 1.0$ atm	All $\beta_k = 1$

Product specifications in each batch, j	$x^*_{j,k} = \quad x^*_{j,1} = 0.90$
	$x^*_{j,2} = 0.85$
	$x^*_{j,3} = 0.85$

Table 7.11. Summary of Results

Batch	Optimal Charging Policy			Optimal Desired Products, mol			Optimal Reflux Ratio	Batch Time, hr
j	$f_{1,j}$	$f_{2,j}$	$f_{3,j}$	$P_{j,1}$	$P_{j,2}$	$P_{j,3}$		
1	0	1	0	31.4	0	15.0	3.98	4.61
2	0.67	0	0	28.1	19.5	19.8	7.50	7.42
3	0.33	0	1	12.0	19.0	30.2	8.23	8.23

Total Products = 175 mol   Total Campaign Time = 19.05 hr
Maximum Production Rate = 9.18 mol/hr

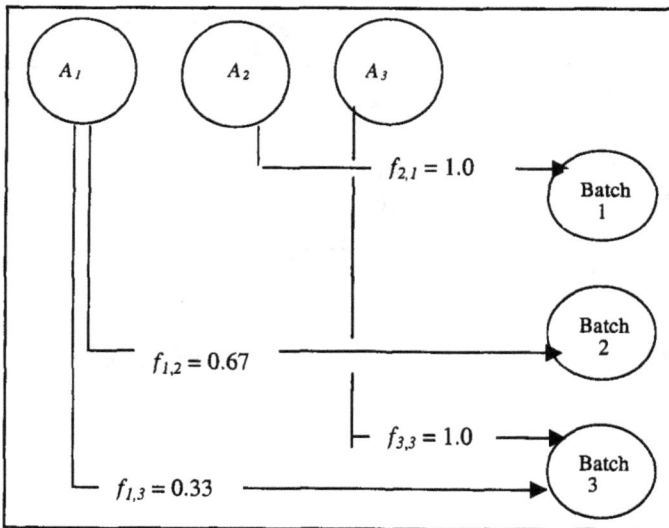

Figure 7.10. Optimal Initial Charging Policy

References

Al-Tuwaim, M.S., and Luyben, W.L., *IEC Res.* **30** (1991), 507.

Bernot, C., Doherty, M.F. and Malone, M.F., *Design And Operating Policies For Multicomponent Batch Distillation. AIChE Annual Meeting*, Los Angeles, USA (1991).

Bernot, C., Doherty, M.F. and Malone, M.F., *IEC Res.* **32** (1993), 293.

Bonny, L., Domenech, S., Floquet, P. and Pibouleau, L., *Chem. Engng. Proc.* **35** (1996), 349.

Diwekar, U.M., *Simulation, Design and Optimisation of Multicomponent Batch Distillation Column.* PhD Thesis, (IIT Bombay, India, 1988).

Diwekar, U.M., Madhavan, K.P. and Swaney, R.E., *IEC. Res.* **28** (1989), 1011.

Diwekar, U.M. and Madhavan, K.P., *IEC. Res.* 30 (1991), 713.

Eliceche, A.M. and Sargent, R.W.H., *IChemE Symp. Ser. No.* **61** (1981), 1.

Farhat, S., Czernicki, M., Pibouleau, L. and Domenech, S., *AIChE J.* **36** (1990), 1349.

Logsdon, J.S., Diwekar, U.M. and Biegler, L.T., *Trans. IChemE*, **68A** (1990), 434.

Logsdon, J.S.and Biegler, L.T., *IEC Res.* **32** (1993), 700.

Mujtaba, I.M., *Trans. IChemE.* **75A** (1997), 609.

Mujtaba, I.M. and Macchietto, S., *Comput. chem. Engng.* **17**(1993), 1191.

Mujtaba, I.M. and Macchietto, S., *J. Proc. Control.* **6** (1996), 27.

Nad, M. and Spiegel, L., *In Proceedings CEF 87* , Sicily, Italy, April (1987), 737.

Sharif, M., *Design of Integrated Batch Processes.* PhD Thesis, (Imperial College, University of London, 1999).

Sharif, M., Shah, N. and Pantelides, C.C., *Comput. chem. Engng.* **22** (1998), s69.

Viswanathan, J. and Grossmann, I.E., *Comput. chem. Engng.* **14** (1990), 769.

CHAPTER 8

8. OFF-CUT RECYCLE

8.1. Introduction - Off-cut Recycle in Binary Separation

When each component in a mixture of n_c components is to be recovered in a separate fraction, there will be in principle (n_c-1) main-cuts, one bottom residue fraction and (n_c-1) intermediate off-cuts. Thus for binary mixtures there is usually one main-cut and one off-cut (Figure 8.1). This intermediate off-cut can either be disposed off safely (meeting environmental constraints) or can be recycled to the next batch and thus a periodic operation (Figure 8.2) can be established (Liles, 1966; Mayur et al., 1970; Nad and Spiegel, 1987; Christensen and Jorgensen, 1987 and Luyben, 1988).

Dynamic optimisation of this type of periodic operation was first attempted and reported in the literature by Mayur et al. (1970), who considered the initial charge to the reboiler as a fresh feed stock mixed with the recycled off-cut material from the previous distillation task. Each batch cycle is then operated in two distillation tasks. During the Task 1, a quantity of overhead distillate meeting the light product specification is collected. The residue is further distilled off in Task 2 until it meets the bottom product specification. The overhead during Task 2 meets neither specifications (but the composition is usually kept close to the that of the initial charge for thermodynamic reasons) and is recycled as part of the charge for the next batch. As the batch cycle is repeated a *quasi-steady state* mode of operation is attained which is characterised by the identical amount and composition of the recycle (from the previous batch) and the off-cut (from the current batch). Luyben (1988) indicates that the *quasi-steady state* mode is achieved after three or four such cycles.

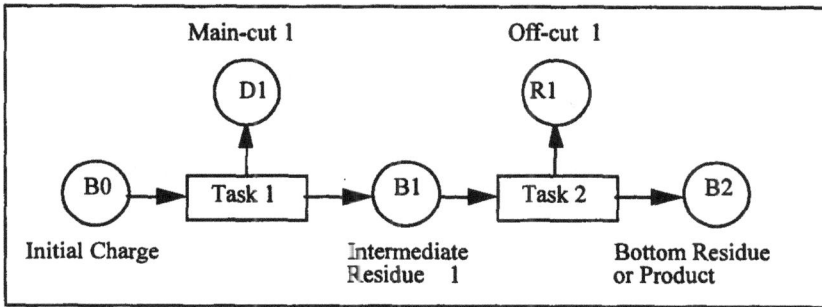

Figure 8.1. STN representation of Binary Batch Distillation with One Main-cut and One Off-cut.

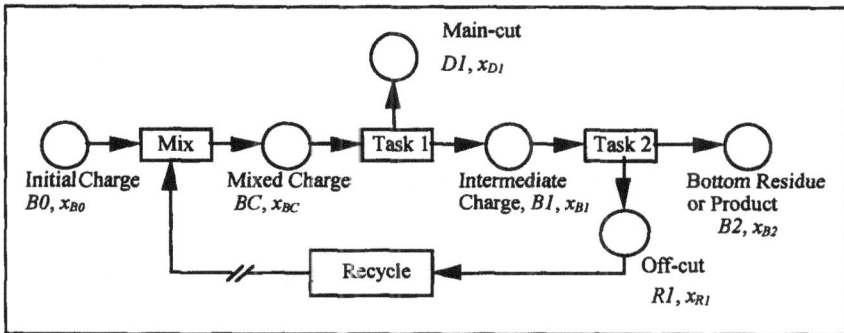

Figure 8.2. Quasi-steady State Off-cut Recycle Strategy in Binary Batch Distillation

Mayur et al. measured the benefits of recycling in terms of a potential reduction in batch time. However the benefits may be measured (Mujtaba, 1989) as follows:

1. Reduction in batch time: For a given fresh feed and a given separation, the column performance is measured in terms of minimum batch time required to achieve a desired separation (specified top product purity (x^*_{D1}) and bottom product purity (x^*_{B2}) for binary mixture). Then an optimal amount and composition of recycle, subject to physical bounds (maximum reboiler capacity, maximum allowable purity of the off-cut) are obtained in an overall minimum time to produce the same separation (identical top and bottom products as in the

case without recycle). The difference in the minimum times obtained with and without recycle shows the benefit of recycling.

2. Increase in productivity: For a fixed reboiler charge it is wished to obtain the optimal amount of fresh feed and the composition of the recycle (note: optimal amount of feed will automatically determine the optimal amount of recycle) to produce a given separation (specified top and bottom product purity for binary mixture) in a minimum batch time. This will give the productivity in terms of fresh feed processed per unit time. For the same given separation and no recycle, the productivity is obtained by processing a charge of fresh feed equal to the reboiler capacity. The difference in these productivities with and without recycle on a fixed reboiler charge basis will show the benefit of recycling.

Mayur et al. (1970) formulated a two level dynamic optimisation problem to obtain optimal amount and composition of the off-cut recycle for the *quasi-steady state* operation which would minimise the overall distillation time for the whole cycle. For a particular choice of the amount of off-cut and its composition $(R1, x_{R1})$ (Figure 8.1) they obtained a solution for the two distillation tasks which minimises the distillation time of the individual tasks by selecting an optimal reflux policy. The optimum reflux ratio policy is described by a function $r_1(t)$ during Task 1 when a mixed charge (BC, x_{BC}) is separated into a distillate $(D1, x^*_{D1})$ and a residue $(B1, x_{B1})$, followed by a function $r_2(t)$ during Task 2, when the residue is separated into an off-cut $(R1, x_{R1})$ and a bottom product $(B2, x^*_{B2})$. Both $r_1(t)$ and $r_2(t)$ are chosen to minimise the time for the respective task. However, these conditions are not sufficient to completely define the operation, because $R1$ and x_{R1} can take many feasible values. Therefore the authors used a sequential simplex method to obtain the optimal values of $R1$ and x_{R1} which minimise the overall distillation time. The authors showed for one example that the inclusion of a recycled off-cut reduced the batch time by 5% compared to the minimum time for a distillation without recycled off-cut.

Christensen and Jorgensen (1987) investigated the possible economic impact of off-cut recycling on some difficult binary separations. For tray column they defined a measure of *the degree of difficulty* (*q*) of a given separation in a given column (described in detail in Chapter 3). With several binary examples they showed that this measure could qualitatively predict the profitability of using an off-cut recycle. They also measured the benefits of recycling in terms of reduction in batch time. They found that the greater the measure of the difficulty of separation, the larger the benefits of using off-cut recycle.

It is quite clear from the previous works that recycling of off-cut material is particularly interesting if a given separation is to be performed in an existing column. The number of trays in the column may not be quite appropriate for the distillation task at hand, and running the column in a conventional manner (without recycling) may need a very long time. In such a case recycling offers a possibility of reducing the distillation time (Mujtaba and Macchietto, 1993). In addition, recycling an off-cut may be used advantageously to reduce capital investment for a given batch separation by allowing a smaller column to be used than would otherwise be necessary with ordinary operation.

The problem of choosing whether and when to recycle each off-cut and the size of the cut is a difficult one. Liles (1966) considered dynamic programming approach and Luyben (1988) considered repetitive simulation approach to tackle this problem. Mayur et al. (1970) and Christensen and Jorgensen (1987) tackled it as a dynamic optimisation problem using Pontryagin's *Maximum Principle* applied to very simplified column models as mentioned in Chapters 4 and 5.

Mujtaba (1989) used the measure of *the degree of difficulty* of separation proposed by Christensen and Jorgensen (1987) to decide whether or not an off-cut is needed. The optimal control algorithm of Morison (1984) was then used to develop operational policies for reflux ratio profiles and amount and timing of off-cuts which minimise the total batch time. A more realistic dynamic column model (type IV as presented in Chapter 4) was used in the optimisation framework.

8.2. Classical Two-Level Optimisation Problem Formulation for Binary Mixtures

Before going to a detailed formulation of the problem the following discussion is worthwhile:

Consider a problem with no recycle as shown in Figure 8.3 where an initial charge to the column is $B0$ with composition x_{B0}. The charge is separated into two fractions, the overhead as $(D1, x_{D1})$ and the bottom as $(B2, x_{B2})$.

The overall mass balance is therefore:

$$B0 = D1 + B2 \tag{8.1}$$

$$B0x_{B0}^1 = D1x_{D1}^1 + B2x_{B2}^1 \tag{8.2}$$

$$\sum x_{D1}^i = 1 \tag{8.3}$$

$$\sum x_{B2}^i = 1 \tag{8.4}$$

where $i = 1,2$ is the component index.

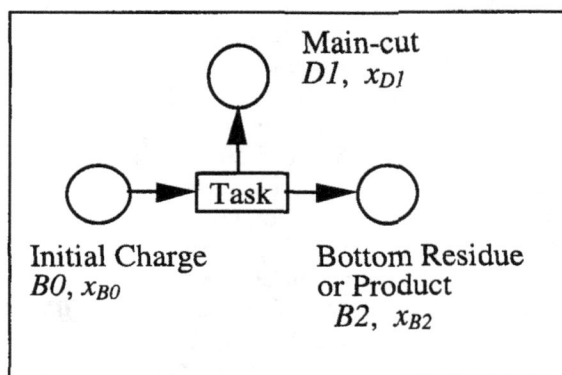

Figure 8.3. A Typical Single Batch Operation

For a given initial charge $(B0, x_{B0})$ the unknown variables in the above system of equations are: $D1$, x^1_{D1}, x^2_{D2}, $B2$, x^1_{B2}, x^2_{B2}. Therefore, the degree of freedom is (DF) = 2. One of the choices of decision variables could therefore be $(x^1_{D1}$ and $x^1_{B2})$. Since we deal with only binaries in this section we drop out the superscripts to indicate the component number. From now on $(x^1_{D1}$ and $x^1_{B2})$ will be expressed as $(x^*_{D1}$ and $x^*_{B2})$ meaning these variables are specified. With these specifications we can now easily formulate a dynamic optimisation (time optimal control) problem for the no recycle case mentioned above. The problem can be stated as:

Starting with a given fresh feed of $(B0, x_{B0})$ select a reflux ratio profile $r(t)$ to achieve an overhead product $(D1, x^*_{D1})$ in a minimum time (t_F) leaving the final bottom product $(B2, x^*_{B2})$. This can be formulated as:

$$\text{Min}_{r(t)} \qquad J = t_F \qquad\qquad (8.5)$$

subject to constraints

$$x_{D1}(t_F) = x^*_{D1}$$
$$x_{B2}(t_F) = x^*_{B2}$$
and the model equations

Now we can remove the 2nd constraint from the above formulation and substitute it by the relation: $D1(t_F) = \underline{D1}$; where, $\underline{D1}$ is the solution of Equations (8.1-8.4) for a given charge $B0$, x_{B0} and specifications x^*_{D1} and x^*_{B2}.

Since the two specifications above are equivalent we end up with the same optimal solution (Mujtaba and Macchietto, 1988; Mujtaba, 1989).

The necessity of the above discussion will now be realised (also see Christensen and Jorgensen, 1987). Figure 8.2 shows a *quasi-steady state* mode of operation, with off-cut recycle. A fresh charge $B0$ of composition x_{B0} is mixed with the off cut ($R1$, x_{R1}) from the previous batch to produce a mixed charge to the reboiler (BC, x_{BC}). The main cut ($D1$, x^*_{D1}) is produced over the time t_1 (Task 1), leaving a residue ($B1$, x_{B1}). At this time the distillate is simply diverted to a second receiver, and further distillation in Task 2 for time t_2 produces the off cut and the final bottom product ($\underline{B2}$, x^*_{B2}) where, $\underline{B2}$ is the solution of Equations (8.1-8.4) as mentioned before.

The following optimisation problem is considered.

given: a batch charge ($B0$, x_{B0}), a desired amount of distillate $\underline{D1}$ of specified purity x^*_{D1} and a final desired bottom product $\underline{B2}$ of specified purity x^*_{B2}

determine: the amount and composition of the off-cut ($R1^*$, x^*_{R1}) and the reflux ratio policy $r(t)$

so as to minimise: the overall distillation time

subject to: any constraints

Writing a mass balance equation around the mixer of Figure 8.2 we can easily show that there are again two degrees of freedom left. So one of the possible choices could be to specify $R1$ and x_{R1} or to optimise these variables as mentioned above. A two level optimisation problem is therefore formulated as:

$P0$ Min$\quad J = t_1 + t_2$
$\quad R1^*, x^*_{R1}$

$$(8.6)$$

subject to $a_R < R1^* < b_R$
$\qquad\qquad a_{xR} < x^*_{R1} < b_{xR}$

where t_1 is obtained from

$P1$ Min$\qquad\qquad t_1$
$\quad r_1(t)$

subject to $D1 = \underline{D1}$
$\qquad\qquad x_{D1} = x^*_{D1}$

$$(8.7)$$

and   bounds on reflux ratio

and t_2 is obtained from-

P2   Min t_2
                $r_2(t)$

        subject to $R1 = R1^*$ (8.8)
                                $x_{R1} = x^*_{R1}$
        and   bounds on reflux ratio

where, a_R, b_R, a_{xR} and b_{xR} are bounds on the amount of recycle cut and on its composition. In addition, all differential and algebraic model equations must be added as equality constraints to problem *P1* and *P2*, with suitable boundary conditions. These are mentioned in detail in earlier chapters.

Thus the multiperiod optimisation problem is formulated as a sequence of two independent dynamic optimisation problems (*P1* and *P2*), with the total time minimised by a proper choice of the off cut variables in an outer problem (*P0*) and the *quasi-steady state* conditions appearing as a constraint in *P2*. The formulation is very similar to those presented by Mujtaba and Macchietto (1993) discussed in Chapter 5. For each iteration of *P0*, a complete solution of *P1* and *P2* is required. Thus, even for an intermediate sub-optimal off cut recycle, a feasible *quasi-steady state* solution is calculated. The gradients of the objective function with respect to each decision variable (*R1* or x_{R1}) in problem *P0* were evaluated by a finite difference scheme (described in previous chapters) which again requires a complete solution of problem *P1* and *P2* for each gradient evaluation (Mujtaba, 1989).

Reflux ratio is the control variable which governs the operation during each period. This is discretised into a small number of intervals (typically 3 to 10) with a constant value during each interval. The optimal values in each interval and the switching times for each of the inner problems were obtained by Mujtaba (1989) using the optimal control algorithm of Morison (1984) with optimisation based on a successive quadratic programming (SQP) as described in chapter 5. The outer problem *P0* was also solved using an SQP based algorithm.

8.2.1. Example Problems - Set 1

Mujtaba and Macchietto (1988) and Mujtaba (1989) used two typical binary mixtures and a variety of separation specifications to demonstrate how optimal recycle policies can be obtained and to assess the validity of *the degree of difficulty* measure. The mixtures were: 1) Benzene-Toluene; 2) Butane-Pentane. For simplicity ideal equilibrium and Antoine's vapour pressure equations were used. The light component was always the first one.

Column configuration, initial charge to the column, and separation requirements for several cases, are presented in Table 8.1. The minimum time required for a given separation ($\underline{D1}$, x^*_{DI}) with off cut recycle is reported in Table 8.1 as t_r (solution of *P0*) and with no recycle as t_{nr} (solution of *P1* with *R1*=0). In both cases the reflux ratio was discretised into 3 control intervals during the main cut period and a single interval during off-cut production. The bounds on the reflux ratio $0.2 < r < 1.0$ were used for all cases. In all the cases 4% of the fresh feed was used as total column holdup, half of it in the condenser, the rest equally distributed on the plates.

Liquid compositions of plates, condenser holdup tank and accumulator (differential variables) at time t=0 are set equal to the fresh charge composition (x_{B0}) to the reboiler. It is also possible to set these values to mixed charge composition (x_{BC}). Reboiler holdup and compositions were initialised to the mixed charge (*BC*, x_{BC}) at each iteration of *P0*. Mujtaba and Macchietto (1988) and Mujtaba (1989) considered Type IV-CMH model for the process and the model was solved at time t=0 to initialise all other variables. The first product (*D1*, x_{D1}) (see Figure 8.2) was drawn off starting from $t = 0$. For the second distillation task no re-initialisation was required. The distillate was simply diverted to a different product accumulator and integration was continued.

The results are summarised in Table 8.1 and Figures 8.4-8.8 which show the optimal reflux ratio profiles and the corresponding accumulated and instant distillate composition curves. The optimal reflux ratio profiles presented in Figures 8.4-8.8 are different from case to case.

Table 8.1. Input Data and the Results.
[Adopted from Mujtaba and Macchietto (1988) and Mujtaba, 1989]

Case	Mixture No.	N_T	x^*_{DI}	q	t_{nr} hr	t_r hr	% Time Saved	Optimal $R1$	x_{RI}
1	1	8	0.960	0.483	3.90	------	-----	------	-----
2	1	4	0.900	0.597	3.10	2.94	5.03	0.50	0.607
3	1	3	0.900	0.746	4.27	3.46	18.97	1.07	0.654
4	1	3	0.912	0.799	5.57	3.91	29.90	1.22	0.676
5	1	3	0.920	0.836	7.22	4.25	41.10	1.50	0.679
6	2	7	0.970	0.400	2.74	------	------	------	-----
7	2	3	0.900	0.500	1.87	------	-----	------	-----
8	2	3	0.940	0.637	2.80	2.51	10.29	0.50	0.571
9	2	3	0.955	0.706	3.88	2.95	23.88	0.99	0.649
10	2	3	0.970	0.799	13.06	3.62	2.26	1.34	0.665

Key:	For all cases, $B0 = 5.0$ kmol,	$x_{B0} = 0.60$ molefraction of component 1
	$\underline{D1} = 3$ kmol,	condenser vapour load = 3.0 kmol/hr
	% Time saved is based on t_{nr}.	

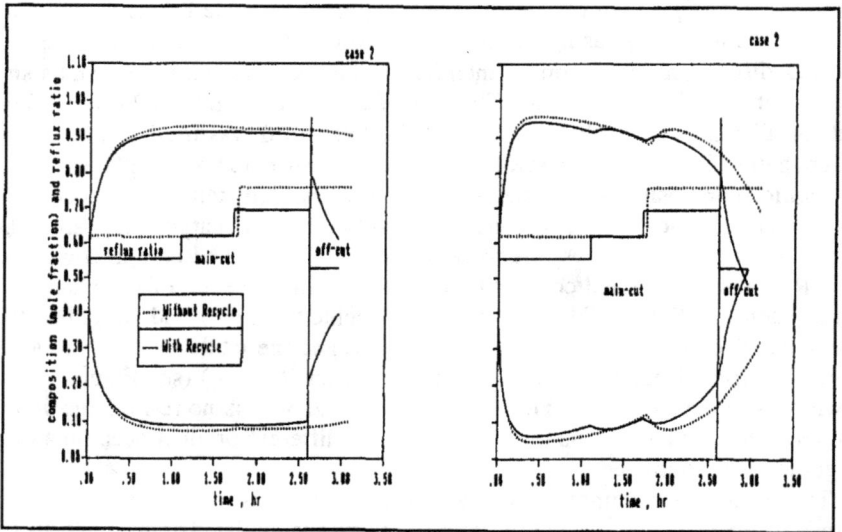

Figure 8.4. Accumulated and Instant Distillate Composition and Optimal Reflux Ratio Profiles (case 2). [Mujtaba, 1989]

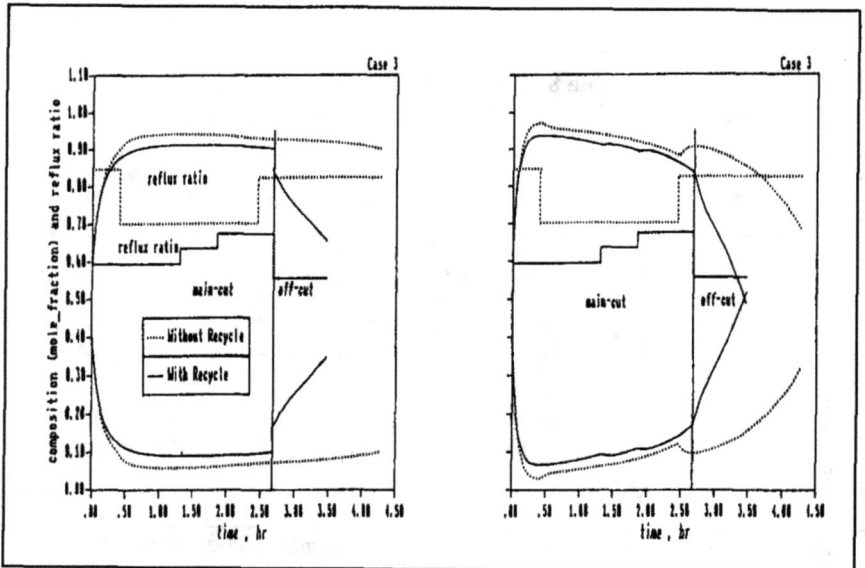

Figure 8.5. Accumulated and Instant Distillate Composition and Optimal Reflux Ratio Profiles (case 3). [Mujtaba, 1989]

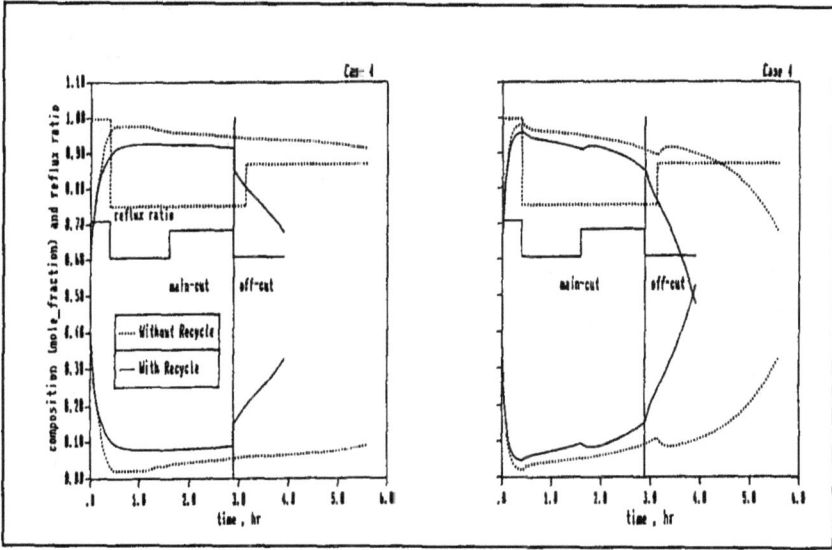

Figure 8.6. Accumulated and Instant Distillate Composition and Optimal Reflux Ratio Profiles (case 4). [Mujtaba, 1989]

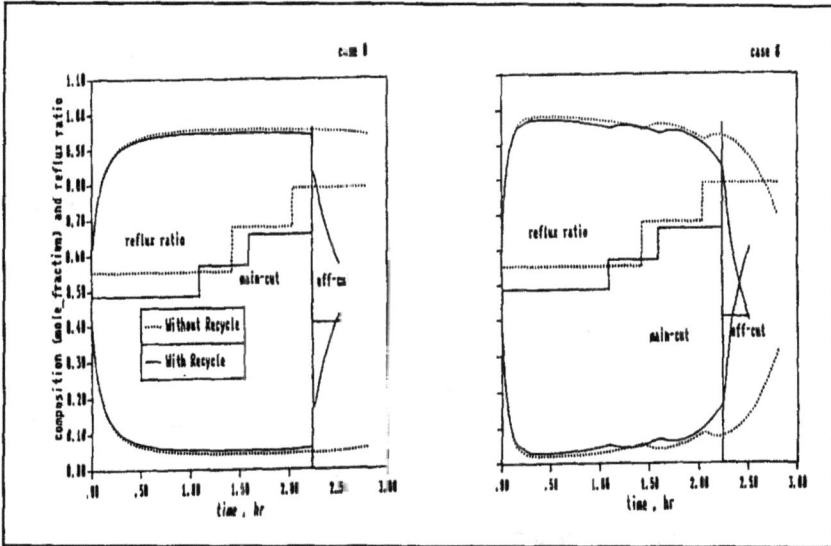

Figure 8.7. Accumulated and Instant Distillate Composition and Optimal Reflux Ratio Profiles (case 8). [Mujtaba, 1989]

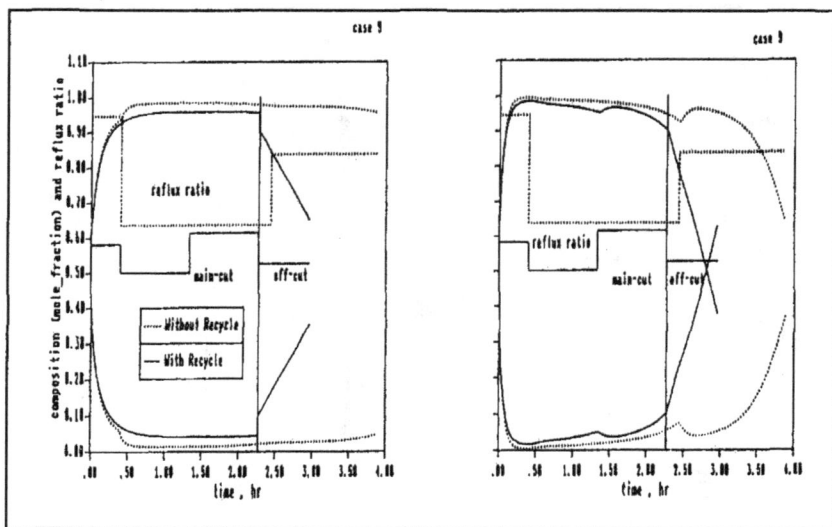

Figure 8.8. Accumulated and Instant Distillate Composition and Optimal Reflux Ratio Profiles (case 9). [Mujtaba, 1989]

With difficult separations (Figures 8.5 and 8.6, no recycle cases) a high value of the reflux ratio is required initially to obtain a high distillate concentration of the desired product. The reflux ratio goes down and up again at the end. A period of total reflux operation is automatically determined, if found to be optimal, which is the case for difficult separations (Figure 8.6, no recycle case). With an easy separation (Figures 8.4, 8.5, 8.7, recycle case) the reflux ratio profile is usually increasing all along as expected. This is because the effect of recycling the off-cut is to make the separation for the main cut easier, for a given column configuration. But for more difficult separations (Figure 8.6 and 8.8) even the recycle cases take an initial high reflux although this is comparatively lower than for cases without recycle (Figure 8.6 and 8.8). The optimal reflux ratio during off cut production is fairly low in all cases and in agreement with practical experience (Rose, 1985).

The benefits of recycling are also quite clear. In some cases the recycling of off-cut materials results in operation time savings of over 70% of the optimal time required without recycle. Also the size of the optimal off-cut increases (and so does its composition) as the separation becomes more difficult (higher q). Table 8.1 shows that recycling is beneficial for q values greater than approximately 0.6. However, it is noticed that the extent of the benefit for similar values of q varies from case to case. Comparison of case 4 and case 10 (have the same and quite high *q*) shows that the time savings are extremely different. Although high *q* values suggest that recycling should be used the extent of time savings depends on various other factors (type of mixture, amount and purity required and vapour boilup rate, etc.).

In all the case studies presented here it is assumed that the amount of fresh feed to be processed in the long production campaign is fixed for every batch cycle, but the reboiler is oversized to some extent. The optimal amount of recycle is obtained within this bound so that maximum benefit can be achieved out of a given column.

8.2.2. Example Problems - Set 2

Christensen and Jorgensen (1987) considered off-cut recycle problem for binary mixtures in tray and packed columns. The two level optimisation problem described earlier was solved using the Pontryagin's *Maximum Principle* and the method of orthogonal collocation was applied to a simple no-holdup column model. Table 8.2 shows a subset of the results presented in Christensen and Jorgensen (1987) for tray columns. The authors determine the optimum values of $(B1, x_{B1})$ to minimise the overall time. The optimum values of $(R1, x_{R1})$ are then determined by using the mass balance equations. The results clearly shows the benefit of off-cut recycle for *q* values greater than 0.6. See the original reference for examples with packed columns.

Table 8.2. Summary of Binary Off-cut Recycle using Tray Columns
[Adopted from Christensen and Jorgensen, 1987]

Case	N_T	α	x_{D1}	*q*	Recycle ?	t_r, hr	% Time Saved	$B1$, kmol	x_{B1}
1	5	2.0	.95	.987	yes	2.72	91.9	116.5	.371
2	5	2.0	.80	.612	yes	1.38	0.07	58.69	.111
3	5	2.0	.75	.543	no	--	--	--	--
4	5	2.1	.85	.650	yes	1.38	0.58	63.51	.126
5	5	2.2	.85	.612	no	--	--	--	--
6	5	1.9	.85	.752	yes	1.89	9.45	75.67	.196
7	5	1.8	.85	.821	yes	2.30	23.2	86.37	.242
8	6	2.0	.85	.597	no	--	--	--	--
9	7	2.0	.85	.522	no	--	--	--	--
10	4	2.0	.85	.835	yes	1.93	26.3	87.64	.248

For all cases: $B0 = 100$ kmol, $x_{E0} = 0.40$ molefraction,
Vapour boilup = 143.10 kmol/hr, $x_{B2} = 0.10$ molefraction

8.3. One Level Optimisation Problem Formulation for Binary Mixtures

It is quite obvious that a two level optimisation formulation can be very expensive in terms of computation time. This is due to the fact that for any particular choice of $R1$ and x_{R1} a complete solution (sub-optimal) of the two distillation tasks are required. The same is true for each gradient evaluation with respect to the decision variables ($R1$ and x_{R1}). Mujtaba (1989) proposed a faster one level dynamic optimisation formulation for the recycle problem which eliminates the requirement to calculate any sub-optimal or intermediate solution. In this formulation the total distillation time is minimised directly satisfying the separation requirements for the first distillation task as interior point constraints and for the second distillation task as final time constraints. It was found that the proposed formulation was much more robust and at least 5 times faster than the classical two level formulation.

Referring to Figure 8.2 and given a batch charge ($B0$, x_{B0}), a desired amount of distillate $\underline{D1}$ of specified purity x^*_{D1} and final bottom product $B2$ of specified purity x^*_{B2} Mujtaba (1989) determined the amount and composition of the off-cut ($R1^*$, x^*_{R1}) and the reflux rate policy $r(t)$ which minimised the overall distillation time. In this formulation instead of optimising ($R1$, x_{R1}) the mixed charge to the reboiler (B_C, x_{BC}) was optimised and at the end of the solution the optimal ($R1^*$, x^*_{R1}) was evaluated from the overall balance around the mixer in Figure 8.2. The dynamic optimisation problem is formulated as:

$P4$ Min $J = t_1 + t_2 = t_F$
         $BC, x_{BC}, r(t)$

subject to interior point constraints
$$D1(t_1) = \underline{D1}$$
$$x_{D1}(t_1) = x^*_{D1} \tag{8.9}$$

and end point constraints

$$B2(t_F) = \underline{B2}$$
$$x_{B2}(t_F) = x^*_{B2}$$

where t_1 is defined by one of the switching times and t_F is the final time.

In addition, all DAE model equations (Type IV-CMH) act as equality constraints in problem $P4$ with suitable boundary conditions as mentioned in the two-level formulation.

The formulation presented in $P4$ automatically establishes the *quasi-steady state* operation of Figure 8.2 where the recycle from the previous batch is identical in amount and composition to those of the current batch.

8.4. Comparison of the Two Level and the One Level Formulations

Mujtaba (1989) considered the separation of a binary mixture (Benzene-Toluene) with off-cut recycle. The optimal operating policy, computation time, etc. were determined using the two level and the one level optimisation problem formulations and the results were compared.

The batch distillation column consisted of 3 internal plates, reboiler and a total condenser. The reboiler was charged with a fresh feed of 5 kmol with Benzene molefraction 0.6. The total column holdup was 4 % of the charge. Half the holdup was in the condenser and the rest was distributed over the plates. The vapour load to the condenser was 3 kmol/hr. The required product purities were $x^*_{D1} = 0.90$ and $x^*_{B2} = 0.15$. The solution of Equations 8.1-8.4 therefore gives $\underline{D1} = 3.0$ kmol and $\underline{B2} = 2$ kmol. This problem is same as case 3 shown in Table 8.1. Three reflux ratio (control) intervals were used to achieve $(\underline{D1}, x^*_{D1})$ and one control interval to achieve $(\underline{B2}, x^*_{B2})$.

The optimal reflux ratio policies together with the switching times and the optimal amount and composition of the recycle obtained by two methods are presented in Table 8.3 and the solution statistics are given in Table 8.4.

The results presented in Table 8.3 are in good agreement. The small differences between the results might be due to the different accuracy set for the optimisation (see Table 8.4). Since the gradients in the two-level formulation were solved by finite difference the inner loop problems (P1 and P2) were to be solved very tightly (accuracy for the optimiser = 1.0E-4). Whereas, the outer loop problem (P0) of the two-level formulation and the one level problem (P4) were solved using the optimiser accuracy = 1.0E-2.

Table 8.3. Comparison of the Results Obtained by Two Problem Formulations [Mujtaba, 1989]

Type	Optimal $R1$ (kmol) x_{R1} (molefraction)	Optimal (Reflux Ratio/End Time (hr)) Intervals			
		1	2	3	4
Two Level Formulation	1.07 0.653	0.594 1.30	0.635 1.85	0.675 2.67	0.554 3.46
One Level Formulation	0.986 0.630	0.600 1.19	0.641 1.91	0.699 2.78	0.589 3.44

Table 8.4. Solution Statistics Using Two Formulations
[Mujtaba, 1989]

	Two-Level	One-Level
Number of Functions	5 (outer loop)	15
Number of Gradients	3 (outer loop)	12
Optimiser Accuracy	1.0E-4 (inner loop) 1.0E-2 (outer loop)	1.0E-2
CPU Time for Functions	10 min	0.4 min
CPU Time for Gradients	10 min	0.7 min
Total Computation Time	81 min	15.5 min

Note: IBM 370 machine was used in this example

The important part of the results by the two methods is the difference in computation time. As presented in Table 8.4, the one level formulation was about 5 times faster than that of the classical two-level formulation. In fact the total time for the one level solution was only 1.5 times the time required to evaluate just one function (or gradient) values for the two-level formulation.

8.5. More Examples using the One Level Optimisation Formulation

Mujtaba (1989) used a typical nonideal binary mixture (Acetone – Ethanol) and a variety of separation specifications to demonstrate how optimal recycle policies can be obtained using one level formulation and to assess the validity of the *degree of difficulty* of separation measure. The SRK equation of state was used to evaluate vapour liquid equilibria. More accurate physical property models (UNIQUAC, Wilson) could also be used for such mixture. However, the use of SRK in this problem will not hamper seriously the conclusions drawn from this example. Column configuration, initial charges to the column and separation requirements for several cases are presented in Table 8.5. For all cases the initial amount of the fresh feed is 5 kmol with acetone concentration of 0.55 molefraction. In all cases 4% of the fresh feed was used as the total column holdup, half of it in the condenser, the rest equally distributed on the plates. The amount of distillate required ($\underline{D1}$) and the vapour load to the condenser are 2.5 kmol and 2.5 kmol/hr respectively in all cases. The minimum time required for a given separation ($\underline{D1}$, x^*_{D1}) with off-cut recycle is reported in Table 8.5 as t_r and with no recycle as t_{nr}. In both cases the reflux ratio was discretised into 3 control intervals during the main cut period and a single interval during the off cut production.

Table 8.5. Input Data and Results (nonideal case)
[Mujtaba, 1989]

Case	N_T	x^*_{DI}	q	t_{nr} hr	t_r hr	% Time Saved	Optimal $R1$(kmol)	x_{RI}
1	5	0.900	0.607	3.77	3.70	1.86	0.50	0.558
2	5	0.910	0.638	4.11	3.98	3.11	0.70	0.608
3	4	0.900	0.728	4.72	4.40	6.68	0.73	0.662
4	3	0.900	0.910	11.32	5.56	50.86	1.91	0.673

Key: % time saved is based on t_{nr}

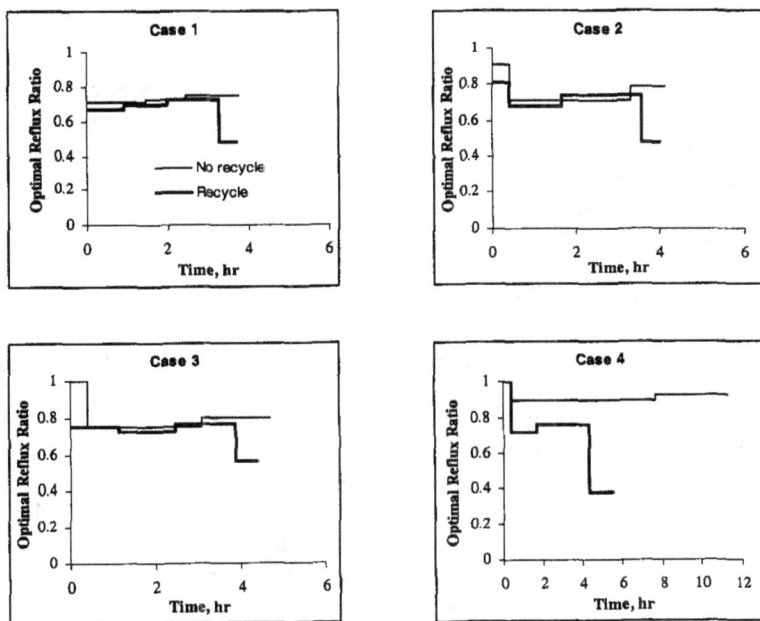

Figure 8.9. Optimal Reflux Ratio Profiles
(One-Level Formulation - Nonideal Case)

The results are summarised in Table 8.5. Figure 8.9 shows the optimal reflux ratio profiles for cases with recycle and without recycle. The benefits of recycling are again quite clear and the discussion presented in section 8.4 equally applies for the cases studied. Table 8.5 shows that recycling is again beneficial for q values greater than approximately 0.60. The solution statistics presented in Mujtaba (1989) shows that the computation time with nonideal mixtures is about 3 times greater compared to the cases with ideal mixtures. This can be explained from the fact that most of the time is consumed in evaluating physical properties. In ideal mixture cases the time spent in physical property calculations was about 28-30% of the total time while for nonideal cases it was about 65-70%.

8.6. Notes on Binary Off-cut Recycle

The one level optimal control formulation proposed by Mujtaba (1989) is found to be much faster than the classical two-level formulation to obtain optimal recycle policies in binary batch distillation. In addition, the one level formulation is also much more robust. The reason for the robustness is that for every function evaluation of the outer loop problem, the two-level method requires to reinitialise the reflux ratio profile for each new value of $(R1, x_{R1})$. This was done automatically in Mujtaba (1989) using the reflux ratio profile calculated at the previous function evaluation in the outer loop so that the inner loop problems (specially problem $P2$) could be solved in a small number of iterations. However, experience has shown that even after this re-initialisation of the reflux profile sometimes no solutions (even sub-optimal) were obtained. This is due to failure to converge within a maximum limit of function evaluations for the inner loop problems. On the other hand the one level formulation does not require such re-initialisation. The reflux profile was set only at the beginning and a solution was always found within the prescribed number of function evaluations.

Recycling of an off-cut is found to be beneficial in most cases for q values greater than about 0.60. On the basis of these results this measure can qualitatively predict the benefit of using off cut recycle and therefore can be used as a decision variable. Optimal values for the amount and composition of the recycle and the reflux policies can then be obtained using the methods presented in this chapter. The methods are capable of showing when and how long an initial total reflux operation is needed for a particular separation. In some cases more than 70% savings of batch time was found when recycle was used. Further improvements on the column performance may be expected from finer control of the reflux ratio, obtained using a larger number of control intervals.

In all the case studies presented in this chapter it was assumed that the amount of fresh feed to be processed in the long production campaign is fixed for every batch cycle. And the reboiler was oversized to some extent so that it could accommodate the extra charge from the off cut recycle. The optimal amount of recycle was obtained within that bound (40% oversized) so that maximum benefit could be achieved out of the given column.

However, in practice it is always desirable to charge the reboiler to its full capacity. In that case, it might be required to determine the optimal amount of fresh feed, the optimal amount of recycle and its composition so as to maximise the amount of fresh feed processed in unit time (productivity). No proper optimisation problem formulation has been attempted so far for this type of problem. However, Mujtaba (1989) scaled the results (amount of fresh feed and recycle) presented in Table 8.1 to a fixed reboiler capacity of 5 kmol in order to asses whether the reductions in operation times do indeed translate into higher productivity. Using the same recycle compositions as those mentioned in Table 8.1, the minimum operation time to separate the mixed charge into distillate and bottom products of the same purity (x^*_{D1}, x^*_{B2}) was calculated. This will automatically produce an off-cut identical in amount and composition to that recycled (scaled) and therefore satisfies the *quasi-steady state* requirement. The productivity thus obtained [fresh feed (scaled) processed in unit time] is then compared with the productivity obtained by processing a fresh fixed reboiler charge (5 kmol) for the same given separation task (x^*_{D1}, x^*_{B2}).

This study was done for case 2 and case 4 of Table 8.1 (for a low and high value of q). The results are summarised in Table 8.6. The purpose of this study is to show that recycling is beneficial even on a fixed reboiler capacity basis. Although these scaled results will in general not be strictly optimal, the benefits of recycling in terms of productivity are quite clear from Table 8.6. Also the benefit increases with q as before.

8.7. Introduction - Off-cut Recycle in Multicomponent Separation

This section presents the dynamic optimisation problem formulation of Mujtaba (1989) and Mujtaba and Macchietto (1992) to obtain optimal recycle policies in multicomponent batch distillation. Some special cases were identified where the methods used for the binary case could be applied fairly easily to multicomponent mixtures. The previously mentioned measure q of *the degree of difficulty* of separation was used to identify those special cases. A new operational strategy regarding the order of off-cuts recycle in a multicomponent environment was discussed. The Benefits of recycling were correlated against the measure q.

Table 8.6. Benefits of Recycling Based on Fixed Reboiler Charge. [Mujtaba, 1989]

Case	B^S_0 kmol	R^S_1 kmol	x^*_{D1}	x^*_{B2}	D^S_1 kmol	q	t_{nr} hr	t_{rS} hr	P_{nr}	P_r	X
2	4.55	.55	.900	.150	2.73	.6	3.10	2.62	1.61	1.73	7.4
4	4.02	.98	.912	.132	2.41	.8	5.57	3.09	0.90	1.30	44.9

Key : BC^* = Fixed Reboiler Charge = 5 kmol
Scale Factor (SF) = BC^* / $(B0 + R1)$ (of Table 8.1)
$B^S_0 = B0$ x SF (fresh feed) $R^S_1= R1$ x SF   (recycle) $BC^* = B^S_0 + R^S_1$
x^*_{D1} , x^*_{B2} are the specifications on the top main product and final bottom product
and are the same as those set in Table 8.1. (note: these specifications yielded
$\underline{D1}$ = 3.0 kmol in Table 8.1)
$\underline{D1}^s$ is the solution of Equations 8.1-8.4 given B^S_0, x^*_{D1}, and x^*_{B2}
t_{nr} is the batch time without recycle using 5 kmol (fixed reboiler charge) of fresh
feed for the given separation x^*_{D1} and x^*_{B2} (this time is same as those reported in
Table 8.1 as t_{nr})
t_{rS} is batch time with recycle using 5 kmol of mixed charge (B^*_c) for the same
separation x^*_{D1} and x^*_{B2}

P_{nr} = Productivity without recycle = BC^* / t_{nr}, kmol/hr
Pr = Productivity with recycle = B^S_0 / t_{rS}, kmol/hr
X = % Productivity increase = $(Pr - P_{nr})$ x100 / P_{nr}

Note: In all cases x_{B0}, V_C and x_{R1} are the same as those used in Table 8.1

As discussed in the previous section, the work of Mayur et al. (1970) and
Christensen and Jorgensen (1987) on the optimal recycle policy was restricted to
binary mixtures. The benefits of recycling were measured in terms of a reduction in
batch time although increase in productivity could be a possible alternative. Luyben
(1988) considered this productivity measure (as defined as "capacity" which
includes both batch time and a constant charging and cleaning time) in a simulation
of multicomponent batch distillation with recycle. Luyben (1988), however, showed
the effect of different parameters (no of plates, relative volatilities, etc.) on the
productivity and did not actually consider the effect of off-cuts recycle on the
productivity.

The mathematical formulation of the dynamic optimisation problem to obtain
optimal off-cut recycle policy was done for a *quasi-steady state* operation using
binary mixtures where there was only one main-cut and one off-cut (Mayur et al.,

Christensen and Jorgensen, Mujtaba and Macchietto). The off-cut was always recycled at the beginning of the batch.

Since in a multicomponent mixture there can be (n_c-1) main-cuts and (n_c-1) off-cuts, there could be a number of operational strategies regarding the way the off-cuts are recycled. Here, these strategies will be discussed first and then a general dynamic optimisation formulation of Mujtaba (1989) and Mujtaba and Macchietto (1992) will be presented to obtain off-cut recycle policy for one of such operational strategies. The possible difficulties for solving the general formulation will also be outlined.

8.7.1. Operational Strategies for Off-cut Recycle

Recycling of all the off-cuts to the beginning of a new batch is quite common (see Figure 8.10). STN representation is avoided for convenience. The main advantage of this strategy is that only one storage vessel is required to store all the off-cuts.

However, this type of recycle policy has the following disadvantages:

a)   For a given reboiler capacity (volumetric), the total amounts recycled increase with the number of off-cuts. This reduces the amount of fresh feed that can be processed (and products produced). For example, Luyben (1988) considered the above operational strategy for off-cut recycle for a ternary separation problem. Analysis of the results of Luyben (1988) show that the amount of fresh feed processed at *quasi-steady state* decreased by about 33% and productivity decreased by 13-16% with respect to those obtained with only fresh feed and no recycle for the same total charge and vapour boilup rate. Also, the batch time for the *quasi-steady state* operation was increased by 10-13% compared to cases with no off-cut recycle (Mujtaba, 1989).

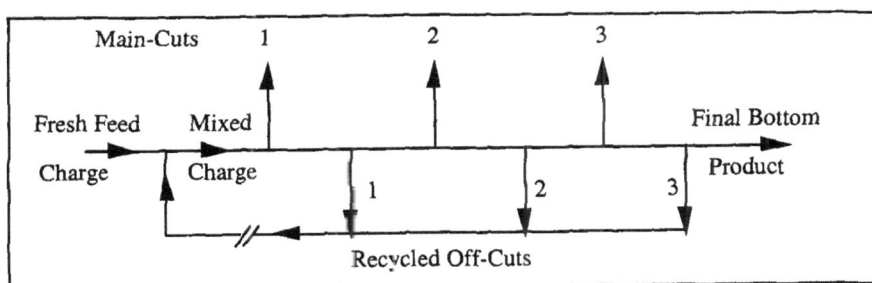

Figure 8.10. Typical Multicomponent Batch Operation with Recycled Off-Cuts

b) Recycling of heavy cuts to the initial charge will dilute the lighter components in the reboiler mixture and will consequently require higher energy. This is not a thermodynamically sound policy since it involves remixing of products which have already been separated (Mujtaba, 1989).

One can overcome the disadvantages of the above recycling policy by collecting and storing the off-cuts separately and recycling each of them to the reboiler in a sequential order (Mujtaba, 1989). Figure 8.11 shows the general sequential recycling strategy for multicomponent system. In practice this can be obtained by adding off-cuts from the previous batch to the reboiler at suitable times rather than at the beginning of the cycle. Quintero-Marmol and Luyben (1990) also discussed a similar strategy.

With this recycling strategy the reduction in the amount of fresh feed processed in a batch for a *quasi-steady state* operation is only equal to the amount of the first off-cut. This approach will also reduce the presence of high boiling point species in the reboiler mixture at the beginning of the process and minimise remixing of products already separated. However this policy will require n_c-1 distinct intermediate tanks. Also a more sophisticated controller will be required to charge the off-cut at the right time.

8.8. Optimisation Problem Formulation for Multicomponent Mixtures

Mujtaba (1989) considered the off-cut recycling strategy shown in Figure 8.11 and formulated an optimisation problem with an objective to minimise the overall batch distillation time. The problem can be stated as:

given: a fresh feed $(B0, x_{B0})$ and the desired purity specification of the main-cuts $(x^1_{D1}, x^2_{D2}, x^3_{D3}, ...)$

determine: the amount and composition of the off-cuts $[(R1, x_{R1}), (R2, x_{R2}), (R3, x_{R3}), ...]$ and the reflux ratio policy $r(t)$

so as to minimise: the overall distillation time

subject to: any constraints

The optimisation problem (MOP1) can be mathematically written as:

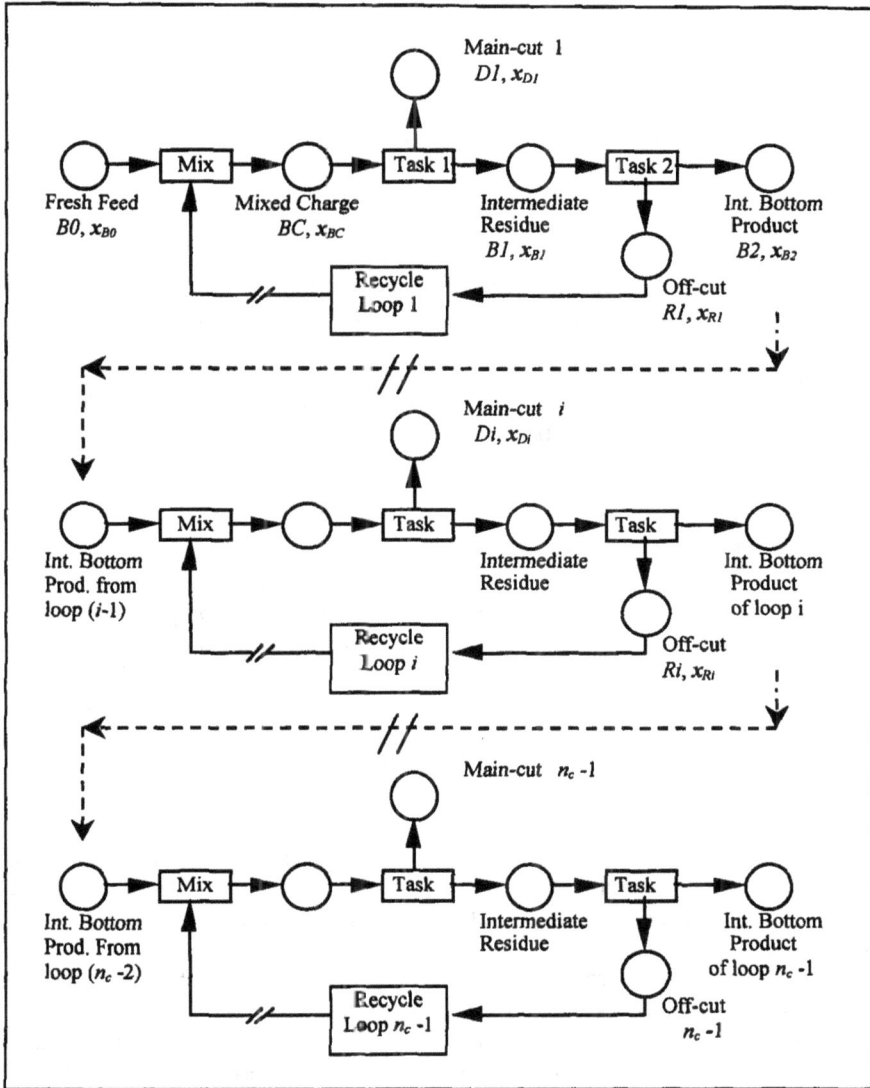

Figure 8.11. General Sequential Off-cut Recycle Strategy in Multicomponent Batch Distillation

Min $J = t_1 + t_2 + t_3 + t_4 + \ldots\ldots = t_F$
$[(R1, x_{R1}), (R2, x_{R2}), (R3, x_{R3}), \ldots\ldots], r(t)$

subject to :   (8.10)

         1. bounds on the amount of off-cuts
         2. bounds on the purity of off-cuts

    and interior point constraints

         1. purity requirements of the main-cuts
         2. requirements of off-cuts to establish *quasi-steady state* operation

where, t_1, t_2, t_3 ... are the batch times for the production of individual cuts (main-cuts and off-cuts). In addition to the constraints mentioned above the DAE process model (Type IV-CMH) equations were added as equality constraints to the optimisation problem.

8.8.1. Solution of the Optimisation Problem

The general formulation presented above results in a very complex optimisation problem and the solution of the problem using existing methods was very difficult. Also as the number of components increase, the above formulation will results in a very big optimisation problem with a large number of optimisation variables. Even for a 3 component mixture, this will result in six decision variables $(R1, x^1_{R1}, x^2_{R1}, R2, x^1_{R2}, x^2_{R2})$ in addition to the variables arising from the number of control intervals and switching times $[r(t)]$ in each of the four cuts (2 main-cuts and 2 off-cuts).

In addition to the above-mentioned problem, numerical difficulties may arise. The system (model equations) describing the multicomponent off-cut recycle operation needs to be reinitialised at the end of each main-cut and off-cut to accommodate the next off-cut to the reboiler. To optimise these initial conditions (new mixed reboiler charge and its composition) it is essential to obtain the objective function gradients with respect to these initial conditions.

The operational strategy shown in Figure 8.11 can be considered as a sequence of processes and the optimal control algorithm given by Morison (1984), Vassiliadis et al. (1993a,b) can be used for such sequential (or multistage) processes. These algorithms provide the initialisation procedure of such sequential processes and provide all the required gradients information. See the original references for further information and implementation strategies of such algorithms.

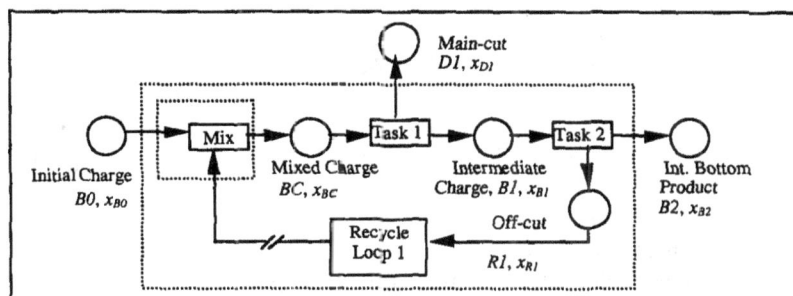

Figure 8.12. Off-cut Recycle after the First Cut

In this chapter, a decomposition strategy considered by Mujtaba (1989) and Mujtaba and Macchietto (1992) is presented in which the whole multiperiod optimisation problem is divided into a series of independent dynamic optimisation problems. This is presented in the next section.

8.9. Decomposition of the Optimisation Problem Formulation for Multicomponent Mixtures

Before discussing how the general optimisation problem mentioned in the previous section is decomposed into a series of independent dynamic optimisation problem and before presenting the mathematical formulation for such problem the following discussions are worthwhile.

Here, a three-component mixture is considered for convenience of discussion. Consider the overall and component mass balances around the outer loop shown in Figure 8.12 (the decomposed optimisation problem for the first cut). The balance equations are the same as those presented in Equations 8.1-8.4. For a given charge ($B0$, x_{B0}) and for a 3-component mixture there will be 5 equations with 8 unknown variables ($D1$, x^1_{D1}, x^2_{D1}, x^3_{D1}, $B2$, x^1_{B2}, x^2_{B2}, x^3_{B2}). Therefore, the degree of freedom (DF) of the steady state mass balances is 3. Several choices of consistent specifications are to specify (x^1_{D1}, x^2_{D1}, x^1_{B2}) or (x^1_{D1}, x^1_{B2}, x^2_{B2}), etc. In all cases, however one needs to specify at least 2 compositions in either the main-cut ($D1$) or the bottom product ($B2$). From the mathematical point of view any one of these specifications can be used to solve the mass balance equations, but it is very difficult to achieve a separation when specifying two components purity in a single product (top or bottom) as discussed in earlier chapters. This is because these purities depend on various other factors (relative volatility, number of plates, etc.).

Usually separation specifications in multicomponent mixtures are imposed on a particular (key) component of the cut, e.g. on component 1 in main-cut 1, on component 2 in main-cut 2 and likewise. Of course, it is possible to set intuitively some of the heavier component compositions to zero during lighter cuts and some of the lighter component compositions to zero during heavier cuts. But it is really difficult to specify independently the compositions of the preceding and few successive component compositions in a particular cut.

The question that immediately comes into mind in such cases is why not specify only one component either on the top or on the bottom product and solve the optimisation problem. The answer is that with such specifications it is very easy to formulate and solve an optimisation problem for multicomponent case without recycle (several examples were presented in chapter 5 and also see Diwekar et al., 1987), but such a problem with off-cut recycles is very difficult. This is due to the fact that it is very difficult to satisfy the conditions for the *quasi-steady state* operation. However, the problem can always be formulated and solved with great ease if the only concern was to produce cuts (main-cut and off-cut) satisfying the required amount and purity on one of the components only (see Mujtaba, 1989) and which did not require to consider a *quasi-steady state* operation.

To determine the optimal *quasi-steady state* off-cut recycle policies Mujtaba (1989) restricted to the special case considered by Luyben (1988), that is, to the case of sharp separation between all components (either using wide range of relative volatilities or a large number of plates). The main assumptions are (with reference to Figure 8.12) as follows:

1. No 3rd component with respect to a particular main-cut component[1] would appear in the main-cut. e.g. $x^3_{D1} = 0$ for main-cut 1 , $x^4_{D2} = 0$ for main-cut 2, etc.

2. Any amount left of a particular *cut component* will be removed totally in the next main-cut. For example the amount left of component 1 after main-cut 1 will be totally removed in main-cut 2 and therefore component 1 will not appear in subsequent cuts.

3. No 3rd component with respect to a particular off-cut component would appear in the off-cut.
    e.g. for off-cut 1 $x^3_{R1} = 0$, for off-cut 2 $x^4_{R2} = 0$ and likewise. This assumption together with the assumption no. 2 implies that each off-cut would only comprise of the main-cut component and the next component in the mixture.

---

[1] *cut component* is the desired (key) component in a cut. For example in main-cut 1, the component 1 is the *cut compoent*.

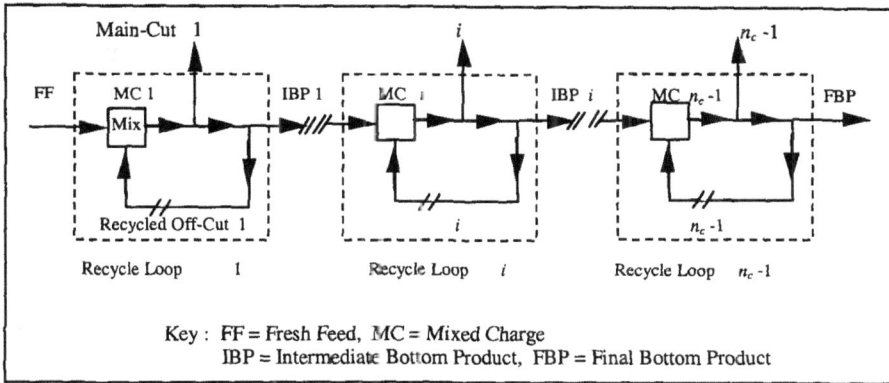

Figure 8:13. Decomposed Multicomponent Batch Operation with Off-Cut Recycle

Now with these assumptions and for given specifications (x^I_{D1}, x^I_{B2}) on component 1 in main-cut 1 and in the intermediate final bottom product (bottom product after 1 main-cut and 1 off-cut), solving the mass balance around the outer loop (Figure 8.12) will give the rest of the variables $(D1, B2,.....\text{etc.})$. Let the solution of $D1$ be given by $D1 = \underline{D}1$, and that of $B2$ be given by $B2 = \underline{B}2$.

Also with these assumptions and specifications the whole multiperiod optimisation problem shown in Figure 8.12 will now be decomposed into a series of independent dynamic optimisation problems (Figure 8.13). Referring to Figure 8.12 and Figure 8.13 the optimisation problem for Recycle Loop 1 may now be described as follows:

given:	a batch charge $B0$, x_{B0}, a desired amount of distillate $D1$ of specified purity x^I_{D1} and final bottom product $B2$ of specified purity x^I_{B2}
determine:	the amount and composition of the off-cut $(R1, x^I_{R1})$ and the reflux rate policy $r(t)$
so as to minimise:	the overall distillation time
subject to:	any constraints

The optimisation problem is now the same as *P4* (Equation 8.9) for binary mixtures. Also because of the assumptions made earlier, the formulation will automatically account for the *quasi-steady state* operation. The optimisation problems for all other recycle loops can be described similarly.

The benefits of recycling can be correlated against the measure of **the degree of difficulty** of separation, **q**. Mujtaba (1989) and Mujtaba and Macchietto (1992) used the same measure to identify the situations when the assumptions mentioned above are valid.

8.10. Measure of "the degree of difficulty" of Separation, q for Multicomponent Mixtures

Referring to Figure 8.12, Mujtaba (1989) applied the same definition of **the degree of difficulty** of separation for binary mixtures (described in Chapter 3) for multicomponent mixtures.

$$q = \int_{x_{B2}^1}^{x_{B0}^1} \frac{(N_{min}+1)dx_B^1}{(x_{B0}^1 - x_{B2}^1)(N_T+1)} \tag{8.11}$$

N_{min} is again evaluated using the Fenske's equation:

$$N_{min}+1 = \frac{\ln\left(\dfrac{x_{D1}^1}{x_{D1}^r}\right)\left(\dfrac{x_B^r}{x_B^1}\right)}{\alpha_{1-r}} \tag{8.12}$$

where, r is any reference component (not to be confused with the reflux ratio). x_B^1 and x_B^r are intermediate compositions of components 1 and r as the reboiler composition changes from x_{B0}^1 to x_{B2}^1. x_{D1}^1 is the specified composition of the cut-component in main-cut 1 and x_{D1}^r is the composition of component r in main-cut 1.

In order to evaluate Equation 8.12 one must obtain the values of all x_B^i (i=1, 2,....n_c) at intermediate points as the reboiler composition changes from x_{B0}^1 to x_{B2}^1. The average relative volatility α_{1-r} at each intermediate point is obtained based on the relative volatilities at the top and bottom of the column which are evaluated using bubble point calculations. Diwekar and Madhavan (1991) obtained these x_B^i with respect to the change in composition of a reference component. Here, say for the first cut, we can use component 1 as a reference component for this purpose (note: this reference component must not be confused with the reference component r mentioned in Equation 8.12). The following equation given by the authors (referring to Figure 8.12) can be used to evaluate x_B^i:

$$x_B^i = x_{D1}^i - \frac{x_{D1}^i - x_{B0}^i}{x_{D1}^1 - x_{B0}^1} \left(x_{D1}^1 - x_B^1 \right) \tag{8.13}$$

The derivation of this equation was simply done by doing a material balance over the entire column ignoring column holdup. Detailed of this derivation was given by the original authors. The same equations (Equations 8.11-8.13) are used to evaluate q values for subsequent cuts. In evaluating Equation 8.11 for a particular cut we use the reference component r as the component next to the *cut component*.

8.11. Example Problems using Multicomponent Mixtures

8.11.1. Example 1

Mujtaba (1989) and Mujtaba and Macchietto (1992) considered a ternary separation using Butane-Pentane-Hexane mixture. Only the optimal operation for the first main-cut and the first off-cut was considered. Table 8.7 lists a variety of separation specification (on CUT 1) and column configurations in each case. A fresh feed of 6 kmol at a composition of <0.15, 0.35, 0.50> (mole fraction) is used in all cases. Also in each case a constant condenser vapour load of 3 kmol/hr is used. For convenience Type IV-CMH model was used with ideal phase equilibrium.

Table 8.7. Input Data and Results for Example 1. [Mujtaba, 1989]

Case	N_T	x^l_{D1}	x^l_{B2}	$D1$ kmol	q	t_{nr} hr	t_r hr	% Time Saved	Optimal $R1$ kmol	x_{R1}
1	4	.850	.01	1.0	.533	3.07	1.99	35.33	.54	.273
2	5	.935	.011	0.9	.561	3.80	2.05	46.04	.64	.276
3	5	.950	.009	0.9	.601	----	2.27	-------	.75	.278
4	3	.850	.010	1.	.667	----	2.29	-------	1.18	.281

Key : N_T = no. of plates t_{nr} = batch time without recycle
        t_r = batch time recycle   % Time saved is based on t_{nr}

The column compositions are initialised to the composition of the mixed reboiler charge and a total of 2% of the fresh feed is used as column holdup. Half of the column holdup is assumed to be in the condenser and the rest is distributed equally over the plates. Piecewise constant reflux ratio was used, with 3 time intervals for the main-cut separation and 1 interval for the off-cut.

The value of q, the optimal batch time with and without recycle, the percentage time savings due to recycle and the optimal amount and composition of the recycle are presented in Table 8.7 for cases wherever it is applicable. The accumulated and instant distillate composition curves with and without recycle cases are shown in Figures 8.14-8.17. These figures also show the optimal reflux ratio profiles for each case.

As can be seen from Table 8.7 the batch time for operations without recycle is not reported for case 3 and 4. A true *quasi-steady state* mode of operation was not obtained for case 4 because of the violation of the sharp separation assumption mentioned in section 8.9. It is clear from Figure 8.17 that a substantial amount of the 3rd component is going to appear in the recycle cut, although the main-cut was free from that component. Therefore, the dynamic optimisation technique used was not suitable for this case. However, this case is an example where the off-cut recycle policies are generated by satisfying the amount and composition of the off-cut component only (i.e. by violating *quasi-steady state* operation mode which requires other component compositions of the off-cut to be identical with those of the recycled off-cut). For case 3 it was not possible to obtain the separation in the given column without recycle. This situation clearly shows the benefit of recycling of the off-cut.

The results presented in Table 8.7 and in the Figures 8.14-8.17 (recycle cases) clearly show that there is an upper limit of q, beyond which the assumptions mentioned in section 8.9 are not valid and the method cannot be applied. The investigation shows that up to q equal to about 0.6, the method is adequately applicable.

As in binary cases, recycling of off-cuts offers a possibility of significant reduction in batch time. In some cases it is found to be more than 45%. The results also show that the measure q can be used to predict qualitatively whether recycle is beneficial or not. As the value of q increases the benefit of recycle increases and so do the amount of and composition of the lighter component in the cut.

An interesting part of the results mentioned so far is that in all cases, an initial total reflux operation was found to be necessary. This was to eliminate the third component from the overheads. The length of that period again depends on the ease of separation. For no recycle cases, the times at total reflux are slightly larger than those for recycle cases. This is due to the fact that the recycle of off-cut eases the separation of the main-cut. After the total reflux period of operation an increasing reflux ratio profile was obtained for the main-cut in all cases. The off-cut was always obtained at low reflux ratio as were the case with binary mixtures (Rose,

1985). Also, the operation with off-cut recycling requires lower energy consumption due to shorter batch time.

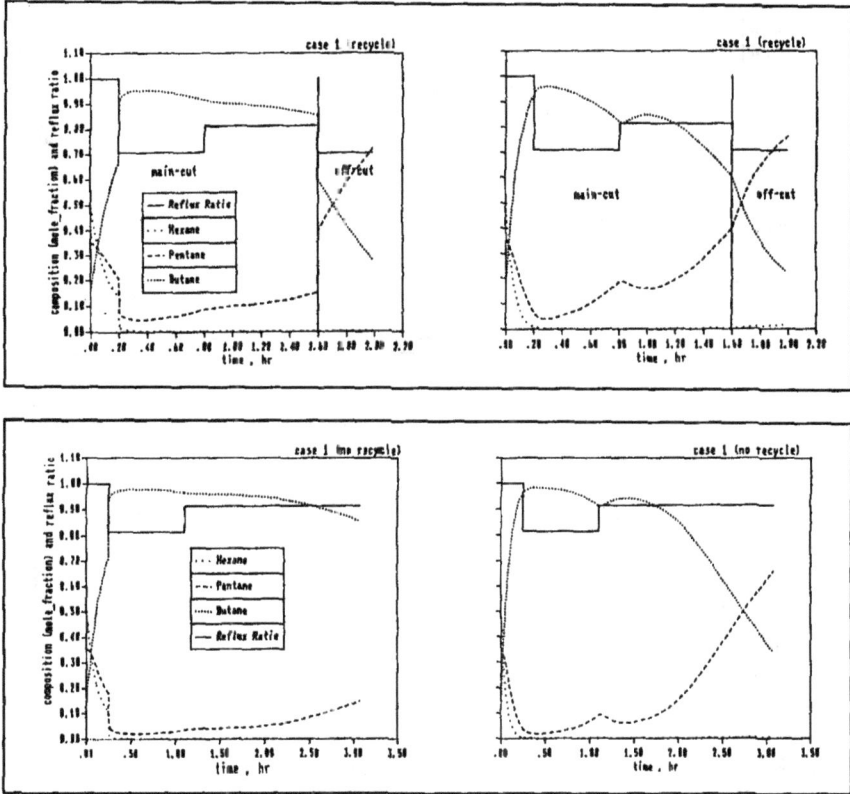

Figure 8.14. Accumulated and Instant Distillate Composition and Optimal Reflux Ratio Profiles (case 1). [Mujtaba and Macchietto, 1992; Mujtaba, 1989][a]

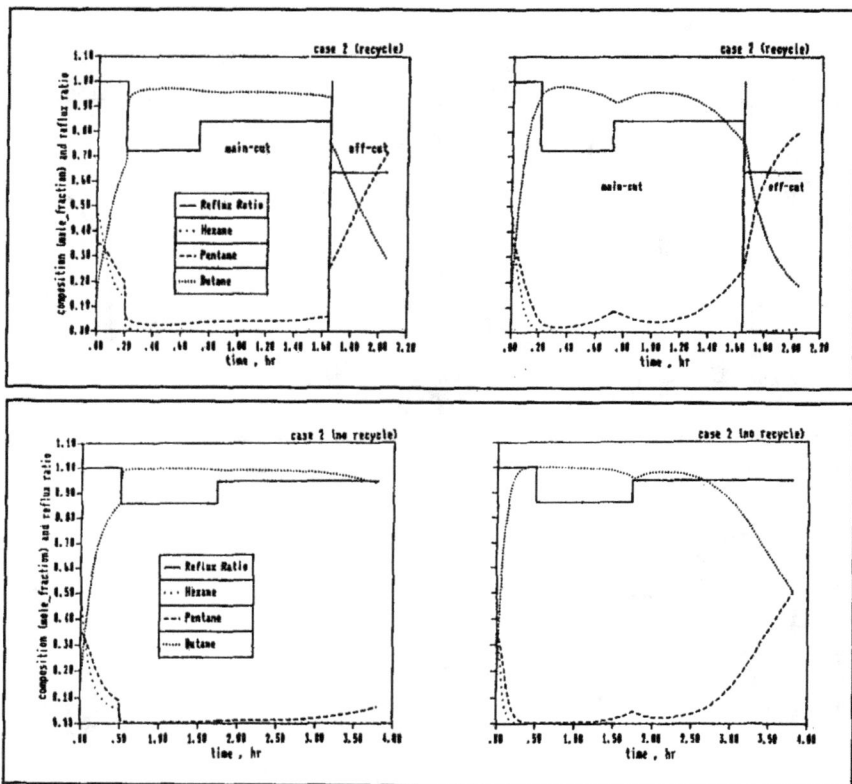

Figure 8.15. Accumulated and Instant Distillate Composition and Optimal Reflux Ratio Profiles (case 2). [Mujtaba and Macchietto, 1992; Mujtaba, 1989][a]

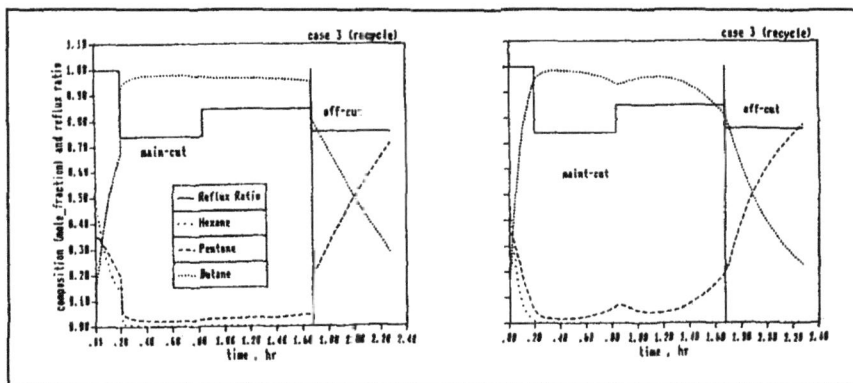

Figure 8.16. Accumulated and Instant Distillate Composition and Optimal Reflux Ratio Profiles (case 3). [Mujtaba and Macchietto, 1992; Mujtaba, 1989][a]

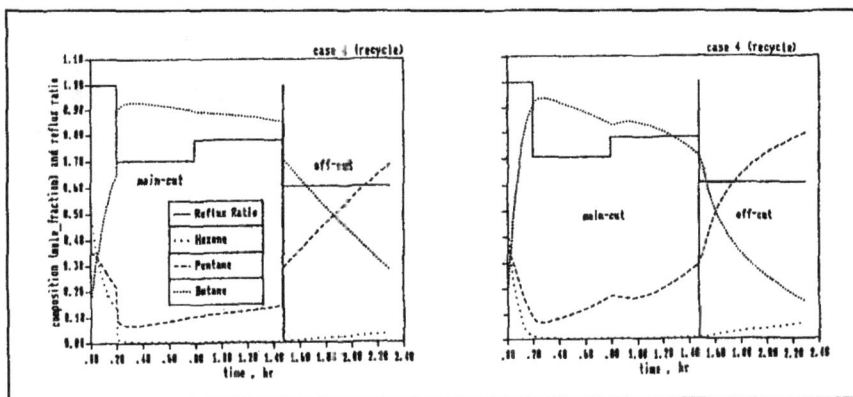

Figure 8.17. Accumulated and Instant Distillate Composition and Optimal Reflux Ratio Profiles (case 4). [Mujtaba, 1989]

Mujtaba and Macchietto (1992) investigated how productivity (kmol of product/hr) is affected by the proposed recycle policy. Consider a fixed vapour load to the condenser (or vapour boilup rate) and fixed product compositions (e.g. x^1_{D1} and x^1_{B2} for recycle loop 1 of Figure 8.13). Now scale the batch times (presented as

---

[a] Reprinted from *Computers & Chemical Engineering*, **16S**, Mujtaba, I.M. and Macchietto, S., An optimal recycle policy for multicomponent batch distillation, s273-280, Copyright (1992), with permission from Elsevier Science .

t_{nr} in Table 8.7) obtained with a fixed fresh feed charge of 6 kmol, to calculate the batch time (with no off-cut production and recycle) required to process any charge. Thus for case 1 of Table 8.7, if the still capacity is fixed at $(B0 + R1)$ $(6.0 + 0.54 = 6.54$ kmol), and with the same x^1_{D1} and x^1_{B2}, $\underline{D_1}$ is 1.09 kmol. For the same column and vapour load to the condenser the batch time t_{nr} becomes 3.346 hr and the productivity is $1.09/(3.346 + 0.5) = 0.283$ kmol of distillate/hr (where the still charging time is assumed to be 0.5 hr). For the same mixed still charge of 6.54 kmol, the productivity for case 1 (with off-cut production and recycle), is $1.0/(1.99 + 0.5) = 0.401$ (assuming 0.5 hr as the simultaneous charging time of the fresh feed and the recycle). Comparison of these productivities shows an improvement of about 42%. Similar calculations for case 2 show an improvement of productivity by 67%.

8.11.2. Example 2

This example is taken from Mujtaba (1989) and Mujtaba and Macchietto (1992) where the same ternary mixture as in example 1 was considered for the whole multiperiod operation which includes 2 main-cuts and 2 intermediate off-cuts. The column consists of 5 (N_T) intermediate plates, a total condenser and a reboiler. The column is charged with the same amount and composition of the fresh feed as was the case in example 1. Column initialisation, holdup distribution and condenser vapour load are also same as those in example 1.

The specifications for the first cut are the same as those presented in Table 8.7 for case 2 which is $x^1_{D1} = 0.935$ and $x^1_{B2} = 0.011$. These results in $\underline{D1} = 0.9$ kmol and $q_1 = 0.561$. For cut 2 $x^2_{D2} = 0.82$ and $x^2_{B2} = 0.13$ were specified which resulted in $\underline{D2} = 2.0$ kmol and $q_2 = 0.415$.

The minimum batch times for the individual cuts and for the whole multiperiod operation are presented in Table 8.8 together with the optimal amount of recycle and its composition for each cut. The percentage time savings using recycle policies are also shown for the individual cuts and also for the whole operation. Figure 8.18 shows the accumulated distillate and composition profile with and without recycle case for the operation. These also show the optimal reflux ratio profiles. Please see Mujtaba (1989) for the solution statistics for this example problem.

It is clear from the Table 8.8 that although the recycling of off-cut was beneficial for the first cut it was not at all beneficial for the second cut. This was quite expected because the q value for the second cut (q_2) was quite small. However based on the total operation time recycling was still beneficial compared to the no recycle case.

It is also clear from the Figure 8.18 that the assumptions mentioned in section 8.9 were not violated in any cuts. Although the cut 1 operation required initial total reflux period to remove the 3rd component from the top of the column, initial total reflux period was not required for cut 2 operation because no 3rd component with respect to the *cut component* was present in cut 2.

8.12. Multicomponent Off-Cut Recycle Policy of Bonny et al.

Bony et al. (1996) proposed a multicomponent off-cut recycle policy that is somewhat in between the policies considered by Luyben (1988) and Mujtaba (1989). Each off-cut generated from a particular batch in a campaign mode operation (described in Chapter 6 and 7) is split into fractions and are recycled to subsequent batches in the campaign mode. The policy is described in Figure 8.19. Here, $s^1_{1,j}$ denotes the fraction of off-cut 1 generated in batch 1 and recycled in batch j. Similarly, $s^m_{j,NB}$ denotes the fraction of off-cut m generated in batch j and recycled in batch NB. Note Bony et al. did not consider *quasi-steady state* operation.

Table 8.8. Summary of the Results for Example 2

t^1_{nr} hr	t^2_{nr} hr	t^1_r hr	t^2_r hr	Optimal				T_{nr} hr	T_r hr	T^1_S hr	T^2_S hr	TS
				$R1$	x^1_{R1}	$R2$	x^2_{R2}					
3.8	1.7	2.05	1.69	.64	.276	.41	.400	5.51	3.74	46	1.17	32

Key :

t^1_{nr}	= optimal batch time without recycle for cut 1.
t^2_{nr}	= optimal batch time without recycle for cut 2
t^1_r	= optimal batch time for recycle loop 1 (off-cut production and recycle).
t^2_r	= optimal batch time for recycle loop 2 (off-cut production and recycle).
$R1$	= amount recycled in off-cut 1, kmol
x^1_{R1}	= butane mole fraction in off-cut 1
$R2$	= amount recycled in off-cut 2, kmol
x^2_{R2}	= pentane mole fraction in off-cut 2
T_{nr}	= total batch time for the whole operation without recycle.
T_r	= total batch time for the whole operation with recycle.
T^1_S	= % time saving based on t^1_{nr} for cut 1.
T^2_S	= % time saving based on t^2_{nr} for cut 2.
TS	= % time saving based on T_{nr} for the whole operation.

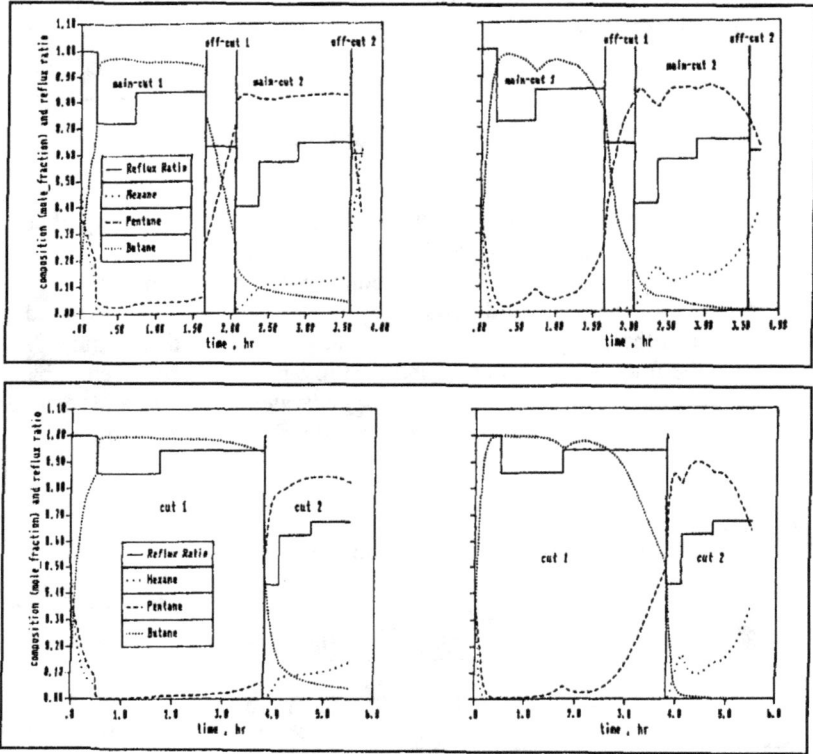

Figure 8.18. Accumulated, Instant Distillate and Optimal Reflux Ratio (example 2)
[Mujtaba and Macchietto, 1992; Mujtaba, 1989][b]

The fractions of each off-cut (such as $s^l_{1,j}$, $s^m_{1,j}$, $s^l_{j,NB}$, etc.) charged to subsequent batches are included in the decision variable list within the optimisation problem formulation by Bony et al. (presented in Chapter 7).

This recycle policy has the option of charging the light off-cuts at the beginning of a fresh batch as well as charging of heavy off-cuts at the beginning of a fresh batch which may not be thermodynamically sensible.

---

[b] Reprinted from *Computers & Chemical Engineering*, **16S**, Mujtaba, I.M. and Macchietto, S., An optimal recycle policy for multicomponent batch distillation, s273-280, Copyright (1992), with permission from Elsevier Science .

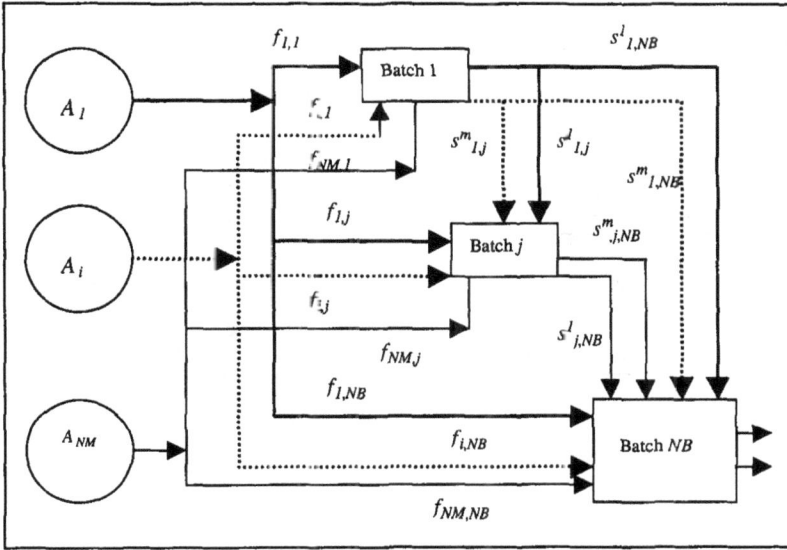

Figure 8.19. Off-cut Recycle Policy of Bonny et al. (1996)

8.12.1. Example

Bonny et al. (1996) considered the separation of 3 multicomponent mixtures composed of same ternary components (Cyclohexane, n-Heptane, Toluene) but with different compositions in 3 batches. The input data for the problem is given in Table 8.9. A single constant reflux ratio is considered in each batch which is optimised together with the batch time. Productions of off-cuts are considered. The objective of the operation is to maximise the productivity. The results without off-cut recycle are summarised in Table 8.10 and Figure 8.20. The results with off-cut recycle are summarised in Table 8.11 and Figure 8.21.

Table 8.9. Input Data

	A_i mol	Composition, molefraction
Mixture 1	140	<0.40, 0.40, 0.20>
Mixture 2	60	<0.50, 0.30, 0.20>
Mixture 3	60	<0.10, 0.50, 0.40>

$B_0 = 100$ mol	V, Vapour Boilup = 91.77 mol/hr
$NB = 3$	N=No of plates = 10
Pressure = 1.0 atm	All $\beta_k = 1$

Product Specifications in each batch, j

$$x^*_{j,k} = \quad x^*_{j,1} = 0.90$$

$$x^*_{j,2} = 0.85$$

$$x^*_{j,3} = 0.85$$

Table 8.10. Summary of Results (without Off-cut Recycle)

Batch j	Optimal Charging Policy $f_{1,j}$	$f_{2,j}$	$f_{3,j}$	Optimal Reflux Ratio	Batch Time, hr
1	.715	0	0	7.50	7.42
2	0	1	0	3.67	2.61
3	.285	0	1	8.37	6.92

Batch j	Optimal Production of Main-cut (P) and Off-cut (R), mol $P_{j,1}$	$R_{j,1}$	$P_{j,2}$	$R_{j,2}$	$P_{j,3}$
1	28.1	24.9	19.8	7.6	19.8
2	19.7	31.6	0	0	8.7
3	9.8	27.5	18.7	11.8	32.2

Total Main-cut Products	= 156.5 mol
Total Campaign Time	= 16.95 hr
Maximum Production Rate	= 9.23 mol/hr

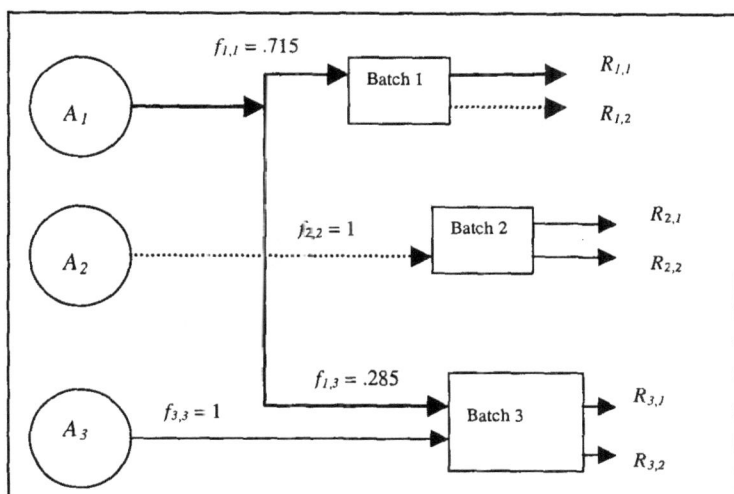

Figure 8.20. Optimal Initial Charge Policy without Off-cut Recycle

Table 8.11. Summary of Results (with Off-cut Recycle)

Batch	Optimal Charging Policy							Optimal Reflux Ratio	Batch Time, hr
j	$f_{1,j}$	$f_{2,j}$	$f_{3,j}$	$s^1_{1,j}$	$s^2_{1,j}$	$s^1_{2,j}$	$s^2_{2,j}$		
1	.715	0	0	-	-	-	-	7.48	7.42
2	.285	1	0	0	0	-	-	6.49	6.62
3	0	0	1	0	1	0	1	7.97	4.92

Batch	Optimal Production of Main-cut (P) and Off-cut (R), mol				
j	$P_{j,1}$	$R_{j,1}$	$P_{j,2}$	$R_{j,2}$	$P_{j,3}$
1	28.2	23.9	20.6	7.7	19.7
2	33.1	27.0	8.66	12.3	18.8
3	0.04	18.5	20.5	11.3	29.6

Total Main-cut Products	= 179.2 mol
Total Campaign Time	= 18.96 hr
Maximum Production Rate	= 9.45 mol/hr

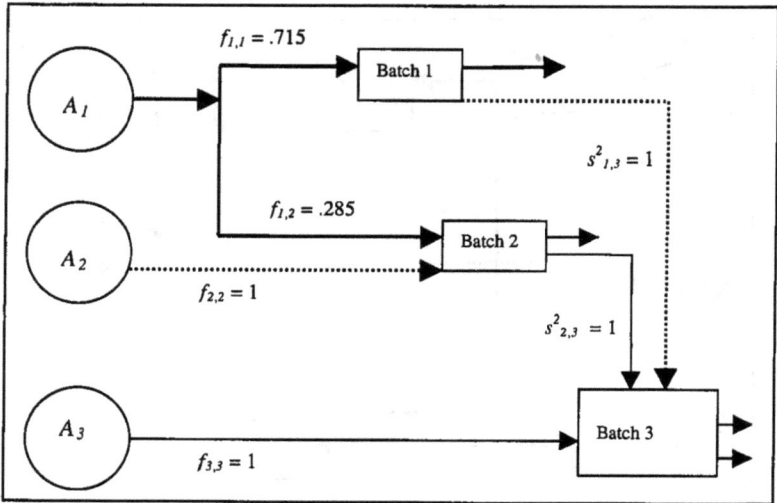

Figure 8.21. Optimal Initial Charge Policy with Off-cut Recycle

The results clearly show the effect of off-cut recycling on the initial charge policy of each charge. Although very little improvement in terms of production rate is observed, the initial charge policies with and without recycle of off-cuts are quite different. Note that with off-cut recycle strategy, heavy off-cuts from batch 1 and 2 are being mixed with fresh charge at the beginning of the 3rd batch. As mentioned earlier this is not a thermodynamically sound policy. For more examples see the original reference.

References

Bonny, L., Domenech, S., Floquet, P. and Pibouleau, L., *Chem. Engng. Proc.* **35** (1996), 349.

Christensen, F.M. and Jorgensen, S.B., *Chem. Eng. J.* **34** (1987), 57.

Diwekar, U.M. and Madhavan, K.P., *IEC. Res.* **30** (1991), 713.

Diwekar, U.M., Malik, R.K. and Madhavan, K.P., *Comput. chem. Engng.* **11** (1987), 629.

Diwekar, U.M., Madhavan, K.P. and Swaney, R.E., *IEC. Res.* **28** (1989), 1011.

Luyben, W.L., *IEC Res.* **27** (1988), 542

Logsdon, J.S., Diwekar, U.M. and Biegler, L.T., *Trans. IChemE*, **68A** (1990), 434.

Logsdon, J.S.and Biegler, L.T., *IEC Res.* **32** (1993), 700.

Liles, J.A., Optimal scheduling of a batch distillation process, M.Sc. Thesis, (Imperial College, University of London, 1966).

Mayur, D.N., May, R.A., and Jackson, R., *Chem. Eng. J.* **1** (1970), 15.

Morison, K.R., *Optimal Control of Processes Described by Systems of Differential and Algebraic Equations.* PhD. Thesis, (Imperial College, University of London, 1984).

Mujtaba, I.M., *Optimal Operational Policies in Batch Distillation.* PhD Thesis, (Imperial College, University of London, 1989).

Mujtaba, I.M. and Macchietto, S. In *Recent Progres en Ginie de Procedes*, eds. Domenech, S. et al. (Lavoisier Techniche et Documentation, Paris, 1988) **2**, 191.

Mujtaba, I.M. and Macchietto, S., *Comput. chem. Engng.* **16S** (1992), S273.

Mujtaba, I.M. and Macchietto, S., *Comput. chem. Engng.* **17** (1993), 1191.

Nad, M. and Spiegel, L., *In Proceedings CEF 87* , Sicily, Italy, April (1987), 737.

Quintero-Marmol, E. and Luyben, W.L., *IEC Res.* **29** (1990), 1915.

Rose, L.M., *Distillation Design in Practice* (Elsevier, New York, 1995).

Vassiliadis, V.S., Sargent, R.W. and Pantelides, C.C., *IEC. Res.* **33** (1994a), 2111.

Vassiliadis, V.S., Sargent, R.W. and Pantelides, C.C., *IEC. Res.* **33** (1994b), 2123.

CHAPTER 9

9. BATCH REACTIVE DISTILLATION (BREAD)

9.1. Introduction

Reactive Distillation processes combine the benefits of traditional unit operations with a substantial progress in reducing capital and operating costs and environmental impact (BP Review, 1997; Taylor and Krishna, 2000).

Traditionally, as in many chemical industries, reaction and separation take place separately (Figure 9.1) in a batch reactor followed by a batch distillation column (Charalambides et al., 1994). Therefore, the distillation of desired species cannot influence the conversion of the reactants in the reactor.

However, conventional batch distillation with chemical reaction (reaction and separation taking place in the same vessel and hence referred to as *Batch REActive Distillation- BREAD*) is particularly suitable when one of the reaction products has a lower boiling point than other products and reactants. The higher volatility of this product results in a decrease in its concentration in the liquid phase, therefore increasing the liquid temperature and hence reaction rate, in the case of irreversible reaction. With reversible reactions, elimination of products by distillation favours the forward reaction. In both cases higher conversion of the reactants is expected than by reaction alone. Therefore, in both cases, higher amount of distillate (proportional to the increase in conversion of the reactant) with desired purity is expected than by distillation alone (as in traditional approach) (Mujtaba and Macchietto, 1997).

An extensive literature survey shows that very little attention has been given to modelling and simulation of batch reactive distillation, let alone optimisation of such process. The published literature deals with the mathematical modelling and numerical integration of the resulting dynamic equations systems, with few presenting computer simulation vs experimental results. Only few authors have discussed the design, control and optimal operational aspects of batch reactive distillation processes.

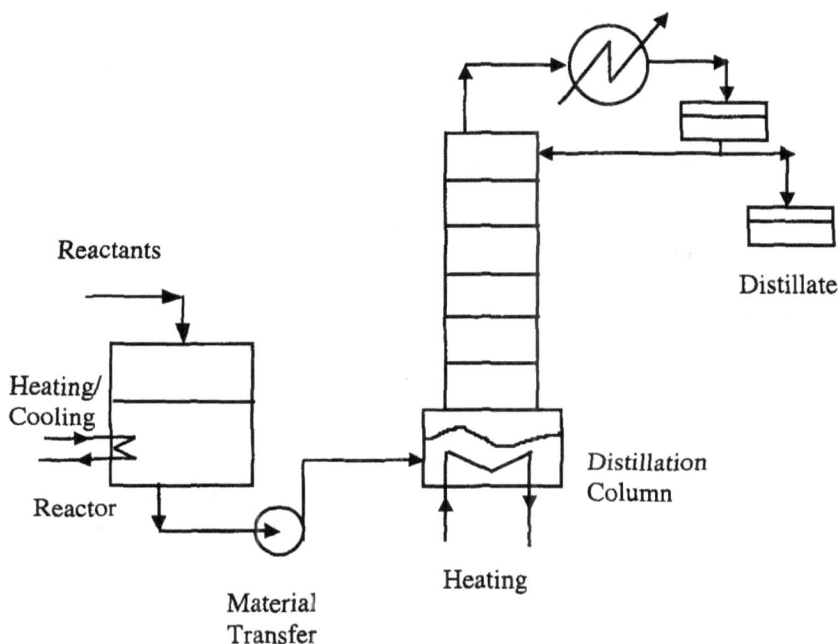

Figure 9.1. Traditional Batch Reaction-Distillation System

9.2. Review

9.2.1. Experimental Studies

Corrigan and Ferris (1969) studied the methanol esterification with acetic acid in an experimental batch distillation column, with emphasis on the design and construction of the equipment. It was noted that until 1963, analytical procedures such as gas phase chromatographic method for the system of methyl alcohol, methyl acetate, acetic acid and water had not been developed. Among the other objectives of Corrigan and Ferris (1969) were to develop an analytical method for such system and to obtain experimental data that could be used to verify the method. The basic pieces of equipment that were designed were a ten gallon jacketed glass lined reactor, a ten gallon glass-lined receiving or feed tank, three sections of Oldershaw sieve tray column, reflux splitter, feed tray, condenser, and a flexopulse timer.

Using appropriate reflux, methyl acetate (the lowest boiling component) at a purity of 92.5% (mol) was possible to produce with a very high recovery rate.

Lehtonen et al. (1998) considered polyesterification of maleic acid with propylene glycol in an experimental batch reactive distillation system. There were two side reactions in addition to the main esterification reaction. The equipment consists of a 4000 ml batch reactor with a one theoretical plate distillation column and a condenser. The reactions took place in the liquid phase of the reactor. By removing the water by distillation, the reaction equilibrium was shifted to the production of more esters. The reaction temperatures were 150-190^0 C and the catalyst concentrations were varied between 0.01 and 0.1 mol%. The kinetic and mass transfer parameters were estimated via the experiments. These were then used to develop a full-scale dynamic process model for the system.

9.2.2. Modelling and Simulation

Egly et al. (1979), Cuille and Reklaitis (1986), Mujtaba (1989), Reuter et al. (1989), Albet et al. (1991), Basualdo and Ruiz (1995) and Wajge and Reklaitis (1999) considered the development of mathematical models to simulate BREAD processes. In most cases, the model was posed as a system of Differential and Algebraic Equations (DAEs) and a stiff solution method was employed for integration.

9.2.3. Design, Control and Optimisation

Egly et al. (1979) considered the *minimum time* optimisation problem using a detailed dynamic process model (Type V) but no details were given regarding the input and kinetic data of the problem.

Wilson (1987) discussed the optimal design of batch distillation processes using a simplified column model for BREAD and using repetitive simulation strategy. For a commercially used complex parallel reaction scheme and using a simple economic model the author showed the benefit of integrating reaction and distillation. A number of plots of process efficiency (in terms of product cost contribution per unit product) for a range of alternative process and design variable choices were generated and an optimal design and operation policy of reactive batch distillation were suggested.

Sorensen and Skogestad (1994) developed control strategies for BREAD processes by repetitive simulation strategy using a simple model in SPEEDUP package. Wilson and Martinez (1997) developed EKF (Extended Kalman Filter) based composition estimator to control BREAD processes. The estimator was found to be quite robust and was able to estimate composition within acceptable accuracy, even in the face of process/model mismatches. Balasubramhanya and Doyle III

(2000) studied nonlinear model based control of a BREAD process using an example of ethanol esterification. The main focus of the work was to develop a reduced order nonlinear process model that could be used efficiently within a model predictive control strategy.

Mujtaba and Macchietto (1992, 1994, 1997) and Wajge and Reklaitis (1999) developed optimisation strategies for BREAD processes. Walsh et al. (1995) included in their work, the effect of uncertain model parameters on the design of operating procedures of BREAD processes.

In this chapter some of the published work in BREAD process optimisation will be presented in detail. Note that although formation of azeotropes is quite common in reactive distillation, such situation is not considered in this book. Readers are directed to Van Dongen and Doherty (1985), Bernot et al. (1991) and Diwekar (1991) regarding this.

9.3. Selecting the Right Column for BREAD

Barbosa and Doherty (1988) listed a number of chemical reaction schemes which were previously used mainly in continuous distillation. The reaction products in the different reaction schemes considered do not always have lower boiling point than the reactants. The use of conventional batch distillation for those reactions would result in removal of reactants as the distillation proceeds thus lowering conversion and yield of product. Therefore, it is very important to select the right batch distillation column for each type of chemical reaction.

For example, if all the reaction products are valuable and have lower boiling temperature than the reactants, then conventional batch distillation would be most suitable. As the reaction proceeds the products will be separated in different main-cuts in sequential order. Conversion and yield can be greatly improved in such cases. If only some of the reaction products have low boiling temperature, then a conventional batch column will only remove those products as distillation proceeds. To separate the rest of the products by conventional distillation would require the removal of unreacted reactants from the column first.

Conventional batch distillation is not suitable when all reaction products have higher boiling temperatures than those of the reactants. Inverted batch distillation is suitable for such situation. If all the reaction products are valuable, as the reaction proceeds, the products will be separated from the bottom of the column in different main bottom cuts in sequential order.

For cases where some of the reaction products have higher and some lower boiling points than those of the reactants, then neither the conventional nor the inverted batch distillation are suitable. For such reaction schemes, the MVC column will be the most suitable one because the light and heavy products can now be withdrawn simultaneously from the column, thus pushing the reaction further to the product side.

However, there are cases where none of these batch distillation columns can be used economically to improve conversion or yield. Also, complications typically arise if there are any azeotropes present in the mixture.

Mujtaba and Macchietto (1992) summarises a list (Table 9.1) of reaction schemes with boiling points of the species involved. For each reaction scheme (with reactants shown on the left and products on the right), the right type of batch distillation column is indicated. The recommendation does not hold when the desired products are those shown on the left hand side of each reaction scheme. For example, acetic anhydride is produced by dehydration of acetic acid (Acetic Acid $\Leftrightarrow$ Acetic Anhydride + Water). In such cases, the use of CBD will be favourable to remove the water or the use of IBD will be favourable to remove the anhydride. In both configurations the equilibrium will shift to the right and the productivity of anhydride will improve. Wajge and Reklaitis (1999) considered such reaction scheme in a CBD column. Further details are in section 9.9.

9.4. Process Modelling and Simulation

Chapter 4 presents models of different complexity for BREAD processes. For other types of models and underlying assumptions, the readers are directed to Cuille and Reklaitis (1986), Albet et al. (1991), Basualdo and Ruiz (1995), Wajge and Reklaitis (1999), Leversund et al. (1993). An example of BREAD process simulation from Greaves (2003) using the model of Mujtaba and Macchietto (1997) is presented in Chapter 4.

Table 9.1. Most Suitable Batch Column Configuration for Several Chemical Reaction Schemes

$$A + B \Leftrightarrow C$$

A (Boiling Point, K)	B (Boiling Point, K)	C (Boiling Point, K)	Recommended Column
Ethylene Oxide (283.5)	Water (373.2)	Ethylene Glycol (470.4)	Inverted

$$A + B \Leftrightarrow 2C$$

Benzene (353.3)	Xylene (417.5)	Toluene (383.8)	N.S. *
Acetic Anhydride (412.0)	Water (373.2)	Acetic Acid (391.1)	N.S.

$$A + B \Leftrightarrow C + D$$

A	B	C	D	
Acetic Acid (391.1)	Methanol (337.8)	Methyl Acetate (330.1)	Water (373.2)	Conventional
Acetic Acid (391.1)	Ethanol (351.5)	Ethyl Acetate (350.3)	Water (373.2)	Conventional
Acetic Acid (391.1)	Propanol (370.4)	Propyl Acetate (374.8)	Water (373.2)	N.S.
Acetic Acid (391.1)	Butanol (390.9)	Butyl Acetate (399.2)	Water (373.2)	MVC or Inverted

*N.S. = Not suitable for any batch distillation

9.5. Dynamic Optimisation

In this section optimal operation problem of BREAD processes is presented as a proper dynamic optimisation problem incorporating a detailed dynamic model (Type V- CMH). The problem formulation and solution exploit the methods developed for non-reactive batch distillation by Mujtaba and Macchietto (1991, 1993, 1998). These methods are also discussed in Chapters 5 and 6.

Mujtaba and Macchietto (1997) have considered a *maximum conversion problem* for BREAD, subject to given product purity constraints. The reflux ratio is selected as the control parameters to be optimised for a fixed batch time so as to maximise the conversion of the limiting reactant. The optimal product amount, condenser and reboiler duties are also calculated. Referring to Figure 4.5 for CBD column the optimisation problem can be stated as:

given: the column configuration, the feed mixture (B_0, x_{B0}), condenser vapour load (V_2) and a separation task (i.e. achieve the product with purity specification for a key distillate component (x^*_D), the batch time (t^*_f)

determine: the optimal reflux ratio profile $r_f(t)$ for the operation

so as to maximise: an objective function defined for instance the conversion

subject to: equality and inequality constraints.

Mathematically the optimisation problem (*P1*) can be written as:

$$P1 \qquad \underset{r(t)}{\text{Max}} \qquad C$$

$$
\begin{aligned}
\text{subject to} \qquad & t = t^*_f \\
& x_D(t_f) \geq x^*_D \qquad && \text{(inequality constraint)} \\
\text{and} \qquad & f(t, \dot{x}, x, u, v) = 0 \qquad && \text{(model equations, equality constraints)} \\
\text{with} \qquad & f(t_0, \dot{x}_0, x_0, u, v) = 0 \qquad && \text{(initial conditions, equality constraints)} \\
& \text{Linear bounds on reflux ratio} \qquad && \text{(inequality constraints)}
\end{aligned}
$$

where C is the conversion of the limiting reactant, $x_D(t_f)$ is the composition of distillate at the end of the operation (t_f), $r(t)$ is the reflux ratio as a function of time (t). Solution method for such optimisation problem has been discussed earlier.

Note that other types of dynamic optimisation problems such as *minimum time*, *maximum profit*, etc. could also be formulated and solved using the algorithms mentioned in this book. However, Mujtaba and Macchietto (1997) used the results of optimisation problem **P1** to develop a computationally efficient technique to solve *maximum profit* problem which will be discussed in the following sections.

9.6. Example: Dynamic Optimisation

Mujtaba and Macchietto (1997) considered the esterification of ethanol and acetic acid to study dynamic optimisation of BREAD. The reaction products are ethyl acetate and water, with ethyl acetate being the main product. The reversible reaction scheme together with the boiling temperatures of the components are shown below:

	Acetic Acid + Ethanol <=> Ethyl Acetate + Water			
Component	(1)	(2)	(3)	(4)
Boiling Points, K	391.1	351.5	350.3	373.2

Ethyl acetate has the lowest boiling temperature in the mixture and therefore has the highest volatility. Controlled removal of ethyl acetate by distillation will improve the conversion of the reactants by shifting the chemical equilibrium to further right. This will also increase the yield proportionately.

The number of plates (defining the column configuration), feed, feed composition, column holdup, etc. for the problem are given in Table 4.9 (Chapter 4). The vapour-liquid equilibrium data and the kinetic data are taken from Simandl and Svrcek (1991) and Bogacki et al. (1989) respectively and are shown in Table 4.10 (Chapter 4). The vapour and liquid enthalpies are calculated using the data from Reid et al. (1977). As mentioned in Chapter 4, these data do not account for detailed VLE calculations and for any azeotropes formed.

Mujtaba and Macchietto (1997) solved a series of optimisation problems (**P1**) for different but fixed batch time t_f (between 5 and 30 hrs) and for two given product purities, $x^*_D = 0.70$ and $x^*_D = 0.80$. Reflux ratio level was optimised over the batch time of operation. Figure 9.2 shows the typical plots of accumulated distillate and reboiler composition profiles for $t^*_f = 15$ hrs and $x^*_D = 0.80$, with the reflux ratio being optimised.

The maximum conversion, the corresponding amount of product, optimal constant reflux ratio and heat load profiles for different batch times are shown in Figures 9.3-9.6. The maximum conversion profile achieved under total reflux operation (where no product is withdrawn) is also shown in Figure 9.3. The latter approximates the conversion which would be achieved in the absence of distillation. Note that if there is a large column holdup, the conversion under total reflux will not approximate the conversion achieved in the absence of distillation.

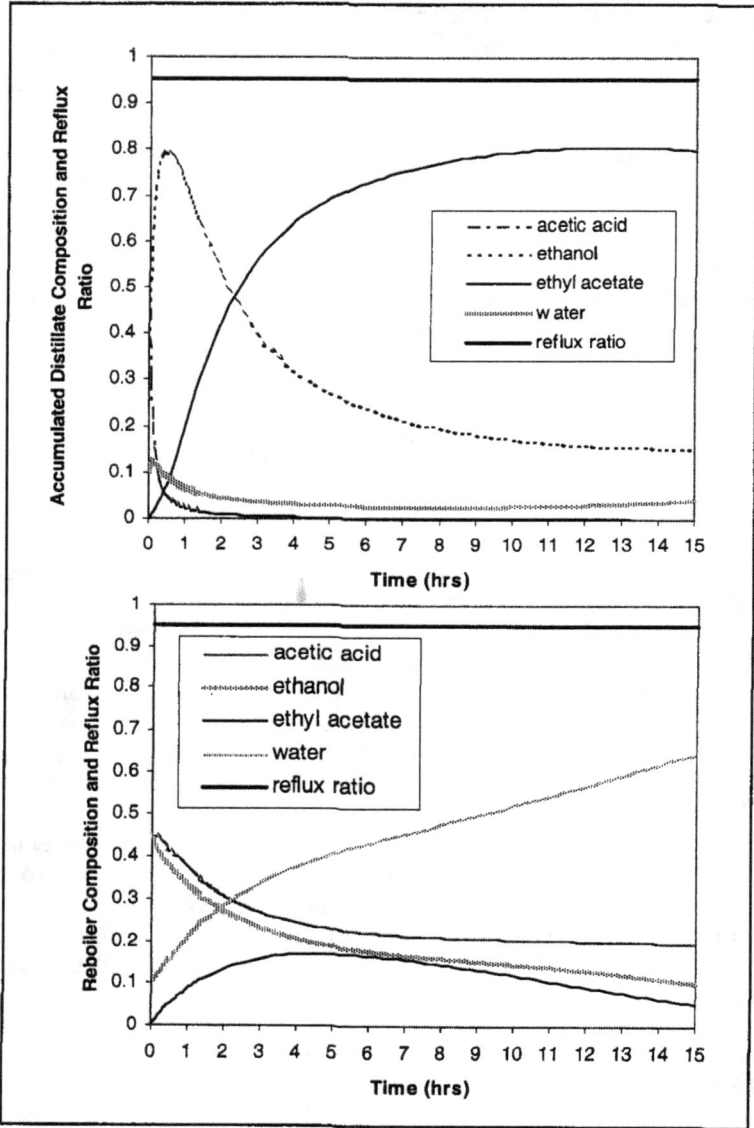

Figure 9.2. Composition and Reflux Ratio Profiles for Ethanol Esterification

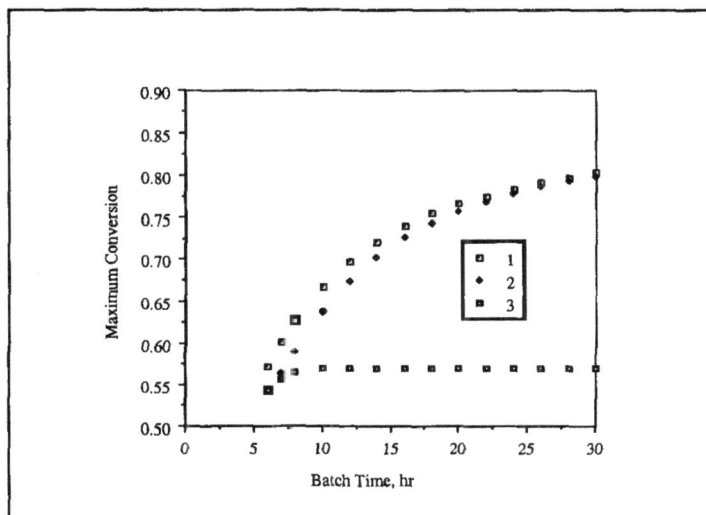

Figure 9.3. Maximum Conversion vs Batch Time. Curve 1- $x^*_D = 0.70$. Curve 2- $x^*_D = 0.80$. Curve 3- total reflux operation[a].

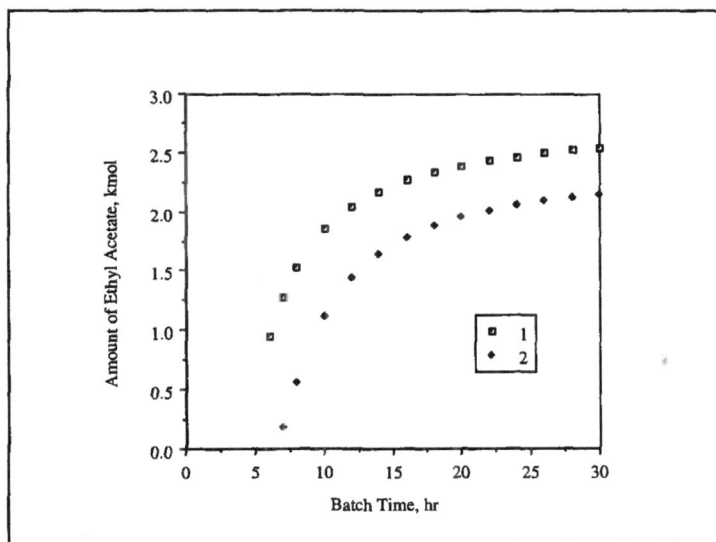

Figure 9.4. Amount of Ethyl Acetate vs Batch Time[a].

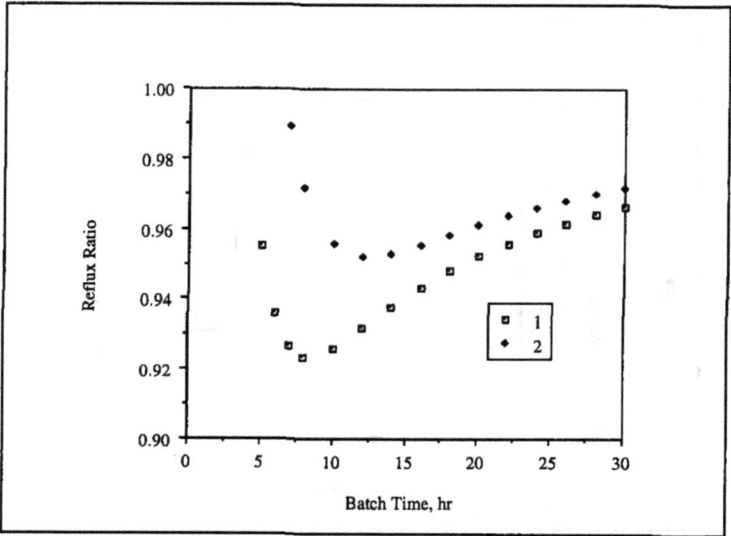

Figure 9.5. Reflux Ratio vs Batch Time[a]. Curve 1- $x^*_D = 0.70$. Curve 2- $x^*_D = 0.80$.

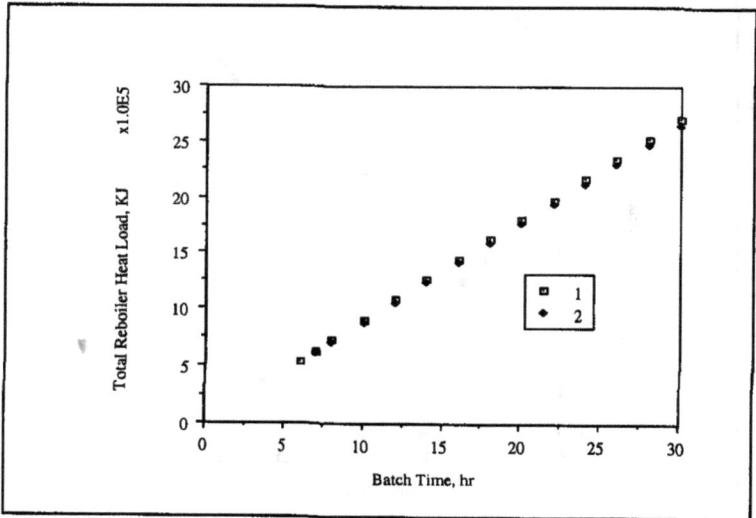

Figure 9.6. Reboiler Heat Duty vs Batch Time[a].

Figure 9.3 shows that the maximum conversion increases with increasing batch time, for each purity specification. This is as expected. As the product species are withdrawn by distillation and there is more time available, the reaction goes further to the right. Since it is easier to shift the equilibrium by eliminating the plentiful product at the given purity at the beginning, this increase is very sharp. It becomes progressively more difficult to remove the product at the given purity. Therefore, the curve becomes flattened near the end of the process. The product amount curve (Figure 9.4) also supports this observation.

Figure 9.5 shows the reflux ratio profile. With only a short available batch time, only a little amount of product is produced by the reaction. Separating this product in the distillate requires a high reflux ratio. With larger batch times, more amount of product is produced by the reaction and separation becomes easier (hence lower reflux ratio). Finally as the batch time is increased further, less and less reactants in the column are available to react to give products. Therefore high reflux ratio is required again to achieve products at the given purity. The total heat load curve (Figure 9.6) follows a linear trend with respect to batch time. This is due to the fact that in practice the maximum heat load is used at all times during a batch.

For each batch time, Figures 9.3 and 9.4 also show that the conversion and the yield for $x^*_D = 0.70$ are higher than those for $x^*_D = 0.80$. A higher reflux ratio (Figure 9.5) is required to produce ethyl acetate product at higher purity with consequently lower rate of reaction and lower conversion.

The percent improvement in conversion (compared to the conversion achieved under total reflux operation) achieved for different fixed batch times is shown in Figure 9.7. The results show that about 40% more conversion is possible when the column is operated optimally compared to total reflux operation.

Computationally, the solution of the dynamic optimisation problem is time consuming and expensive. Mujtaba and Macchietto (1997) reported that the number of "Function" and "Gradient Evaluations" for each *maximum conversion problem* is between 7-9. A fresh solution would require approximately 600 cpu sec in a SPARC-1 Workstation. However, subsequent solutions for different but close values of t_f could take advantage of the good initialisation values available from the previous solutions.

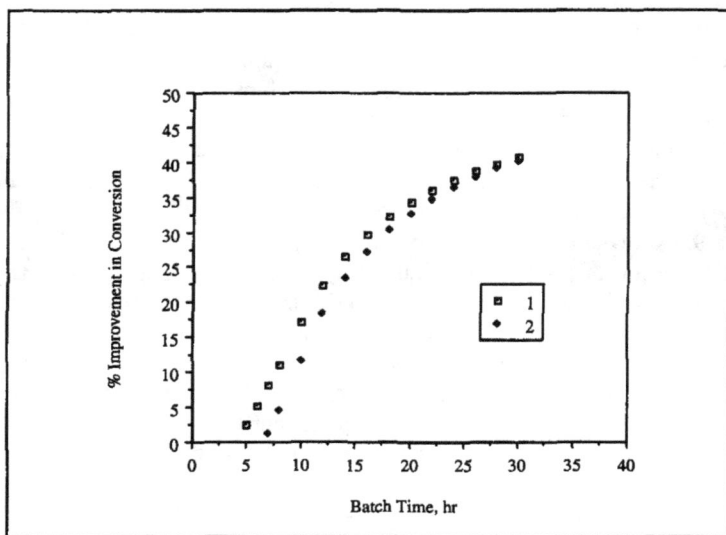

Figure 9.7. Percent Improvement in Conversion vs Batch Time[a].
Curve 1- $x^*_D = 0.70$. Curve 2- $x^*_D = 0.80$.

9.7. Profit Maximisation via Maximum Conversion Optimisation

A *maximum profit* problem can be formulated as follows:

P2   Max P
                 $t_f, C, r(t), D_1$

        subject to: product purity constraints
                                       DAE model equations
                                       etc.

Mujtaba and Macchietto (1997) defined a typical profit function P as:

$$P = \frac{C_{D1}D_1 - C_{B0}B_0 - C_hQ_R}{t} \qquad (9.1)$$

---

[a] Reprinted with permission from (Mujtaba, I.M. and Macchietto, S., *Ind. Eng. Chem. Res.* **36** (6), 2287-2295). Copyright (1997). American Chemical Society.

where D_1 is the amount of product (kmol), B_0 is the amount of raw material (kmol), Q_R is the total heat input (kJ), t is the batch time (hr), C_{D1} is the product price ($/kmol), C_{B0} ($/kmol) is the raw materials cost and C_h ($/kJ) is the energy cost for heating (operating cost).

Using the above profit function, the solution of problem **P2** will automatically determine the optimum batch time (t_f), conversion (C), reflux ratio (r) and the amount of product (D_1). However, as the cost parameters $(C_{D1}, C_{B0},$ etc.) can change from time to time, it will require a new solution of the dynamic optimisation problem **P2** (as outlined in Mujtaba and Macchietto, 1993, 1996), to give the optimal amount of product, optimal batch time and optimal reflux ratio. And this is computationally expensive. To overcome this problem Mujtaba and Macchietto (1997) calculated the profit of the operation using the results of the *maximum conversion* problem (**P1**) which were obtained independent of the cost parameters.

Note that for a fixed operation time, t in Equation 9.1, the profit will increase with the increase in the distillate amount and a *maximum profit* optimisation problem will translate into a *maximum distillate* optimisation problem (Mujtaba and Macchietto, 1993; Diwekar, 1992). However, for any reaction scheme (some presented in Table 9.1) where one of the reaction products is the lightest in the mixture (and therefore suitable for distillation) the maximum conversion of the limiting reactant will always produce the highest achievable amount of distillate for a given purity and vice versa. This is true for reversible or irreversible reaction scheme and is already explained in the introduction section. Note for batch reactive distillation the *maximum conversion* problem and the *maximum distillate* problem can be interchangeably used in the *maximum profit* problem for fixed batch time. For non-reactive distillation system, of course, the *maximum distillate* problem has to be solved.

9.7.1. Example

Using the above argument Mujtaba and Macchietto (1997) used the results (in terms of the amount of distillate, recovery, energy consumption, etc.) of the *maximum conversion* optimisation problem to evaluate the maximum achievable profit (for a given operation time) using the Equation 9.1. The maximum achievable profit at different batch time is shown in Figure 9.8. The cost parameters are given in Table 9.2. Figure 9.8 shows that the optimum batch time (for $x^*_D = 0.70$) lies between 12-14 hrs where the profit is the maximum for the entire range of operation.

Table 9.2. Cost Parameters for Ethanol Esterification[*]
[Adopted from Mujtaba and Macchietto, 1997]

C_1 = Cost of Acetic Acid, \$/kmol	= 32.01
C_2 = Cost of Ethanol, \$/kmol	= 17.87
$C_3 = C_{DI}$ = Price of Ethyl Acetate at $x^*_D = 0.70$, \$/kmol	= 80.0[*1]
$C_3 = C_{DI}$ = Price of Ethyl Acetate at $x^*_D = 0.80$, \$/kmol	= 85.0[*1]
C_4 = Cost of water, \$/kmol	= 0.0[*2]
C_{BO} = Raw material cost at the feed composition, \$/kmol	= 22.45
C_h = Steam cost (at 100 psig), \$/kJ	= 0.32[*3]

* Prices taken from : Chemical Marketing Reporter, October' 1992
*1 Assumption based on quoted price for product purity $x_D = 0.85$
*2 Assumption
*3 Taken from Peters and Timmerhaus (1980) and adjusted for inflation

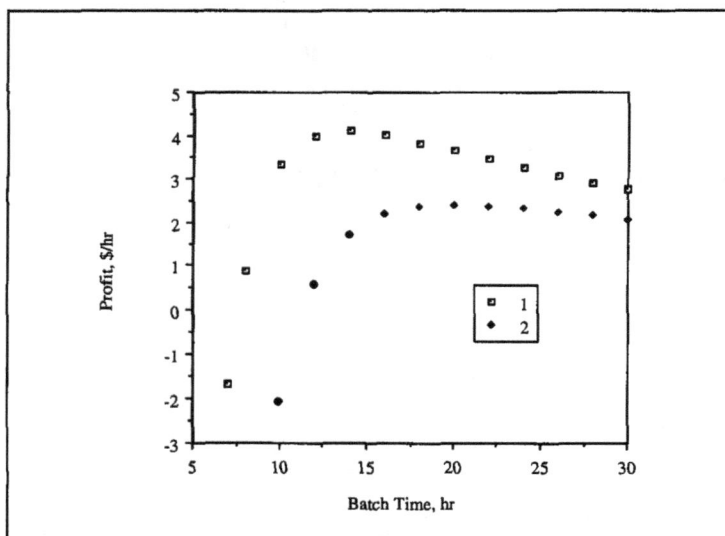

Figure 9.8. Maximum Profit vs Batch Time[b].
Curve 1- $x^*_D = 0.70$. Curve 2- $x^*_D = 0.80$.

---

[b] Reprinted with permission from (Mujtaba, I.M. and Macchietto, S., *Ind. Eng. Chem. Res.* **36** (6), 2287-2295). Copyright (1997). American Chemical Society.

9.8. Polynomial Based Optimisation Framework - A New Approach

The solution of dynamic optimisation problems with detailed dynamic model by sophisticated numerical techniques, as is required above, is computationally very expensive (Mujtaba and Macchietto, 1993, 1996) and therefore is not suitable for on-line optimisation where a quick and reliable solution is necessary. While optimising the operation of a batch distillation plant it is always desirable to operate at conditions which maximises the profit. However, the profit is usually a function of products prices, raw material cost, utility costs, etc. Mujtaba and Macchietto (1993) showed that changes in these costs significantly change the operating condition of the plant. This consequently means that an expensive dynamic optimisation problem has to be solved each time there is a price change to obtain optimal operating conditions. A further complication in the model by including chemical reaction would make the solution even more expensive. Therefore, there is a potential to develop techniques for solving dynamic optimisation problems cheaply for reactive or non-reactive batch distillation while taking into consideration of a detailed dynamic model.

Mujtaba and Macchietto (1997) proposed a new and an alternative technique that permits very efficient solution of the *maximum profit* problem using the solutions of the *maximum conversion* problem already calculated. This is detailed and explained in the following using again the ethanol esterification example presented in the previous section.

For a given product purity, the results presented in Figures 9.3-9.6 are only dependent on the batch time (t) but are independent of the cost function P and cost parameters used to compute the profit. These results can be easily represented by mathematical functions (as the results are quite well behaved with respect to batch time) leading to algebraic equations as follows:

$$\text{Maximum Conversion: } C = g_1(t) \tag{9.2}$$
$$\text{Optimum Amount of Distillate: } D_1 = g_2(t) \tag{9.3}$$
$$\text{Optimum Reflux Ratio: } r = g_3(t) \tag{9.4}$$
$$\text{Total Reboiler Heat Load: } Q_R = g_4(t) \tag{9.5}$$

where, $g_1(t)$, $g_2(t)$, etc. are polynomial functions.

The profit function in Equation 9.1 can now be written as:

$$P = \frac{C_{D1}g_2(t) - C_{B0}B_0 - C_h g_4(t)}{t} \tag{9.6}$$

which is a function of only one variable (t) for a give set of values of (C_{D1}, C_{B0}, C_h).

The dynamic optimisation problem *P2* now results in a single variable algebraic optimisation problem. The only variable to be optimised is the batch time t. The solution of the problem does no longer require full integration of the model equations. This method will solve the *maximum profit* problem very cheaply under frequently changing market prices of (C_{DI}, C_{B0}, C_h) and will thus determine new optimum batch time for the plant. The optimal values of C, D_1, r, Q_R, etc. can now be determined using the functions represented by Equations 9.2-9.5.

The new technique is illustrated below with the same example (ethanol esterification) and using the results of the *maximum conversion* problem. For $x^*_D = 0.70$, Mujtaba and Macchietto (1997) used 5th order polynomials to fit the data presented in Figures 9.3 and 9.5; a 3rd order polynomial to fit the data in Figure 9.4 and a 1st order polynomial to fit the data in Figure 9.6 respectively. The resulting curves and the polynomial equations are shown in Figures 9.9-9.12.

Now, the terms D_1 and Q_R in Equation 9.1 (same as the functions $g_2(t)$ and $g_4(t)$ in Equation 9.6) can be simply replaced by the time-dependent polynomial equations presented in Figures 9.10 and 9.12. For a given set of cost parameters, batch time is the only remaining optimisation variable in Problem *P2*, which can be obtained with extremely little computational effort.

Once the optimal batch time is obtained, the corresponding optimal reflux ratio and the maximum achievable conversion can be calculated directly using the polynomial equations presented in Figure 9.9 and 9.11, respectively. For frequently changed market prices, the same methodology can be applied efficiently without any difficulties to find the optimal batch time, amount of products, energy costs, etc. to calculate the maximum attainable profit. However, note the same exercise will have to be carried out for each product specification (x^*_D).

For a given product purity of $x^*_D = 0.70$, Mujtaba and Macchietto (1997) solved the *maximum profit* problem for a number of cost parameters using the method described above. The results are presented in Table 9.3. For each case, Table 9.3 also shows the optimal batch time, amount of product, reflux ratio, total reboiler duty and maximum conversion (calculated using the polynomial equations).

Case 1 of Table 9.3 is the base case. It shows the optimisation results using the cost parameters presented in Table 9.2. The maximum profit and optimal batch time obtained by optimisation shows very good agreement to those shown in Figure 9.8. The maximum profit shown in Figure 9.8 is between 3.99-4.13 ($/hr) with an optimum batch time between 12-14 hr. Each of the optimisation problems (i.e. solution of *P2* with Equation 9.6) presented in Table 9.3 requires approximately 3- 4 iterations and about 3- 4 cpu sec using a SPARC-1 Workstation (Mujtaba and Macchietto, 1997).

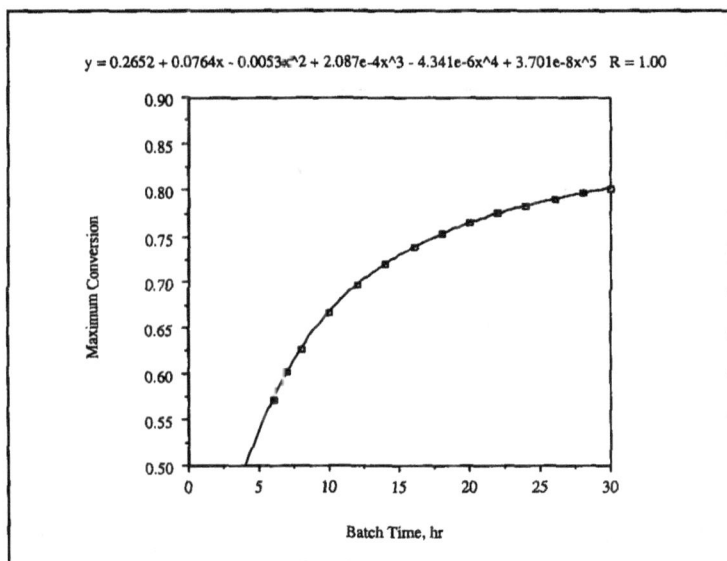

Figure 9.9. Curve Fitting for Maximum Conversion Profile $(x^*_D = 0.70)^c$

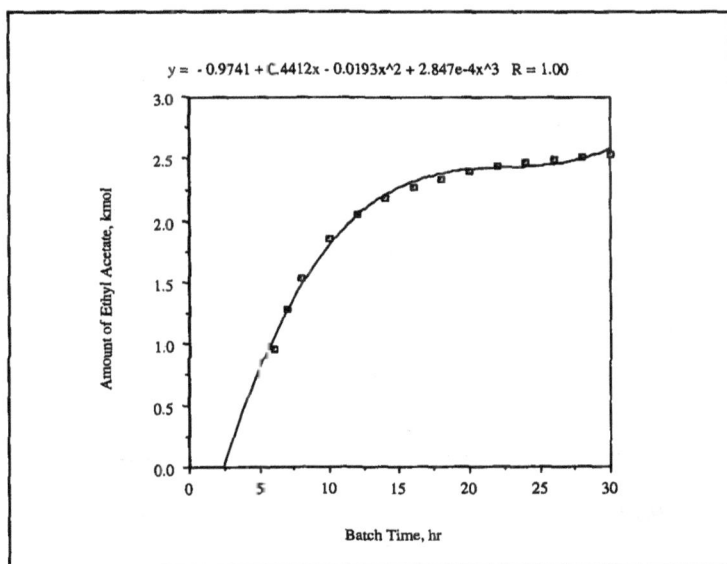

Figure 9.10. Curve Fitting for Product Profile $(x^*_D = 0.70)^c$

$$y = 1.2061 - 0.0932x + 0.0113x^2 - 6.258e\text{-}4x^3 + 1.653e\text{-}5x^4 - 1.681e\text{-}7x^5 \quad R = 1.00$$

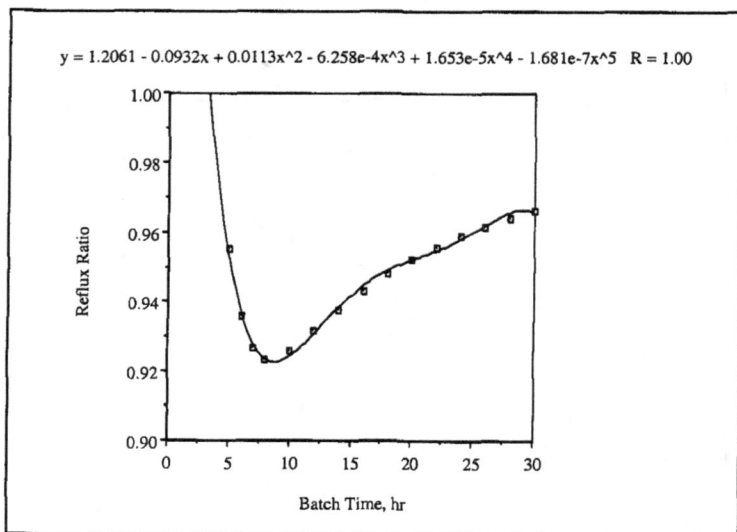

Figure 9.11. Curve Fitting for Reflux Ratio Profile $(x^*_D = = 0.70)^c$

$$y = 0.016 + 0.9027x \quad R \approx 1.00$$

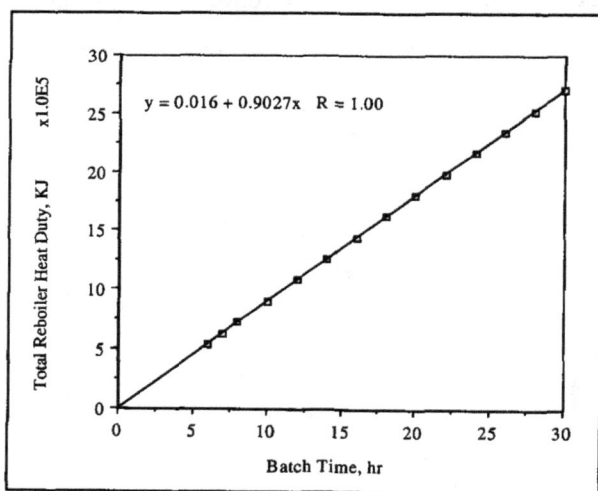

Figure 9.12. Curve Fitting for Reboiler Heat Duty Profile $(x^*_D = 0.70)^c$

---

c Reprinted with permission from (Mujtaba, I.M. and Macchietto, S., *Ind. Eng. Chem. Res.* **36** (6), 2287-2295). Copyright (1997). American Chemical Society.

Table 9.3. Summary of the Results for *Maximum Profit* Problem.
[Adopted from Mujtaba and Macchietto, 1997]

Case	C_{B0} $/kmol	C_{D1} $/kmol	C_h $/kJ	Profit $/hr	Batch Time, hr	D_1 kmol	Reflux Ratio	Max. Conv	Q_R kJ
1	22.45	80.0	0.32	4.29	14.79	2.25	0.936	.729	13.36
2	22.45	150.0	0.32	15.78	11.66	1.99	0.927	.694	10.54
3	22.45	70.0	0.32	2.80	15.83	2.30	0.939	.739	14.31
4	15.00	80.0	0.32	7.03	12.53	2.08	0.923	.705	11.33
5	35.00	80.0	0.32	0.59	19.96	2.40	0.945	.766	18.04
6	22.45	80.0	0.45	4.13	14.79	2.25	0.936	.729	13.36

Note that the results of the maximum profit problem obtained using the techniques presented above will be close to those determined by rigorous optimisation method (using the techniques presented in Mujtaba and Macchietto, 1993, 1996) only if the polynomial approximations are very good as were the case for the example presented here. Mujtaba et al. (2004) presented Neural Network based approximations of these functions.

9.9. Campaign Mode Operation Optimisation

Wajge and Reklaitis (1999) developed an optimisation framework RBDOPT allowing campaign optimisation of BREAD processes with minimal input from the users. The framework facilitates several operational features such as: continuous flow of one of the reactants, intermediate feed, inverted batch distillation, etc. The background process model is based on rigorous mass, energy and thermophysical properties resulting in a set of Differential and Algebraic Equations (DAEs) similar to Type V model.

RBDOPT is coded in an object oriented environment using (C^{++}) and uses distributed computing techniques to speed up the calculations. The thermophysical properties are estimated using Physical Property Data Service (PPDS). PPDS supports over 900 components and 36 physical property routes including NRTL, SRK, UNIFAC and UNIQUAC. The general structure of RBDOPT is shown in Figure 9.13. RBDOPT defines a batch distillation column as an object. Procedures and data are partitioned into *public* and *private* categories, depending on the importance of the level of details by the user. For example, a user might not be interested in the total reflux conditions, intermediate enthalpy values, etc. Therefore, these information can be classified as private data. On the other hand the data related to instant distillate composition, accumulated product compositions, cut time, etc. are important to a user and therefore can be classified as public data. For optimising *multiple separation duties* in a single sequence or in a multiple sequence (campaign mode with several mixtures processed in a single column) operations, the main module can easily define multiple duties as different *objects* of the distillation *class* and keep the simulation details of those duties active simultaneously for the eventual processing by an optimiser. See the original reference for further details.

9.9.1. Example: Hydrolysis of Acetic Anhydride

Wajge and Reklaitis (1999) considered a simple single duty multiperiod operation using the hydrolysis reaction of acetic anhydride (Ullmann, 1985) to demonstrate the RBDOPT framework. Acetic anhydride is manufactured by dehydration of acetic acid. The by-product of the process is a mixture of acetic anhydride, water and acetic acid with boiling points 412 K, 373.2 K and 391.1 K respectively for each component in the mixture. The main concern is that the acetic anhydride and water in the mixture may hydrolyse quite quickly to produce back the acetic acid.

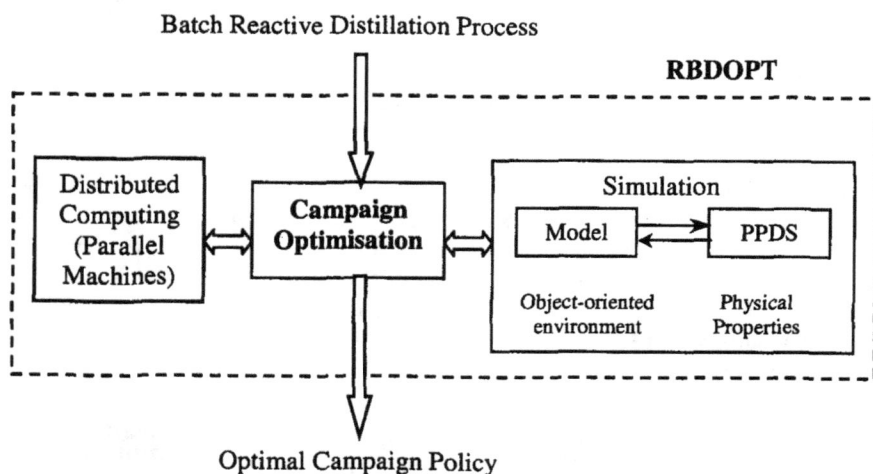

Figure 9.13. General Structure of RBDOPT

Wajge and Reklaitis (1999) considered the following operation sequence using a CBD column:

1. CUT 1 – Recover most of the water (the lowest boiling component)
2. CUT 2 – Maximise the amount of acetic acid recovery, thus leaving acetic anhydride as the final bottom product in the reboiler.

The optimisation problem was presented as:

$$\underset{R}{\text{Max}} \quad \text{(amount of acetic anhydride} + wt_{end})$$

subject to:

DAE model equations	(Equality Constraints)
$x^*_{acetic\ acid} \geq 0.9$ in CUT 2	(inequality constraints)
$x^*_{acetic\ anhydride} \geq 0.92$ in CUT 2	(inequality constraints)

where, R is the external reflux ratio to be optimised, w is the weighting factor and t_{end} is the batch time.

The chemical reaction scheme, kinetic data and other input data are given in Table 9.4. There are 2 control variables in each cut. These are the reflux ratio and the duration of the cut. The results of the optimisation are shown in Table 9.5 and Figure 9.14. The distillation column needs to be operated initially at low optimal reflux ratio to remove most of the water.

Table 9.4. Input Data for Acetic Anhydride Hydrolysis

acetic anhydride + water <=> acetic acid	
$r_{acetic\ anhydride} = 1.05e^5 C_{acetic\ anhydride} C_{water}$	$(\dfrac{gmol}{l.s})$
No. of Plates	10
Initial Charge, B_0 kmol	2 kmol
Initial Composition: x_{B0} = <water, acid, anhydride> molefraction	<0.2, 0.6, 0.2>
Holdup, kmol:	
Per Plate	0.01
Condenser	0.03
Pressure, bar	0.527
Condenser Vapour Load	20 kmol/h

Table 9.5. Summary of the Results of Campaign Optimisation using RBDOPT

Amount of Distillate in CUT 1, kmol	0.446
Composition of CUT1, molefraction	<0.455, 0.517, 0.028>
Amount of Distillate in CUT 2, kmol	1.29
Composition of CUT2, molefraction	<0.068, 0.90, 0.032>
Amount of Bottom Product in CUT2, kmol	0.2621
Composition of Bottom Product, molefraction	<0.0, 0.08, 0.92>
Total Batch Time, hr	0.414

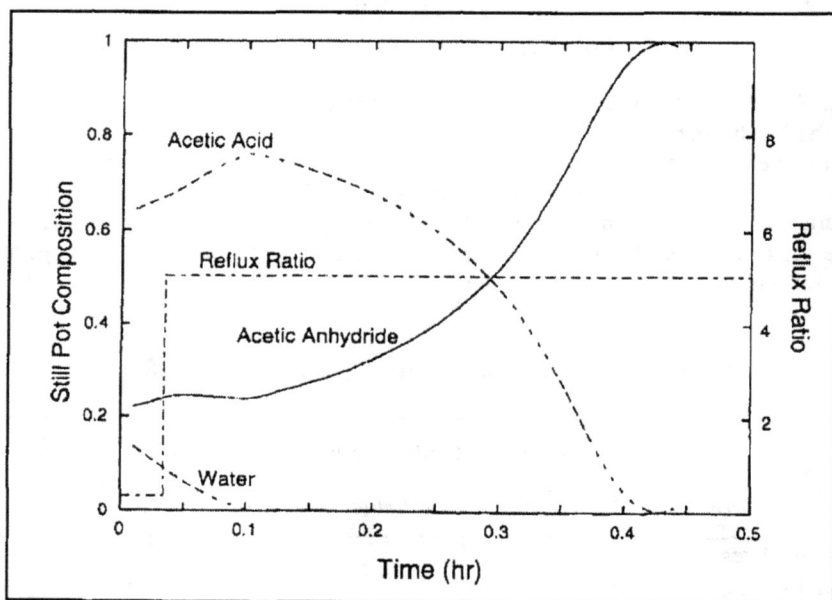

Figure 9.14. Reflux Ratio and Instant Distillate Composition Profile. [Wajge and Reklaitis, 1999][d]

[d] Reprinted from *Chemical Engineering Journal*, **75**, Wajge, R.M. and Reklaitis, G.V., RBDOPT: a general-purpose object-oriented module for distributed campaign optimisation of reactive batch distillation, 57-68, Copyright (1999), with permission from Elsevier Science .

9.10. Optimal Design of Operating Procedures with Parametric Uncertainty

Process dynamics and uncertainties are important issues for optimal design and operation of any processes. All process models are subject to uncertainties, e.g. in physical properties, kinetics, etc.

Walsh et al. (1995) examined the effect of uncertain parameters on the optimal operation of BREAD processes and the use of control to reduce the effect of uncertainties. The study provided an important step towards the implementation of optimal operating policies on real (uncertain) processes.

As presented in the earlier chapters, the operating policy for a batch distillation column can be determined in terms of reflux ratio, product recoveries and vapour boilup rate as a function of time (open-loop control). Under nominal conditions, the optimal operating policy may be specified equivalently in terms of a set-point trajectory for controllers manipulating these inputs. In the presence of uncertainty, these specifications for the optimal operating policy are no longer equivalent and it is important to evaluate and compare their performance.

The effect of uncertainty on the performance of an operating policy can be evaluated by repetitive simulation (Leversund et al., 1993). However, if the uncertainty can be characterised by a finite set of parameterised conditions, then a multi-period approach to design and operation is appropriate (Chapter 7 and Mujtaba and Macchietto, 1996). This approach is well suited to design problems in which the key variability in the process is both known and discrete, e.g. a number of specified feed compositions. If the variability / uncertainty to be accommodated is continuous rather than discrete then the multi-period design must be augmented with a method for finding the worst-case scenarios of the uncertainty, i.e. worst-case design (Grossmann et al., 1983).

Walsh et al. (1995) used a simplified BREAD process model in conjunction with a parameterised uncertainty description to predict the performance. Worst-case design algorithm of Walsh (1993) was used to systematically determine the optimal operating policies that allow performance constraints to be met over a bounded set of uncertain parameters.

9.10.1. The Worst-case Design Algorithm

Walsh (1993) used approximate global maximisation of the maximum constraint violation to identify the worst-case scenarios for consideration in design. Control Vector Parameterisation (CVP) described in earlier chapters is used in solving the underlying dynamic optimisation problems. The general approach of worst-case design is shown in Figure 9.15. See the original reference for further details.

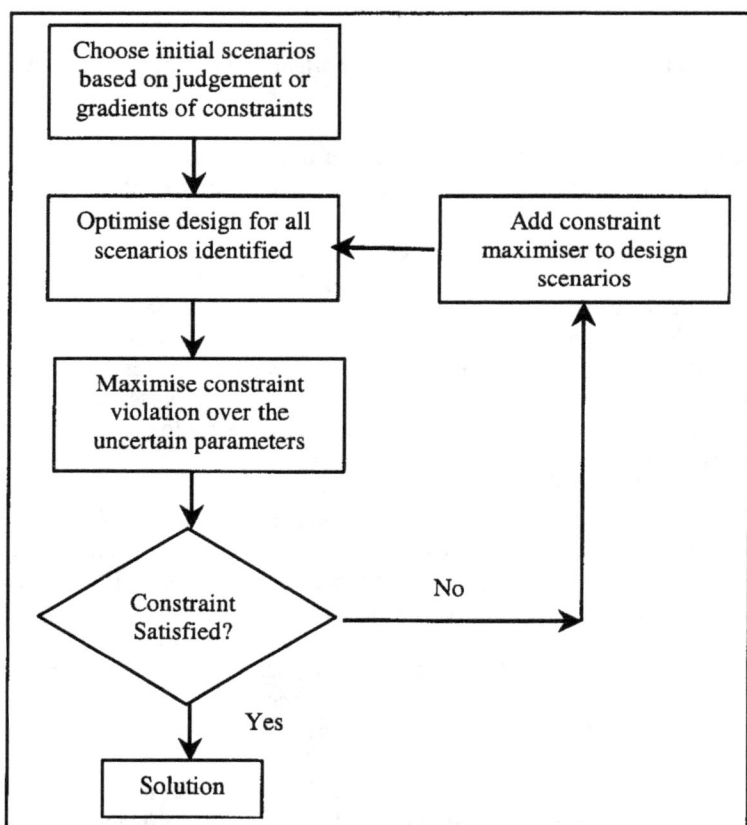

Figure 9.15. Worst-case Design using Outer Approximation Algorithm.

9.10.2. Case Study

9.10.2.1. Process Description

Walsh et al. (1995) considered an industrial batch reactive distillation problem originally presented by Leversund et al. (1993) as a case study. A condensation polymerisation reaction between a dibasic aromatic acid (*R1*) and two glycols (*R2*, *R3*) was considered. The reaction products were a polymer product (*P*) and water

(W). Chemical reaction occurs in the reboiler only. High conversion is achieved by removing the water in the distillate.

The process model is given in Leversund et al. (1993). As the polymer product forms a separate solid phase, this was not included in the process model. The model includes mass and energy balances, column holdup and phase equilibria and results in a set of DAEs. All thermodynamic, physical properties are calculated from library subroutines. The column description and data for the process are given in Table 9.6.

Table 9.6. Input Data for the Case Study

Column:	Accumulator + Total Condenser + 5 Plates + Reboiler
Initial Holdups:	Reboiler = 20.6 kmol Plates = 0.1 kmol
	Condenser = 1.7 kmol Accumulator = 0
Initial Charge Composition:	x^0 (W, R_2, R_3, R_1) = [0.0097, 0.3444, 0.1505, 0.4854] molefraction
Initial Plate Composition:	x^0 (W, R_2, R_3, R_1) = [0.999999, 0.000001, 0.0, 0.0] molefraction
Reaction:	$1.0R_1 + 0.7R_2 + 0.3R_3 \leftrightarrow P(s) + 2.0W$
Reaction Rate:	$r = k_1 c_{R1}(c_{R2} + c_{R3}) - k_2 c_W$
Rate Constants:	$$k_1 = k_{1,0}\ \exp\left(-\frac{7.5 \times 10^4}{R}\left(\frac{1}{T_N} - \frac{1}{T_0}\right)\right)$$
	$$k_2 = k_{2,0}\ \exp\left(-\frac{7.5 \times 10^4}{R}\left(\frac{1}{T_N} - \frac{1}{T_0}\right)\right)$$
	$k_{1,0} = 0.03 \qquad k_{2,0} = 0.0075$
Reference Temperature:	T_0 = 500 K
Column Pressure:	1.013 bar
Value of Product P:	C_{prod} = 50 $/kmol
Cost of Steam:	C_{heat} = 2.844x10^{-6} $/kJ

9.10.2.2. Uncertainty Description

In this example, uncertainties in the effective reboiler heat duty Q_R, in the reaction coefficients, $k_{1,0}$ and $k_{2,0}$, are considered. It is assumed that the reaction coefficients ramp between steady initial and final levels over time. For convenience in formulating the optimisation, the ramp was assumed to start immediately and finish after a typical run time of 20 hrs. The initial and final steady state values (bounds) are: $k_{1,0} \in [0.02, 0.05]$ and $k_{2,0} \in [0.005, 0.02]$. The perturbation to Q_R is allowed to make one step between any two values within the bounds at any time up to 20 hrs from the start of the batch. Within the given bounds, there are seven uncertain parameters as shown below and in Figure 9.16.

- P_1 and P_2 : Initial and Final Steady State Values for $k_{1,0}$
- P_3 and P_4 : Initial and Final Steady State Values for $k_{2,0}$
- P_5 and P_6 : Initial and Final Steady State Values for Q_R
- P_7 : Time when Q_R is perturbed

It is assumed that the nominal Q_R and the internal reflux ratio $r = L/V$ are available for manipulation, and that all column temperatures and the liquid holdups in the reboiler and accumulator, are available as measurements.

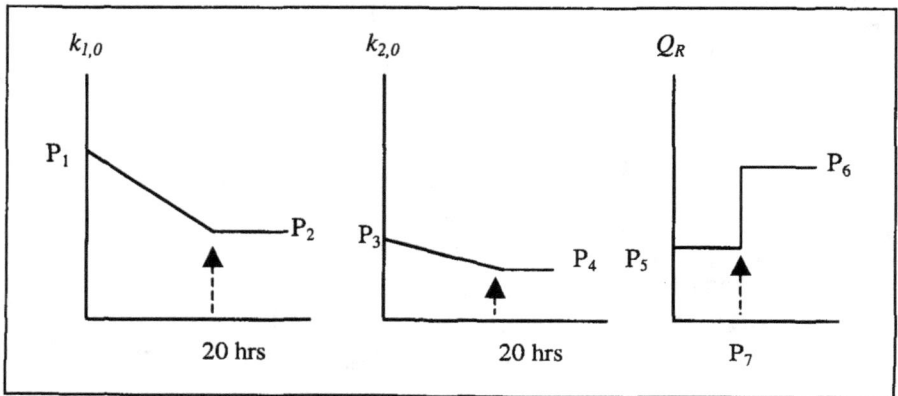

Figure 9.16. Uncertainty Models for Different Parameters

Table 9.7. Comparison of Nominal and Worst-case Algorithm Based Optimum Design of Operating Policies (Open Loop Strategy)

Case	r	Q_R, GJ/hr	$H_{threshold}$, kmol (%)	Profit
1 (nominal)	0.949	0.7473	0.287 (1.3)	20.2
2 (worst-case)	0.880	0.5040	19.70 (88)	15.0

9.10.2.3. Optimisation

The objective of this study is to maximise the profit which is the difference between the mean value of polymer product and the mean cost of steam (based on actual energy used) over the operating time of the plant subject to:

1. reboiler temperature < 513.15 K at all times to avoid product degradation
2. final composition of water in the accumulated distillate > 0.999 mole fraction to avoid any loss of reactants.

It should be noted that the objective imposes no penalty for incomplete conversion of reactants or for changeover between batches and therefore may lead to solutions with unrealistically low conversion to products and short batch times.

The aim of the optimisation is to optimise the operating parameters so as to give *good performance* under uncertainty. By *good performance* it is meant that the constraints are always satisfied and that the average profitability is approximately maximised. The estimate of the average performance is obtained by a weighted average of

- the nominal case with $\Delta Q_R = 0$, $k_{1,0} = 0.03$, $k_{2,0} = 0.0075$
  and
- the least profitable case , identified by the worst-case design algorithm.

9.10.2.4. Results

Open loop strategy: A simple control strategy is adopted: operate the column at constant reflux ratio (r), reboiler heat duty (Q_R) and terminate when the reboiler liquid level falls below a threshold level ($H_{threshold}$) to avoid reboiler running dry. This gives three operating parameters to be chosen.

Table 9.7 shows the results of base case (Case 1) optimisation with nominal values of operating parameters and those obtained by using worst-case design algorithm (Case 2). The profit in Case 2 is based on the weighted average of 18 nominal and 12 worst-case scenarios. Walsh et al. (1995) observed that maximising the constraint violations for the nominal design resulted in both the reboiler temperature constraint and the accumulator composition constraint being violated. Hence the nominal design cannot be used with the level of uncertainty assumed.

The operating policy of Case 2 is a radically different policy to the nominal optimum (Case 1) which failed under uncertainty. Case 2 requires less steam and reflux and terminating after a few hours with most charge (88% of the initial charge in the reboiler) unreacted (compared to about 20 hrs and almost complete conversion for the nominal case).

This highlights the inadequacy of sensitivity analysis of nominal optima as a basis for choosing a control scheme in the face of uncertainty. Uncertainty gives substantial economic penalty with this scheme, which incorporates no feedback.

Closed loop strategy: Walsh et al. (1995) observed that the temperature on the third (middle) plate showed the greatest variability over time. The temperature of the third plate is therefore controlled using reflux rate only, with reboil rate adjusted to a constant nominal value. The target trajectory is approximated as a ramp from the initial plate temperature, 373.15 K, to a new steady value. This results in two new design parameters: the final temperature setpoint and the time to complete the ramp.

Measurement dynamics and tray hydraulics are neglected at this stage of the analysis but should be included in any detailed control analysis of successful operating policy. Actuation limits are included with reflux ratio (r) being limited to between 0.8 and 0.999. The initial controller output (reflux ratio) is also made an optimisation variable. The controller is coarsely tuned by trial and error to give good tracking of the specified trajectory without excessive oscillation. The controller gain used is −0.06/K and the integral time is 6 minutes.

Table 9.8 shows the results of base case (case 1) optimisation with nominal values of operating parameters and those obtained by using worst-case design algorithm (case 2). The profit in Case 2 is based on weighted average of 20.5 nominal and 13.5 worst-case scenarios. Figure 9.17 shows nominal (optimum) and worst-case (optimum) temperature profile.

The nominal profit (case 1) is somewhat higher than that in Table 9.7. This is due to variation of reflux ratio over the batch time. Also in contrast to previous scheme (case 2 of Table 9.7) little loss of nominal performance is entailed in robustly satisfying the constraints. The temperature trajectory of case 2 shown in Figure 9.17 is qualitatively similar to the nominal optimum policy, but has been modified slightly to ensure constraint satisfaction.

Comparison of the results obtained with the open and closed loop strategies shows the importance of considering uncertainty systematically, as there is a substantial difference in the economic penalty for accommodating uncertainty. Please refer to the original reference for further details.

Table 9.8. Comparison of Nominal and Worst-case Algorithm based Optimum Design of Operating Policies (Feedback Strategy)

Case	initial r	Q_R, GJ/hr	$H_{threshold}$, kmol (set)	Profit
1 (nominal)	0.8	0.778	0.2	20.6
2 (worst-case)	0.8	0.644	0.2	17.0

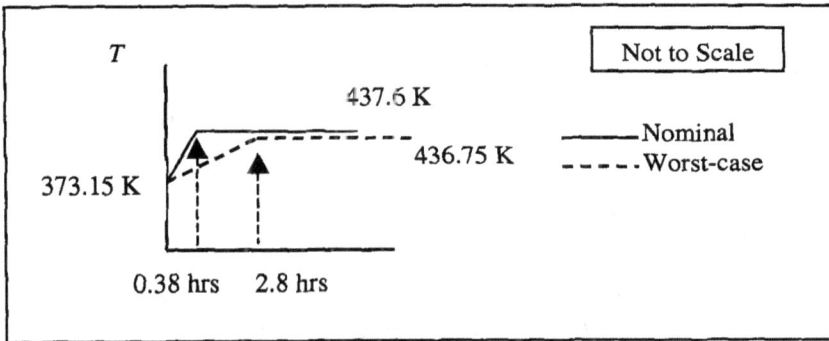

Figure 9.17. Optimum Temperature Trajectories (Feedback Strategies)

References

Albet, J., Le Lann, J.M., Joulia, X. and Koehret, B., *In Proceedings Computer Oriented Process Engineering* (Elsevier Science Publishers B.V., Amsterdam, 1991), 75.

Balasubramhanya, L.S. and Doyle III, F.J., *J. Proc. Cont.* **10** (2000), 209.

Basualdo, M.S. and Ruiz, C.A., *Lat. Ame. Appl. Res.* **25-S** (1995), 29.

Barbosa, D. and Doherty, M.F.. *Chem. Eng. Sci.* **43** (1988), 529.

Bernot, C., Doherty, M.F. and Malone, M.F., *Chem. Eng. Sci.* **46** (1991), 1311.

Bogacki, M.B.; Alejski, K. and Szymanowski, J., *Comput. chem. Engng.* **13** (1989), 1081.

Bosley, J.R. Jr. and Edgar, T.F. *In Proceedings of 5th International Seminar on Process Systems Engineering, Kyongju, Korea, 30 May-3 June,* **1** (1994), 477.

BP Review, UK, October-December, 1997, 15-16.

Charalambides, M.S., Shah, N. and Pantelides, C.C. *In Proceedings IChemE Research Event, London,* 5-6 January, **2** (1994), 908.

Corrigan, T.E. and Ferris, W.R., *Can. J. Chem. Eng.* **47** (1969), 334.

Cuille, P.E. and Reklaitis, G.V., 1986, *Comput. chem. Engng.* **10** (1986), 389.

Diwekar, U.M., *AIChE J.* **37** (1991), 760.

Diwekar, U.M., *AIChE J.* **38** (1992), 1571.

Egly, H. Ruby, V. and Seid, B., *Comput. chem. Engng.* **3** (1979), 169.

Grossman, I.E., Haleman, K.P. and R.E. Swaney, R.E., *Comput. chem. Engng.* **7** (1983), 439.

Greaves, M. A., *Hybrid Modelling, Simulation and Optimisation of Batch Distillation Using Neural Network Techniques.* Ph.D. Thesis, (University of Bradford, Bradford, UK, 2003).

Lehtonen, J., Salmi, T., Harju, T., Immonen, K., Paatero, E. and Nyholm, P., *Chem. Eng. Sci.* **53** (1998), 113.

Leversund, E.S., Macchietto, S., Stuart, G. and Skogestad, S., *Comput. chem. Engng.* **18** (1993), S391.

Mujtaba, I.M., *Optimal Operational Policies in Batch Distillation.* PhD Thesis, (Imperial College, University of London, 1989).

Mujtaba, I.M., Greaves, M.A. and Hussain, M.A., *In Proceedings of the 8th International Symposium on Process Systems Engineering,* Kunming, China, 5 – 10 January, (2004).

Mujtaba, I.M. and Macchietto, S., in *Proceedings PSE'91 - 4th International Symposium on Process Systems Engineering,* **1**: Design, (Quebec, Canada, 1991), 19.1

Mujtaba, I.M. and Macchietto, S., *Comput. chem. Engng.* **17** (1993), 1191.

Mujtaba, I.M. and Macchietto, S., Optimal operation of reactive batch distillation. *Presented at the AIChE Annual Meeting,* Miami Beach, USA, Nov. 1-6, paper no. 135g, 1992.

Mujtaba, I.M. and Macchietto, S., *In Proceedings of IFAC Symposium ADCHEM'94*, Kyoto, Japan, 25 - 27 May, (1994), 415.

Mujtaba, I.M. and Macchietto, S., *J. Proc. Control.* **6** (1996), 27.

Mujtaba, I.M. and Macchietto, S., *IEC Res.* **36** (1997), 2287.

Mujtaba, I.M. and Macchietto, S., *Chem. Eng. Sci.* **53** (1998), 2519.

Peters, M. S. and Timmerhaus, K.D. *Plant design and economics for Chemical Engineers* (3rd ed., McGraw-Hill Book Company. New York, 1980).

Reid, R.C., Prausnitz, J.M. and Sherwood, T.K., *The Properties of Gases and Liquids* (3rd ed., Mc-Graw Hill Book Company, 1977).

Reuter, E., Wozny, G. and Jeromin, L., *Comput. chem. Engng.* **13** (1989), 499.

Simandl, J. and Svrcek, W.Y., *Comput. chem. Engng.* **15** (1991), 337.

Sorensen, E. and Skogestad, S., *J. Proc. Cont.* **4** (1994), 205.

Taylor, R. and Krishna, R., *Chem. Eng. Sci.* **55** (2000), 5183.

Ullmann, F. (ed.), *Ullmann's Encyclopedia of Industrial Chemistry*, **A1**, (Weinham, Germany ,1985).

Van Dongen, D.B. and Doherty, M.F., *Chem Eng. Sci.* **40** (1985), 2087.

Walsh, S.P. *Integrated Design of Chemical Waste Water Treatment Systems*. PhD Thesis, (Imperial College, University of London, 1993).

Walsh, S.P., Mujtaba, I.M. and S. Macchietto, S., *Acta Chimica Solvenica*, **42** (1995), 69.

Wajge, R.M. and Reklaitis, G.V., *Chem. Eng. J.* **75** (1999), 57.

Wilson, J.A., In *IChemE Symposium Series* No. 100 (1987), 163.

Wilson, J.A. and Martinez, E.C., *Chem. Eng. Res. Des.* **75** (1997), 603.

CHAPTER 10

10. BATCH EXTRACTIVE DISTILLATION (BED)

10.1. Introduction

The separation of azeotropic or close boiling mixtures is often the most difficult task in many chemical industries. Under normal operating conditions high recovery of species with high purity is not possible. The separation of these mixtures using a conventional distillation column is not an economically feasible option as a very large column or a very high reflux ratio operation would be required to separate these mixtures. Pressure swing distillation is used sometimes to break the azeotropic composition by changing the pressure but is not an economic option for large swing of the pressure (Towler, 1997). However, extractive distillation provides a potentially very attractive alternative to conventional columns for separating close boiling and /or azeotropic mixtures by introducing a third component (*solvent*) into the system in a considerable quantity. The solvent helps to break the azeotropic composition by changing the liquid phase activity coefficient (Towler, 1997).

In continuous mode, the operation of extractive distillation for binary mixtures requires at least two columns in sequence, one for the separation step and the other for the solvent recovery step when the solvent is to be reused because of economic and/or environmental constraints. On the other hand advantages of batch distillation are well known. A single mixture can be separated into several products (*single separation duty*) and multiple mixtures can be processed, each producing a number of products (*multiple separation duties*) using only one column (Logsdon et al., 1990; Mujtaba and Macchietto, 1996; Sharif et al., 1998). Although batch azeotropic distillation is frequently applied (e.g. for making absolute alcohol) few attempts have been made for realising extractive distillation in a batch system. The reason for that is probably the very complex operation of batch distillation by itself and the further complication of the operation on introduction of a solvent (resulting in a greater reboiler volume). The location of the solvent feed, the solvent flow rate, number of plates in the column, reflux ratio, etc. all influence the separation achieved.

Lang et al. (1994, 1995) studied the effects of the operational parameters on BED processes using binary mixtures. Three distinct time intervals were considered for the first distillation step where the most volatile component was being separated. The column was operating under total reflux condition for the first two intervals without and with the solvent feed, the length of each interval being fixed. The solvent feed rate, reflux ratio, feed mixture charge, etc. were all varied and the column operation was simulated repeatedly to see the effects of these parameters on the performance of BED. Note that pre-setting the total reflux operation period and solvent rate can seriously affect the productivity (amount of product/time) and/or the profit of the operation (see sections 5.10.2 and 7.3.4). Also note that these studies do not exactly quantify the effect of the parameters on the operation as no *performance measure* of the column was clearly defined and the products were not achieved with specified purities. In distillation, whether batch, continuous or extractive, purity of the main products must be specified as it is driven by the customer demand. The amount of product and the operation time can be dictated by the economics (*maximum profit* or *maximum productivity* problem) or one of them can be fixed and the other obtained by solving a *minimum time* or a *maximum distillate* problem (Mujtaba, 1997).

Dussel and Stichlmair (1995), Ahmad and Barton (1996), Safrit and Westerberg (1997) discussed the theoretical aspects of the separation of azeotropic mixtures in conventional and non-conventional BED column configurations using simulation techniques. Refer to the original references for further details.

Tran and Mujtaba (1997) and Mujtaba et al. (1997) highlighted the operating features and limitations of BED processes for close boiling and azeotropic mixtures. However, the works were limited to the separation of only one key component in the distillate without due regard to the recovery of solvent or the separation of other components in the feed mixture.

Mujtaba (1999) considered the conventional configuration of BED processes for the separation of binary close boiling and azeotropic mixtures. Dynamic optimisation technique was used for quantitative assessment of the effectiveness of BED processes. Two distinct solvent feeding modes were considered and their implications on the optimisation problem formulation, solution and on the performance of BED processes were discussed. A general Multiperiod Dynamic Optimisation (MDO) problem formulation was presented to obtain optimal separation of all the components in the feed mixture and the recovery of solvent while maximising the overall profitability of the operation.

Low and Sorensen (2002) and Low (2003) considered optimal operation of BED processes in CBD and MVC columns using azeotropic mixtures. They used *profit* as the *measure* to compare the performances of such columns. For a given separation task, an MVC column was found to be superior to a CBD column. However, for maximum profitability most of the initial feed needed to be charged in the reboiler rather than in the middle vessel of the MVC column. See the original references for further details.

In this chapter some of the work reported by Mujtaba and co-workers in the published literature will be discussed and presented for conventional BED processes.

10.2. Comparison Between a CBD and a BED Process

Figure 10.1 shows typical distillate composition profiles for close boiling mixtures (binary) and implications of using CBD column for such mixtures.

The accumulator 1 is rich in component 1 and the accumulator 2 is rich in component 2. However, due to close boiling points of the components, the components cannot be separated at high purity unless a large column is used with high reflux ratio.

Figure 10.2 shows typical distillate composition profiles for close boiling mixtures using a solvent in a CBD column. The CBD process becomes a conventional BED process with the addition of the solvent. Due to the addition of solvent, the components can be separated at high purity using a small column with low reflux ratio. See Safrit and Westerberg (1997) and Low and Sorensen (2002) for unconventional BED processes.

10.3. Solvent Feeding Modes and Operating Constraints

Mujtaba (1999) discussed two modes of solvent feeding for conventional BED processes.

10.3.1. Batch Mode

In this mode the solvent is charged in the reboiler with the feed mixture at the beginning of the process. As the reboiler has a limited capacity, it, therefore, limits the amount of feed mixture that can be processed. This increases the number of batches to be processed in a *campaign mode* operation. In this mode, finding the optimum feed charge to solvent ratio is an important factor to maximise the productivity.

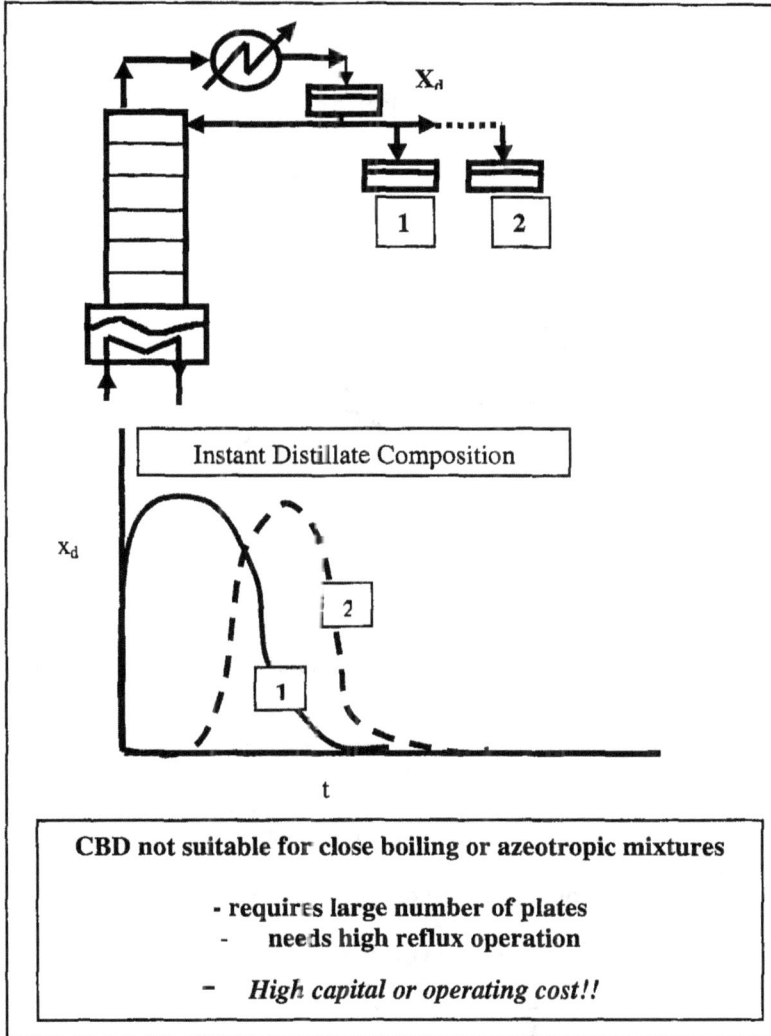

Figure 10.1. CBD Column Processing Close Boiling Mixtures

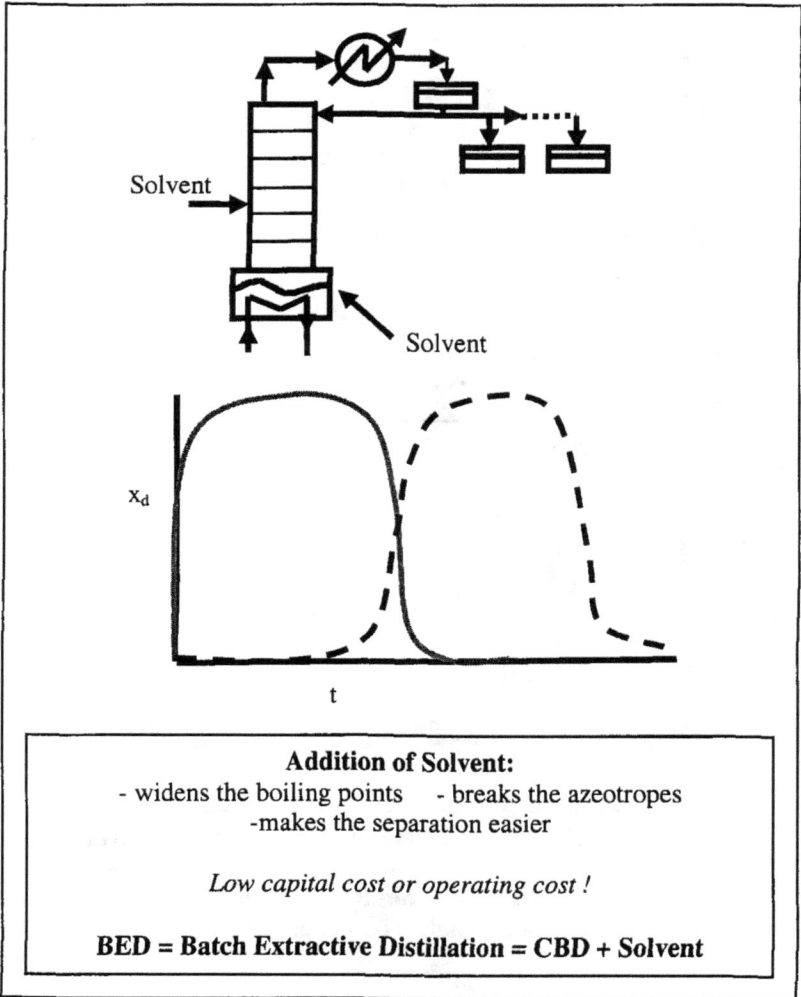

Figure 10.2. BED Process

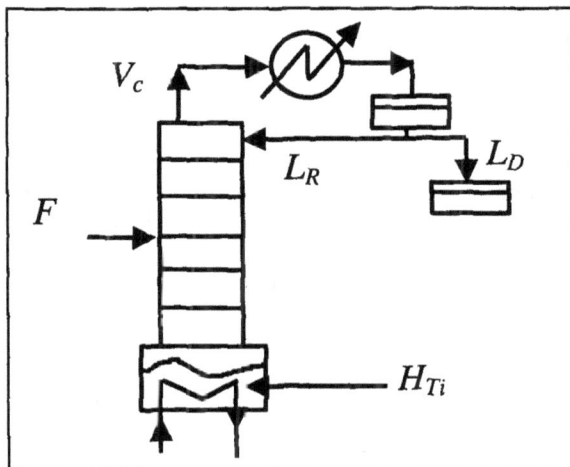

Figure 10.3. Batch Extractive Distillation (BED) Process.
[Mujtaba, 1999][a]

10.3.2. Semi-continuous Mode

In this mode the solvent is fed to the column in a *semi-continuous* fashion at some point of the column (Figure 10.3). Mujtaba (1999) noted two strategies in this mode of operation as far as charging of the initial feed mixture is concerned.

10.3.2.1. Full Reboiler Charge

In this strategy the feed mixture is charged in the reboiler (at the beginning of the process) to its maximum capacity. For a given condenser vapour load V_c, if the reflux ratio R (which governs the distillate rate, L_D, kmol/hr) and the solvent feed rate F (kmol/hr) are not carefully controlled the column will be flooded. To avoid column flooding Tran and Mujtaba (1997) developed the following necessary and sufficient condition:

$$L_D \geq F \tag{10.1}$$

where, $L_D = V_c(1-R)$

---

[a] Reprinted with permission from IChemE, UK. Full reference is at the end of the chapter.

This leads to: $R \le 1 - (F / V_c)$
and

$$R_{max} = 1 - (F / V_c) \qquad (10.2)$$

For a given value of F and assuming that the column design (V_c) is fixed the above inequality constraint on R (Equation 10.2) must be satisfied to avoid flooding. However, if the column operates using a reflux ratio below R_{max} for some time, the reboiler will be able to accommodate extra solvent and the recovery or the productivity can be further improved. Mujtaba (1999) had considered time dependent (piecewise constant) solvent feed rate within a distillation Task.

10.3.2.2. Fractional Reboiler Charge

For many azeotropic mixtures, *full feed charge* is not a suitable option to achieve the required product specifications in a given sized column, operating even at R_{max} (given by Equation 10.2). The amount of solvent required in such separations is more than what the column can accommodate without flooding. Mujtaba et al. (1997) proposed *fractional feed charge* strategy for this type of mixtures. In this strategy the feed mixture is charged to a certain fraction of the maximum capacity. The column can operate at a reflux ratio greater than R_{max} for some period until the reboiler level reaches to its maximum capacity, C.

For a given initial feed charge (B_0), and operating at constant values (which can be optimised) of reflux ratio and solvent feed rate, the length of this period can be calculated using:

$$t_{flood} = \frac{C - B_0}{F - L_D} = \frac{C - B_0}{F - V_c(1 - R)} \qquad (10.3)$$

Beyond this time period the column operation should satisfy the condition given by Equation 10.2 to avoid flooding.

Note that for both strategies the reboiler holdup must not exceed the maximum capacity (*constraint*) at any time (*path*) within the entire operation period to avoid column flooding and this imposes a *path constraint* in the optimisation problem as discussed below.

10.4. Operational Constraint (Path Constraint)

Mujtaba (1999) explained the BED operational constraint (*path constraint*) using two time intervals in a separation Task (producing a cut). To ensure a safe operation Mujtaba (1999) proposed the following treatment to the *path constraint*. The reboiler holdup profile (H_{Ti}, Figure 10.3) during any distillation Task is the solution of the total mass balance equation and can be expressed as:

$$H_{Ti} = B_{i-1} + [F - V_i(1-R)]t_{Ti} \qquad [i = 1, 2, 3] \qquad (10.4)$$

where t_{Ti} is the distillation Task time.

10.4.1. One Time Interval

For fractional charge strategy, typical reboiler holdup profiles (H_{Ti} vs. t) for a distillation Task with one time interval are shown in Figure 10.4. For a given value of V_c, the values of R and F will decide the slope of the straight lines shown in the figure. Figure 10.4 shows that the BED operation dictated by line 1 violate the *path constraint* as the reboiler holdup goes above the maximum reboiler capacity (C). The operations dictated by other lines are desirable, however, the values of R and F are to be optimised to maximise the productivity.

10.4.2. Two Time Intervals

For fractional charge strategy, Figure 10.5 shows typical reboiler holdup profiles for two time intervals. Again the operation dictated by line 1 is not desirable although the holdup at the end of the operation satisfies the constraint (Equation 10.5).

$$B(t_{Ti}) \leq C \qquad (10.5)$$

The situation when H_{Ti} crosses C can be easily avoided by setting an upper limit on t_1, calculated based on the feasible bounds on R and F. The column will flood in a shortest possible time when it operates at R^u and F^u (upper limits) in the first time interval. Therefore, the maximum allowable length for the first interval (for any Task i) can be calculated using:

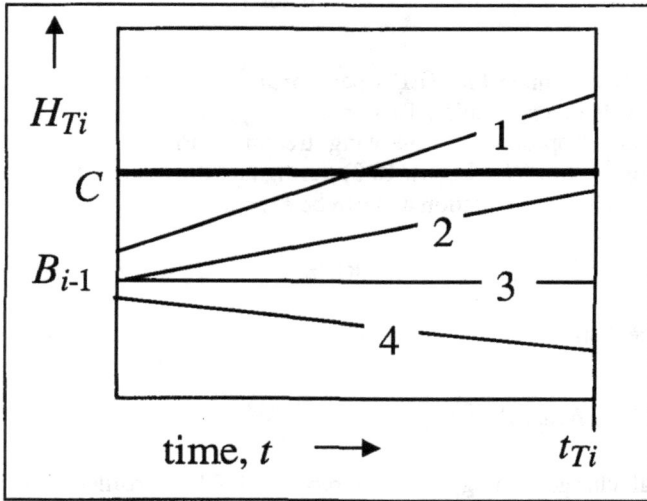

Figure 10.4. Typical Reboiler Holdup Profile with One Time Interval.
[Mujtaba, 1999][b]

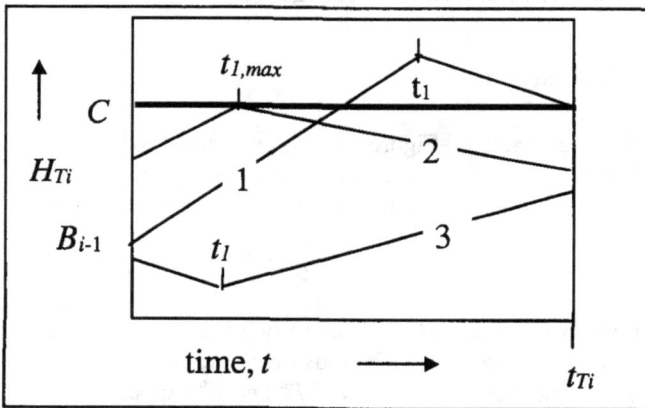

Figure 10.5. Typical Reboiler Holdup Profile with Two Time Intervals.
[Mujtaba, 1999][b]

---

[b] Reprinted with permission from IChemE, UK. Full reference is at the end of the chapter.

$$t_{1,max} = \frac{C - B_{i-1}}{F^u - V_c(1 - R^u)} \qquad (10.6)$$

Use of other values for F and R (smaller than F^u and R^u) for a period $t_1 \le t_{1,max}$ will always give $H_{Ti}(t_1) < C$ (line 3 in Figure 10.5).

10.5. General Multiperiod Dynamic Optimisation (MDO) Problem Formulation

Mujtaba (1999) described a general optimisation problem formulation referring to Figure 10.6. The separation of binary close boiling or azeotropic mixtures was considered. The solvent used was the highest boiling component in the mixture and therefore was recovered as the final bottom product in the process. A total of three distillation Tasks thus constituting multiperiod operation were considered (represented by a STN in Figure 10.6):

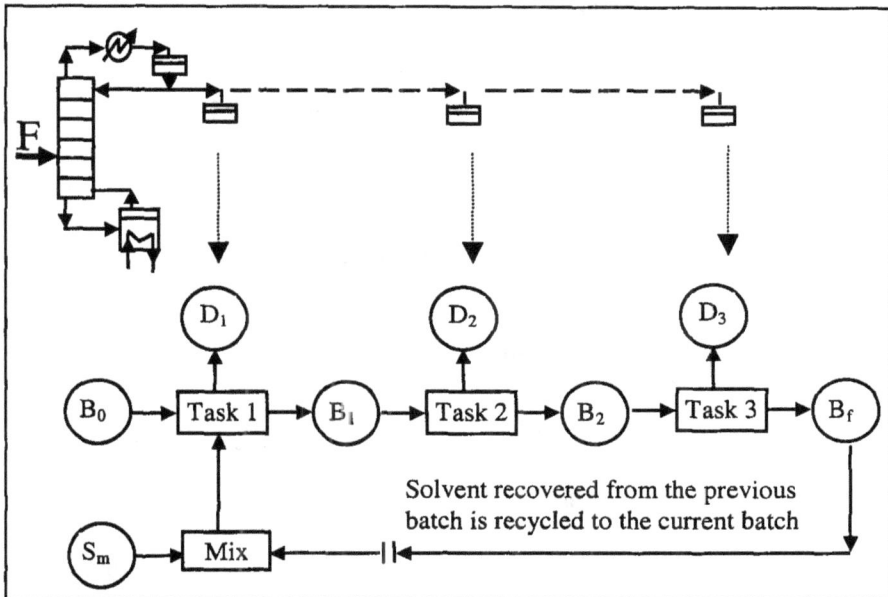

Figure 10.6. STN representation for binary BED process with solvent recovery.

(a)   From the binary feed mixture (represented by state B_0) and the solvent, the distillation Task 1 produces a distillate (D_1) rich in highly volatile component (component 1) leaving the intermediate residue (B_1) in the still.

(b)   The distillation Task 2 produces an off-cut (D_2) and recovers most of the remaining component 1, the composition being same as that of the original feed mixture (so that it can be stored for further processing if needed).

(c)   The distillation Task 3 produces a distillate (D_3) recovering component 2 and leaving the solvent in the still.

Note here, that if the solvent is the mid-boiling component in the mixture, then the solvent is to be recovered in Task 2 and Figure 10.6 can be modified according to new operating strategy. Also the production of off-cut (D_2) may not always be necessary.

Mujtaba (1999) used *profit* as the *performance measure* of BED processes. However, other measures such as *productivity* could also be considered (Tran and Mujtaba, 1997; Mujtaba et al., 1997). With *profit* as the *performance measure*, the MDO problem (*maximum profit* problem) can be formally stated as:

given: the column configuration, solvent feed location, the feed mixture, vapour boilup rate and separation tasks (e.g. achieve products at specified purity)

determine: the optimal reflux ratio, solvent feed rate profiles, recovery of key component and operation time for each Task.

so as to maximise: the profit.

subject to: equality and inequality constraints.

Mathematically the optimisation problem (*OP*) can be written as:

OP   Max   P
                $\{Re_{Ti}, t_{Ti}, R(t_{Ti}), F(t_{Ti})\}$
                (for all Tasks)

          s.t.   Model Equations   (equality constraints)
                    Product Specifications   (inequality constraints)
                    $B(t) \leq C$ (path constraint)
                    $R^l \leq R \leq R^u$ (inequality constraints)
                    $F^l \leq F \leq F^u$ (inequality constraints)

where Re_{Ti} is the recovery of the key component, $R(t_{Ti})$ is the reflux ratio, $F(t_{Ti})$ is the solvent feed rate profiles and t_{Ti} is the Task time for each Task T_i; $B(t)$ is the reboiler holdup at any time t $[0, \Sigma t_{Ti}]$; C is the maximum capacity of the reboiler; R^l and R^u are the lower and upper bounds on the reflux ratio within which it is optimised. Similarly, F^l and F^u are the lower and upper bounds on the solvent feed rate. The profit P is formulated using the cost of raw materials and utilities and prices of the products (Mujtaba and Macchietto, 1993):

$$P = \frac{C_{D1}D_1 + C_{D3}D_3 - C_{B0}B_0 - C_S S_m}{t_{T1} + t_{T2} + t_{T3}} - C_{fc} \qquad (\$/h) \qquad (10.7)$$

C_{D1}, C_{D3} are the prices of the desired products (\$/kmol), C_{B0} is the raw material cost (\$/kmol), C_S is the cost of solvent (\$/kmol), C_{fc} is a fixed operating cost (\$/hr). D_1, D_3, B_0, S_m are the amount of desired products, binary feed mixture and the make-up solvent respectively. t_{T1}, t_{T2}, t_{T3} are the batch times in Task 1, 2 and 3 respectively. Mujtaba (1999) assumes that there is no saleable value for the off-cut (D_2) and it is being stored free of charge. Note D_1, D_3, D_2, B_0, S_m, B_f represent the amount of materials in states (Figure 10.6) D_1, D_3, D_2, B_0, S_m, B_f respectively.

A two-level optimisation solution technique as presented in Chapter 6 and 7 can be used for a similar optimisation problem. For a given product specifications (in terms of purity of key component in each Task) and considering Re_{Ti} as the only outer level optimisation variable, the above MDO problem (OP) can be decomposed into a series of independent *minimum time* problem (Single-period Dynamic Optimisation (SDO) problem) in the inner level. For each iteration of the outer level optimisation, the inner-level problems are to be solved. As mentioned in the earlier chapters, the method is efficient for simultaneous design and operation optimisation especially with **multiple separation duties**.

However, Mujtaba (1999) shows that using only the product specifications and a few logical and sensible assumptions the above MDO problem can be decomposed into a series of independent SDO problems in one level which is computationally less expensive than the method proposed by Mujtaba and Macchietto (1993, 1996).

10.6. Product Specifications and Decomposition of MDO Problem into Single-period Dynamic Optimisation (SDO) Problems

Today's market is customer oriented. To satisfy customer demands all products must be produced on-specification. This means that all the components (or at least the desired component) must be recovered at given specifications. For binary mixtures Mujtaba (1999) assumes that both the components in the feed mixture are recovered at specified purities. Also the author assumes that the solvent recovered has the same composition as the original solvent fed to the column (see Figure

10.6). This allows easy and efficient recycling of the solvent in subsequent batches. In summary, the distillate purity for each Task and final bottom product are imposed as constraints in the optimisation problem (as shown in *OP* above).

However, the purity demand of component 2 in the distillate achieved in Task 3 will dictate the recovery of component 1 in Task 2. Mujtaba and Macchietto (1993) argues that in most cases more than 95% recovery of the component 1 is necessary in this Task and there is a very little economic benefit to optimise the recovery in that Task. Therefore, minimisation of batch time will maximise the profit in this Task (Task 2).

The specification on the final bottom product (solvent) composition determines the amount of distillate and the amount of solvent to be recovered in Task 3 (mass balance). Therefore, with fixed amount of distillate and bottom products, minimisation of the Task time will maximise the profit. The recovered solvent from any batch (B_f) is recycled to the next batch (Figure 10.6). A fresh make-up solvent (S_m) is used with the recycled solvent if needed. Note there will be no recycled solvent for the first batch in a campaign mode operation.

Neither the minimisation of batch time (by fixing the recovery of component 1) nor maximisation of the distillate (by fixing the batch time) will result in the maximisation of the profit for the Task 1. The recovery of component 1 together with the batch time must be optimised to maximise the profit. This results in a *maximum productivity* (amount of distillate over time) optimisation problem for this Task.

The operation strategies and the arguments presented above decomposes MDO problem (*OP*) into a series of three independent SDO problems as: (1) *Maximum productivity* problem for Task 1 and (2) *Minimum time* problems for Tasks 2 and 3.

10.6.1. *Maximum Productivity Problem*

The optimisation problem can be stated as –

given:	the column configuration, solvent feed location, the feed mixture, vapour boilup rate and a separation task (e.g. achieve a product at specified purity)
determine:	the optimal reflux ratio, solvent feed rate profiles for the operation and the optimum amount of product and Task time
so as to maximise:	the productivity
subject to:	equality and inequality constraints.

Mathematically the problem (*OP1*) can be written as:

OP1   Max $Prod = D_i/t_{Ti}$
$R(t_{Ti}), F(t_{Ti})$

s.t.   Model Equations   (equality constraints)

$x_{Di}^k \geq \underline{x}_{Di}^k$ (inequality constraints)

$B(t_{Ti}) \leq C$ (path constraint)
$R^l \leq R \leq R^u$ (inequality constraints)
$F^l \leq F \leq F^u$ (inequality constraints)

where, x_{Di}^k is the composition of the accumulated distillate achieved (molefraction) at the end of operation time t_{Ti} for Task i and $\underline{x}_{Dk}^k$ is the specified product purity for component k.

10.6.2. Minimum Time Problem

The optimisation problem can be stated as –

given: the column configuration, the feed mixture, vapour boilup rate, a separation task in terms of product purity, recovery or amount of product

determine: optimal reflux ratio, solvent feed rate profiles for the operation

so as to minimise: the Task time

subject to: equality and inequality constraints (e.g. model equations)

Mathematically the problem (*OP2*) can be written as:

OP2   Min t_{Ti}
$R(t_{Ti}), F(t_{Ti})$

s.t.   Model Equations   (equality constraints)

$x_{Di}^k \geq \underline{x}_{Di}^k$ (inequality constraints)

$D_i \geq D_i^*$ (inequality constraints)
$B(t) \leq C$ (path constraint)
$R^l \leq R \leq R^u$ (inequality constraints)
$F^l \leq F \leq F^u$ (inequality constraints)

where D_i is the total amount of distillate produced (kmol) over the operation time t_{Ti} for Task i and D_i^* is the specified amount of distillate to be produced in Task i.

Note with only one time interval, the optimisation problem (*OP1* or *OP2*) can be simplified by replacing the *path constraint* with the end time constraint (Equation 10.5). At the optimal solution the reboiler holdup profile will be on or below the thick line shown in Figure 10.4 depending on the strategy for the initial feed charge. For full charge strategy either of the constraints given by Equation 10.2 or 10.5 can be used to prevent column flooding.

With two time intervals, if the constraint (Equation 10.6) is imposed together with the Equation 10.5 in the optimisation problem, the reboiler holdup profile will be on or below the thick line shown in Figure 10.5. For full charge strategy, however, the constraints given by Equations 10.2 and 10.5 must be satisfied.

The conditions given by Equations 10.2, 10.5 and 10.6 greatly ease the solution of the dynamic optimisation problems (*OP* or *OP1* and *OP2*). However, note that more rigorous treatment of the *path constraint* will be required when F and R are continuous function of time or there are more than two time intervals for a distillation Task. See Vassiliadis (1993), Logsdon and Biegler (1993) and Vassiliadis et al. (1994a,b) for further details on dealing with path constraints (interior point constraints).

10.7. Process Model and the Solution Method

Tran and Mujtaba (1997), Mujtaba et al. (1997) and Mujtaba (1999) have used an extension of the Type IV- CMH model described in Chapter 4 and in Mujtaba and Macchietto (1998) in which few extra equations related to the solvent feed plate are added. The model accounts for detailed mass and energy balances with rigorous thermophysical properties calculations and results to a system of Differential and Algebraic Equations (DAEs). For the solution of the optimisation problem the method outlined in Chapter 5 is used which uses CVP techniques. Mujtaba (1999) used both reflux ratio and solvent feed rate (in semi-continuous feeding mode) as the optimisation variables. Piecewise constant values of these variables over the time intervals concerned are assumed. Both the values of these variables and the interval switching times (including the final time) are optimised in all the SDO problems mentioned in the previous section.

Table 10.1. Input Data for Example 1 and 2. [Adopted from Mujtaba, 1999]

N, number of plates including condenser and reboiler	$= 20$
n_C, number of components including solvent	$= 3$
Feed Composition (excluding the solvent), molefraction	$= < 0.5, 0.5>$
Reboiler maximum capacity	$= 10$ kmol
The condenser vapour load (V_C)	$= 4.0$ kmol/hr
Condenser holdup	$= 0.1$ kmol
Total plate holdup	$= 0.5$ kmol
Column pressure	$= 1.013$ bar

10.8. Case Studies

First, two examples using both problems *OP1* and *OP2* are presented to explain the effects of different solvent feeding modes and *path constraint* on the operation. In these examples only Task 1 of Figure 10.6 is carried out where component 1 is recovered at a given purity. Then, example 3 using Multiperiod Optimisation Problem (*OP*) is presented, where all three Tasks of Figure 10.6 are carried out.

The example problems are taken from Tran and Mujtaba (1997) and Mujtaba (1999) and only binary feed mixtures are considered in all three examples.

Note Lang et al. (1994, 1995) provided case studies to show the effect of reflux ratio and solvent flow on the separation of azeotropic mixtures using repetitive simulation. The readers are directed to the original references for more case studies.

10.8.1. Example 1: Minimum Time Problem - Close Boiling Mixture

10.8.1.1. Batch Solvent Feeding Mode

Here, the separation of *Heptane* is considered from a feed mixture of *Heptane-Toluene* using *Phenol* as the solvent. The objective is to recover 75% of the Heptane in the distillate with 95% purity ($x_{D1}^1 = 0.95$ molefraction) in *minimum time*. Therefore, the optimisation problem *OP2* is considered. The input data are given in Table 10.1. Vapour liquid equilibrium is calculated using NRTL-3 model (Lang et al., 1994).

The amount of solvent (F, kmol) charged in the reboiler is varied and the amount of feed (B_0, kmol) is adjusted to have the reboiler charged to its maximum capacity. This shows the effect of using a sub-optimal amount of solvent on the batch time. In fact F could be used as a decision variable and could be optimised. Two time intervals are used for the entire operation period (Task 1) and in each interval a constant reflux ratio is used, the value of which is optimised. Also the length of each interval is optimised. The effect of *batch mode* solvent feeding on the minimum operation time (t_{T1}, hr) and productivity (*Prod* = Amount of distillate/t_{T1}) is summarised in Table 10.2. Given the amount of feed (B_0) and the composition of Heptane (x_{B0}^1) and the Task (75% recovery (Re_{T1}) of the Heptane in the distillate with $\underline{x}_{D1}^1 = 0.95$), the amount of distillate to be achieved (D_1^*) for each case is calculated using Equation 10.8. The value of D_1^* is then used as the constraint bound in the optimisation problem *OP2*.

$$D_1^* = \frac{Re_{T1}(B_0 \cdot x_{B0}^1)}{\underline{x}_{D1}^1} \tag{10.8}$$

Note that the *path constraint* does not appear in the optimisation problem for batch mode solvent feeding.

For each operation, the amount of distillate (D_1, kmol) achieved and the optimum reflux ratio profile are also presented in Table 10.2. Note in all cases the constraints on the distillate amount and purity are satisfied. Table 10.2 clearly shows that t_{T1} decreases and P increases with F. Equation 10.8 shows that for a given recovery, feed composition and distillate purity; D_1 is directly proportional to B_0.

The operation time without any solvent t_{ns} will also be proportional to B_0 and can be calculated by:

$$t_{ns} = (B_0 t_b) / B_b \tag{10.9}$$

where B_b is the amount of base case feed and t_b is the base case operation time (case 1 in Table 10.2 is the base case).

The exact contribution of the solvent in reducing the operation time is calculated by t_{re} (= t_{ns} - t_{T1}). The value of t_{re} increases up to a maximum value of 0.58 hr (for case 4) with the increase in F. If there were no addition of solvent, the productivity in all cases of Table 10.2 would remain the same (*Prod* = D_1/t_{ns} = 0.417) as the base case (case 1). The increase in *Prod* with F is due to decrease in batch time by t_{re}. The productivity improves by more than 10% in case 5 compared to that for case 1. However, beyond a critical amount of solvent (4 kmol, case 5) the increase in F decreases the productivity. For a given vapour load, the amount of solvent vapour travelling up the column will increase with decreasing B_0. This consequently means higher reflux ratio and therefore longer operation time to maintain the distillate purity. This will also mean a drop in t_{re} beyond a critical value (cases 5 and 6).

10.8.1.2. Campaign Mode Operation with Batch Solvent Feeding

The column is to process a total of 100 kmol of the feed mixture. Since the reboiler capacity is smaller than the total amount of feed to be processed, a *campaign mode* operation will be required processing a number of batches in sequence. Again the objective is to recover 75% Heptane with 95% purity. A total of 39.5 kmol of distillate will be achieved from the campaign operation (from Equation 10.8, using $B_0 = 100$ kmol).

The total number of batches and the total campaign time for different solvent and feed proportion are presented in Table 10.3. Each case of Table 10.3 corresponds to that of Table 10.2 in terms of solvent and feed proportion. Therefore, for each case the results in Table 10.2 are used to calculate the total number of batches (N_B) and the total campaign time (T_{cam}) using:

Table 10.2. Summary of the Results using *Batch Mode* Solvent Feeding.
[Adopted from Mujtaba, 1999]

Case	F kmol	B_0 kmol	x_M molefraction	D_1 kmol	t_{T1} hr
1	0.0	10.0	<.50, .50, .00>	3.95	9.48
2	1.0	9.0	<.45, .45, .10>	3.55	8.18
3	2.0	8.0	<.40, .40, .20>	3.15	7.03
4	3.0	7.0	<.35, .35, .30>	2.76	6.06
5	4.0	6.0	<.30, .30, .40>	2.38	5.15
6	5.0	5.0	<.25, .25, .50>	1.97	4.31

Case	t_{ns} hr	t_{re} hr	Prod kmol/hr	(R_1, t_1)	(R_2, t_2)
1	9.48	0.00	0.417	0.885, 5.53	0.912, 3.95
2	8.53	0.35	0.434	0.882, 4.76	0.905, 3.42
3	7.58	0.55	0.448	0.881, 4.33	0.898, 2.70
4	6.64	0.58	0.456	0.885, 4.32	0.892, 1.74
5	5.69	0.54	0.462	0.886, 2.87	0.884, 2.28
6	4.74	0.43	0.457	0.893, 3.81	0.829, 0.50

Key: x_M = composition of mixed charge (solvent + feed), molefraction
Prod = D/t_{T1}
(R_i, t_i) = (reflux ratio in interval i, interval length)
t_{ns} = minimum batch time without solvent operation, hr

Table 10.3. Results of Campaign Mode Operation (Example 1).
[Adopted from Mujtaba, 1999]

Case	F kmol	B_0 kmol	N_B	T_{cam} hr	Total D kmol
1	0.0	10.0	10.00	94.80	39.5
2	1.0	9.0	11.11	90.92	39.5
3	2.0	8.0	12.50	87.85	39.5
4	3.0	7.0	14.28	86.54	39.5
5	4.0	6.0	16.67	85.83	39.5
6	5.0	5.0	20.00	86.18	39.5

$$N_B = \frac{100}{B_0} \qquad (10.10)$$

$$T_{cam} = N_B \times t_{T1} \qquad (10.11)$$

It is clear from Table 10.3 that although the increase in F increases N_B, it permits a significant saving in the total campaign time (more than 10%) for the same total amount of distillate.

10.8.1.3. Effect of F - Semi-Continuous Solvent Feeding Mode (Full Charge)

Tran and Mujtaba (1997) investigated the effect of solvent feed rate (F) using $V_C = 10$ kmol/hr and N_F (solvent feed location) =18 (counted from the top) under semi-continuous solvent feeding mode with full charge strategy. The column is run in two steps. In STEP 1, the column is run at total reflux for 2 hrs without introducing any solvent. In STEP 2 the solvent is introduced to the column. The flooding constraint given by Equation 10.2 has been added for STEP 2 operation.

For different values of F, R_{max}, optimum reflux ratio (R_2), minimum operation time, productivity are shown in Table 10.4. In all cases the total amount of distillate is 3.95 kmol with 95% purity in Heptane. The productivity $(Prod)$ is calculated using total operation time (t_{total}) which includes 2 hrs of total reflux operation time in STEP 1.

As seen in Table 10.4, the operation time of STEP 2 decreases with the increase in F. The operation time saved is about 7% and 16% in cases 2 and 3 respectively when compared to the base case (case 1). The improvement in productivity is about

7% for case 2 and 12% for case 3 compared to the base case. Cases 4 and 5 yield no solution despite a total reflux operation in STEP 1. Although solutions were found for $R > R_{max}$ meaning flooding of the column, no optimal solutions were available within the allowable R_{max}. That means that the value of V_C is to be increased to push R_{max} higher (according to Equation 10.2). This demonstrates the importance of choosing an appropriate feed rate for a fixed condenser vapour load (i.e. for a fixed column design).

10.8.1.4. Effect of Feed Plate Location (N_F) - Semi-Continuous Solvent Feeding Mode (Full Charge)

Tran and Mujtaba (1997) studied the effect of N_F using $F = 1.0$ kmol/hr and $V_C = 10$ kmol/hr. This gives $R_{max} = 0.90$. Again a two-step operation as in section 10.8.1.3 was considered. The results are presented in Table 10.5. It is evident from Table 10.5 that, the higher the feed plate location (smaller N_F), the shorter the operation time for STEP 2 is and the better the productivity is. Moving the feed plate up the column allows more contact of the solvent with the vapour travelling up the column and makes the separation of Heptane easier (cases 1-4). The improvements in operation time and productivity of case 4 are more than 8% compared to case 1. However, if the feed plate is very near to the top of the column it is possible that some solvent vapour will travel up the condenser and will make the separation of Heptane difficult. In that case a higher reflux operation meaning a longer operation time might be required.

Table 10.4. Effect of F. [Adopted from Tran and Mujtaba, 1997]

Case	F kmol/hr	R_{max}	R_2, t_2	t_{total} hr	*Prod* kmol/hr
1	0.0	1.00	0.889, 3.55	5.55	0.71
2	0.5	0.95	0.876, 3.18	5.18	0.76
3	1.0	0.90	0.867, 2.95	4.95	0.80
4	1.5	0.85	no solution found		
5	2.0	0.80	no solution found		

Key: R_2, t_2 = optimum reflux ratio, minimum time for STEP 2

Table 10.5. Effect of N_F. [Adopted Tran and Mujtaba, 1997]

Case	N_F	R_2, t_2	t_{total} hr	*Prod* kmol/hr
1	18	.867, 2.95	4.95	0.795
2	10	.851, 2.64	4.64	0.849
3	5	.847, 2.57	4.57	0.862
4	3	.846, 2.56	4.56	0.864

Key: R_2, t_2 = optimum reflux ratio, minimum time for STEP 2

10.8.1.5. Effect of V_C - Semi-Continuous Solvent Feeding Mode (Full Charge)

Tran and Mujtaba (1997) investigated the effect of V_C using $F = 1.5$ kmol/hr and N_F = 18. Again a two-step operation as in section 10.8.1.3 was considered. The operation time, reflux ratio profile and productivity using four different values of V_C are summarised under **Scenario 2** in Table 10.6. For each case the column was also run with $F = 0$ for comparison (**Scenario 1**). For **Scenario 1**, $R_{max} = 1.0$. The productivity (*Prod$_1$* or *Prod$_2$*) is calculated using the total operation time ($t_{total,1}$ or $t_{total,2}$) which includes 2 hrs of total reflux operation in STEP 1.

The results clearly show that the operation improves in terms of operation time and productivity with increasing value of V_C for both **Scenario 1** and **2** operations. However at higher value of V_C the **Scenario 1** operation becomes closer to **Scenario 2** operation. For instance, for $V_C = 25$ only 5% improvement in the productivity is noted when the operations of **Scenario 1** and **2** are compared but for $V_C = 15$ the improvement in productivity is about 10%. No solution was found for case 1 (and **Scenario 2**) as the value of $R_{max,2} = 0.85$ does not permit the required separation. Either the solvent flow rate has to be decreased or the column design has to be changed to increase V_C to push $R_{max,2}$ to a higher value to permit the desired separation. Note, however, that high value of V_C will increase the operating cost which must be balanced against the improvement in operation time or productivity.

Table 10.6. Effect of V_C. [Adopted from Tran and Mujtaba, 1999]

| Case | Without Solvent (*Scenario 1*) | | | With Solvent (*Scenario 2*) | | | |
	V_C	$(R_2, t_2)_1$	$t_{total,1}$	$Prod_1$	$R_{max,2}$	$(R_2, t_2)_2$	$t_{total,2}$	$Prod_2$
1	10	.889, 3.55	5.55	0.71	.85	no solution	-	
2	15	.889, 2.37	4.37	0.90	.90	.867, 1.98	3.98	0.99
3	20	.889, 1.79	3.79	1.04	.925	.872, 1.54	3.54	1.11
4	25	.890, 1.44	3.44	1.15	.940	.875, 1.27	3.27	1.21

Key: (R_2, t_2) = optimum reflux ratio, minimum time for STEP 2

10.8.1.6. Effect of Recovery of Heptane (Re_H) - Semi-Continuous Solvent Feeding Mode (Full Charge)

The recovery Re_k of any component k in a mixture can be defined as:

$$Re_k = (D_1 \cdot x_D^k)/(B_0 \cdot x_{B0}^k) \tag{10.12}$$

where D_1 is the amount of distillate collected over batch time t_{T1}, B_0 is the amount of solvent free feed and x_{B0}^k is the composition of component k in B_0.

Tran and Mujtaba (1997) investigated the effect of recovery of Heptane using $F = 0.5$ kmol/hr, $N_F = 5$ and $V_C = 10$ kmol/hr. The values of F and V_C give $R_{max,2} = 0.95$ for the operation with solvent (**Scenario 2**). Again two steps operation as mentioned in section 10.8.1.3 is considered here. For different Re_H values the amount of distillate (calculated using Equation 10.11), operation time, reflux ratio profile and productivity are summarised under **Scenario 2** in Table 10.7. For each case the column is also run with $F = 0$ for comparison (**Scenario 1**). For **Scenario 1**, $R_{max} = 1.0$. As in section 10.8.1.5, the productivity ($Prod_1$ or $Prod_2$) is calculated using the total operation time ($t_{total,1}$ or $t_{total,2}$).

For a given column design and product purity, high recovery makes the separation difficult and therefore requires high reflux ratio and long operation time which gives low productivity. These are clear from Table 10.7. The last column in Table 10.7 gives the percent improvement (P_I) in productivity for **Scenario 2** compared to **Scenario 1**. The values of P_I reveals that as the recovery goes higher the benefit of using BED operation (**Scenario 2**) becomes clearer compared to CBD operation (**Scenario 1**).

Table 10.7. Effect of Heptane recovery. [Adopted from Tran and Mujtaba, 1997]

Case	Re_H	D_I kmol	Without Solvent (**Scenario 1**)			With Solvent (**Scenario 2**)			P_I %
			$(R_2, t_2)_1$	$t_{total,1}$	$Prod_1$	$(R_2, t_2)_2$	$t_{total,2}$	$Prod_2$	
1	75	3.95	.889, 3.55	5.55	.710	.865, 2.92	4.92	.802	12.9
2	85	4.47	.914, 5.22	7.22	.620	.891, 4.09	6.09	.734	18.4
3	90	4.74	.932, 6.95	8.95	.529	.908, 5.19	7.19	.658	24.4
4	95	5.00	.957, 11.74	13.74	.364	.936, 7.84	9.84	.508	39.6

key: P_I = percent improvement in productivity = $100.0 * (Prod_2 - Prod_1) / Prod_1$

                 (R_2, t_2) = (optimum reflux ratio, minimum time) for STEP 2

10.8.2. Example 2: Maximum Productivity Problem - Close Boiling Mixture

Here, the same mixture used for example 1 is considered. Semi-continuous solvent feeding mode with *full charge* strategy is opted in this example. The objective is to maximise the productivity of Task 1 of the STN shown in Figure 10.6. The specification on the distillate composition is 0.95 molefraction in Heptane. The optimisation problem (*OP1*) is considered and both the reflux ratio and solvent rate profiles are optimised. Again two time intervals are used for the entire operation period (Task 1). In each interval, constant reflux ratio and solvent feed rate are used, the values of which are optimised. The input data are the same as those in Table 10.1 except that the maximum reboiler capacity is 25 kmol. The solvent is introduced in plate 6 (N_F).

Since the column is operating in *full charge* mode the constraints given by Equations 10.2 and 10.5 must be satisfied for the first and second intervals respectively. A number of cases were studied using different maximum allowable values for the solvent feed rate (F^u) in the first interval. These give values for R_{max} for the first interval. For each case, Table 10.8 shows the optimal reflux ratio, the solvent feed rate, the lengths of the time intervals and the productivity. The case 1 represents the base case where no solvent was introduced into the column.

In Table 10.8, R_{max} decreases with the increase in F^u according to Equation 10.2. The operation time decreases and the productivity increases with the increase in F^u. These are as expected. Initial total reflux operation was not needed for any case. Significant improvement in productivity (about 77%) is noted for case 4 compared to the base case. This clearly reveals the importance of extractive distillation for close boiling mixtures. The case 5 yields no solution. Although a solution was found for $R > R_{max}$ for the first interval meaning flooding of the column, no optimal solution was obtained with $R \leq R_{max}$. This means that the value of V_C is to be increased to push R_{max} to a higher value according to Equation 10.2. This demonstrates the importance of choosing an appropriate feed rate for a fixed condenser vapour load (i.e. for a fixed column design).

For cases 2-4, the optimal values of R in both intervals were less than R_{max} meaning the column was never flooded during the operation. This is also evident from the reboiler holdup values at the end of each interval. Also for cases 2-4 the constraint given by Equation 10.5 was not active. Mujtaba (1999) noted that cases 2-4 were re-run with only one time interval for the distillate Task. The constraint given by Equation 10.5 was imposed on the optimisation problem. For cases 2-3 the maximum productivity obtained was about 0.55 and for case 4 no solution was obtained. This shows the importance of having time-sequenced operation.

Table 10.8. Summary of the Results for Example 2. [Adopted from Mujtaba, 1999]

Case	F^u kmol/hr	R_{max}	Optimal (R_1, R_2)	Optimal (F_1, F_2) kmol/hr	(H_1, H_2)	t_1, t_{T1} hr	Prod kmol/hr
1	0.00	1.0	0.859, 0.915	0.0, 0.0	------	9.95, 22.49	0.44
2	0.25	0.9375	0.845, 0.826	0.25, 0.22	23.21, 21.99	3.24, 5.81	0.65
3	0.50	0.875	0.833, 0.803	0.50, 0.48	24.06, 23.15	2.02, 4.99	0.74
4	0.75	0.8125	0.812, 0.800	0.62, 0.64	24.13, 23.69	2.01, 4.83	0.78
5	1.0	No Solution					

Key:	R	$= L_R/V_c$ (Figure 10.3), at total reflux $R = 1.0$
	R_1, R_2	= reflux ratio in time interval 1 and 2
	F_1, F_2	= solvent feed rate in time interval 1 and 2
	t_1	= length of interval 1, hr
	t_{T1}	= total Task time, hr
	t_2	= length of interval 2, hr $= t_{T1} - t_1$
	H_1, H_2	= reboiler holdup at the end of t_1 and t_{T1}.

10.8.3. Example 3: Multiperiod Optimisation with Azeotropic Mixture

Using semi-continuous solvent feeding mode with *fractional charge* strategy Mujtaba (1999) considered the separation of Acetone from a feed mixture of Acetone-Methanol with 0.5 molefraction of Acetone and 0.5 molefraction of Methanol. The mixture forms an azeotrope with 0.78-0.80 molefraction of Acetone (Lang et al., 1994; Safrit and Westerberg, 1997). A recycled solvent containing 99.9% Water and 0.1% Methanol was used for the separation of this azeotropic mixture. The objective was to maximise the profit of the operation and therefore the optimisation problem (*OP*) was considered. The input data for the problem are given in Table 10.9. With the given product specifications (Table 10.9) the problem *OP* results in a series of three independent optimisation problems (as described before). These are problem *OP1* for Task 1 and problem *OP2* for Task 2 and 3 respectively (of Figure 10.6).

Table 10.9. Input Data for Example 3. [Adopted Mujtaba, 1999]

N, no. of plates $= 20$ N_F, solvent feed location	$= 6$
n_c, no. of components including solvent	$= 3$
C, reboiler maximum capacity , kmol	$= 25$
Holdup, kmol: Condenser	$= 0.1$
Plate (total)	$= 0.5$
The condenser vapour load (V_c), kmol/hr	$= 4.0$
Column pressure, bar	$= 1.013$

Task Specifications:
- Task 1: Distillate composition = 0.95 molefraction Acetone
- Task 2: 98% recovery of the remaining Acetone with composition = 0.5 molefraction
- Task 3: Distillate composition = 0.95 molefraction Methanol Bottom composition = 0.999 molefraction Water

Cost and Prices:
Acetone = \$18.08/l; Methanol = \$14.56/l; Solvent = \$0.2/l
Raw material = \$5.0/l; Operating cost = \$5.0/hr (fixed)

For Tasks 1 and 3 two time intervals are used. For Task 2 only one time interval is used. Within each interval the reflux ratio, the solvent feed rate and the length of the interval are optimised. Vapour liquid equilibrium is calculated using UNIQUAC model. A number of cases were studied for different amount of initial feed charge (B_0) to the reboiler. For each case, the optimal reflux and solvent feed profiles for each Task, percentage of Acetone and solvent recovered and the overall profit of the operation are shown in Table 10.10.

In Task 1, an initial total reflux operation was required for all cases. The optimal length of the total reflux operation period with B_0 = 15 kmol was longer compared to that with the other values of B_0. The large values of B_0 leave little room for the solvent (as can be seen from the solvent rate profiles). Therefore, a longer total reflux operation helps to break the azeotrope allowing the column profile to be developed to an optimal stage before the distillate withdrawal is initiated in the second time interval. For the same reason, an increasingly higher reflux ratio for the second time interval was necessary to maximise the productivity for each case. This consequently requires a longer operation time. Since, for all the cases (and for all Tasks), the optimal solvent feed rate for the first interval was smaller than F^u and t_1 was smaller (equal in one case) than $t_{1,max}$ the reboiler holdup at time t_1 was lower than the maximum reboiler capacity, C. Also all the cases (and all the Tasks) satisfy the constraint given by Equation 10.5 at time t_{Ti}. For Task 1, it is found that up to B_0 = 10 kmol, the productivity is controlled by the amount of initial feed charged to the reboiler (the productivity increases with B_0). Beyond B_0 = 10 kmol, the productivity is controlled by the available room for the solvent in the reboiler (the productivity decreases with B_0). This phenomenon is clearly explained in Mujtaba et al. (1997).

The initial state of Task 2 is the final state of Task 1. Therefore, H_{T1} of Task 1 becomes the initial feed amount B_1 for Task 2 (Figure 10.6). The distillate compositions for all cases of Task 2 are close to that of the feed mixture and therefore the distillate can be stored, recycled and reprocessed in subsequent batches. In this Task, in fact, the distillate could be obtained at purity higher than 0.5 molefraction in acetone. But re-mixing of this Task with a fresh feed mixture in the next batch would not be a thermodynamically sensible option (Mujtaba, 1989; Quintero-Marmol and Luyben, 1990; Mujtaba and Macchietto, 1992) as explained in Chapter 8. The recovery of Acetone (can be calculated using Equation 10.11) decreases with increasing B_0 in Task 1 (Table 10.10). Therefore the amount of distillate in Task 2 increases with B_0 for a given recovery (98%) of Acetone in Task 2 and so does the operation time.

The final state of Task 2 becomes the initial state of Task 3. Therefore, H_{T2} of Task 2 becomes the initial feed amount B_2 for Task 2 (Figure 10.6). In Task 3 the increase in B_0 increases the amount of distillate. This is due to fixed distillate and bottom compositions. However, since the initial composition of the mixture at the beginning of Task 3 was not identical, the minimum batch time for Task 3 does not necessarily increase with the increase in B_0 and D_3.

Table 10.10. Summary of the Results for Example 3.
[Adopted from Mujtaba, 1999]

Task 1: maximum productivity problem

B_0	R_1, R_2	F_1, F_2	H_1, H_{T1}	$t_{1,max}$	t_1, t_{T1}	D_1	Prod
7	1.0, .776	4.20, 4.45	9.59, 19.30	3.0	.62, 3.52	2.60	.74
10	1.0, .769	4.01, 4.01	12.50, 23.13	2.5	.62, 4.26	3.36	.79
15	1.0, .783	3.80, 2.69	17.5, 25.0	1.67	.67, 5.14	3.88	.76

Task 2: minimum time problem

B_1	R_1	F_1	D_2	x_{D2}	H_{T2}	t_{T2}
19.30	.784	1.0	2.01	<.5, .485, .015>	19.62	2.33
23.13	.732	1.0	3.53	<.5, .483, .017>	22.89	3.29
25.00	.614	1.5	7.47	<.5, .472, .028>	24.79	4.84

Task 3: minimum time problem

B_2	R_1, R_2	F_1, F_2	H_1, H_{T3}	D_3	$t_{1,max}$	t_1, t_{T3}
19.62	.771, .949	0.11, 0.09	18.09, 17.60	2.62	3.59	1.89, 6.20
22.89	.569, .939	0.12, 0.095	20.72, 20.04	3.43	1.40	1.35, 5.92
24.79	.356, .860	0.098, 0.11	22.31, 19.28	4.13	0.14	0.14, 6.87

Summary:

B_0	% Acetone Recovery	Total Solvent Fed, kmol	% Solvent Recovery	Profit, P $/hr
7	70.6	18.44	98.7	245.9
10	63.8	20.96	98.5	288.1
15	49.1	20.16	98.6	248.9

Key: B_0 = amount of initial feed charge in the reboiler, kmol
R_1, R_2 = reflux ratio values in interval 1 and 2 (for each Task)
F_1, F_2 = solvent rate in interval 1 and 2, kmol/hr (For each task)
H_1, H_{Ti} = reboiler holdup in interval 1 and 2, kmol (For each task)
t_1 = length of interval 1, hr (in each Task)
D_i = amount of distillate in Task i, kmol
x_{D2} = distillate composition in Task 2, molefraction
Note: $<F^l, F^u>$ = <0.5, 6.0> kmol for Task 1 and 2 and = <0.05, 1.5> for Task 3
$<R^l, R^u>$ = <0.3, 1.0> for all Tasks

At the end of Task 3, H_{T3} together with the condenser and the column holdup constitute the total amount of solvent recovered at the specified purity. About 98% solvent was recovered which can be reused in subsequent batches. Overall, $B_0 = 10$ kmol is noted to be the optimum amount of feed mixture for the given reboiler capacity and products specifications and this makes the most profitable operation. However, the actual optimum values for B_0 could be between 7-10 or 10-15 kmol.

Mujtaba (1999) noted that for the given separation task for this azeotropic mixture, *full feed charge* strategy was not suitable. This is due to the requirement of $R > R_{max}$ to achieve high purity of Acetone. However, this violates the condition given by Equations and 10.1 and 10.2.

Further study on the separation of Acetone-Methanol using BED processes in CBD and MVC column configurations can be found in Milani (1999), Low and Sorensen (2002) and Low (2003). An example of BED process optimisation for Acetone-Methanol-Water mixture using an MVC column is presented in Chapter 11.

References

Ahmad, B.S. and Barton, P.I., *AIChE J.* **42** (1996), 3419.

Dussel, R. and Stichlmair, J., *Comput. chem. Engng.* **19** (1995), s113.

Lang, P., H, Yatim, P. Moszkowicz and M. Otterbein, *Comput. chem. Engng.* **18** (1994), 1057.

Lang, P., Lelkes, Z., Moszkowicz, P., Otterbein, M. and Yatim, H., *Comput. chem. Engng.* **19** (1995), s645.

Logsdon, J.S. and Biegler, L.T., *IEC Res.* **32** (1993), 700.

Logsdon, J.S., Diwekar, U.M. and Biegler, L.T., *Trans. IChemE*, **68A** (1990), 434.

Low, K.H., *Optimal Configuration, Design and Operation of Batch Distillation Processes*, PhD Thesis, (University of London, 2003).

Low, K.H. and Sorensen, E., *AIChE J.* **48** (2002), 1034.

Milani, S.M., *Trans. IChemE*, **77A** (1999), 469.

Mujtaba, I.M., *Optimal Operational Policies in Batch Distillation*. PhD Thesis, (Imperial College, University of London, 1989).

Mujtaba, I.M., *Trans. IChemE*, **75A** (1997), 609.

Mujtaba, I.M., *Trans. IChemE*, **77A** (1999), 588.

Mujtaba, I.M. and Macchietto, S., *Comput. chem. Engng.* **16S** (1992), S273.

Mujtaba, I.M. and Macchietto, S., *Comput. chem. Engng.* **17** (1993), 1191.

Mujtaba, I.M. and Macchietto, S., *J. Proc. Control.* **6** (1996), 27.

Mujtaba, I.M., Woesten, D. and Avignon, L., In *Proceedings of Sep. Sci. Tech.*, AIChE Annual Meeting, LA, USA, 16-21 November, (1997), paper no. 34u, 157.

Mujtaba, I.M. and Macchietto, S., *Chem. Eng. Sci.* **53** (1998), 2519.

Quintero-Marmol, E. and Luyben, W.L., *IEC Res.* **29** (1990), 1915.

Safrit, B.T. and Westerberg, A.W., *IEC Res.* **36** (1997), 436.

Towler, G., *Lecture Notes on Azeotropic Distillation*, UMIST, UK, 7-8 May (1997).

Tran, S. and Mujtaba, I.M., In *Proceedings of IChemE Jubilee Research Event*, Nottingham, UK, April 8-9, **2** (1997), 689.

Vassiliadis, V.S., *Computational Solution of Dynamic Optimization Problems with General Differential-Algebraic Constraints*. PhD Thesis, (Imperial College, London, 1993).

Vassiliadis, V.S., Sargent, R.W. and Pantelides, C.C., *IEC. Res.* **33** (1994a), 2111.

Vassiliadis, V.S., Sargent, R.W. and Pantelides, C.C., *IEC. Res.* **33** (1994b), 2123.

CHAPTER 11

11. UNCONVENTIONAL BATCH DISTILLATION

11.1. Introduction

Over the last few decades many alternative column configurations and operation have been suggested for batch distillation. The features and characteristics of such configurations have been briefly presented in Chapter 2 with process models of different degree of complexity in Chapter 4. The flexibility and operational issues of some of these configurations will be presented in this chapter.

11.2. Use of Continuous Columns for Batch Distillation

11.2.1. Introduction

In the mid Eighties, 99 batch processes within 74 UK companies were identified (Parakrama, 1985). In the last decade or more a continuous shift towards batch processes has been noticed and many small-scale companies are using their existing continuous columns for batch distillation (Willet, 1995) without much realising the implications of such practice. Also R&D sections of many multi-national chemical companies do the pilot plant study in CBD columns for their continuous distillation columns which are in operation in the plant (Chen, 1995; Jenkins, 2000; Greaves, 2003). Because of confidentiality the results of such studies are not available in public.

Mujtaba (1997) explored the potentials of using continuous columns for batch distillation in detail.

In CBD operation (Figure 11.1) each cut (*one pass*) produces either a desired product (main-cut) or an off-specification product (off-cut). The recovery of key component in each cut can be maximised by controlling reflux ratio optimally and by having a number of time sequenced reflux ratio (Chapter 5, 6). The process model for CBD operation may result in a simple or a complex set of differential and algebraic equations (DAEs) depending on the underlying assumptions and the accuracy of calculations required (Chapter 4). The optimisation problem using such model is however always dynamic. And the computation time for the solution of such problem is always high but varies depending on the type of model used and on the total number of variables to be optimised (Chapter 5, 6).

However, the use of continuous columns Figure 11.2 for batch distillation has several advantages:

(a) modelling task becomes similar to that of steady state continuous distillation,

(b) model results in a set of non-linear algebraic equations rather than a set of non-linear DAEs,

(c) optimisation task becomes easier which results in a steady state optimisation problem as opposed to a dynamic optimisation problem and therefore the computation becomes cheaper.

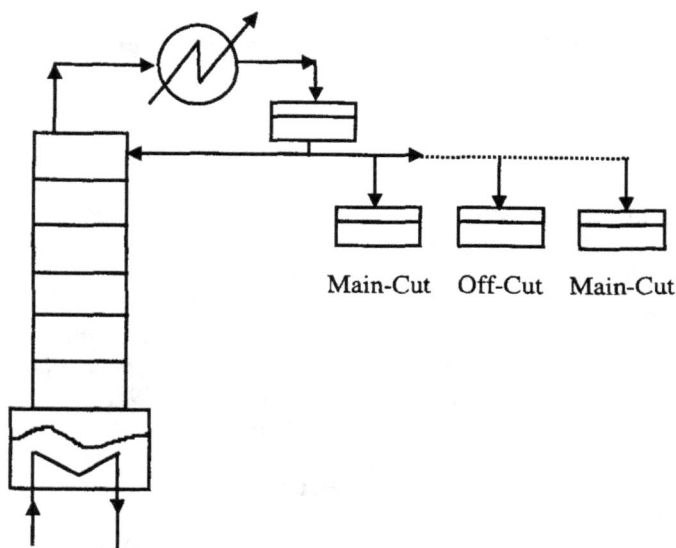

Main-Cut   Off-Cut   Main-Cut

Figure 11.1. Conventional Batch Distillation (CBD)

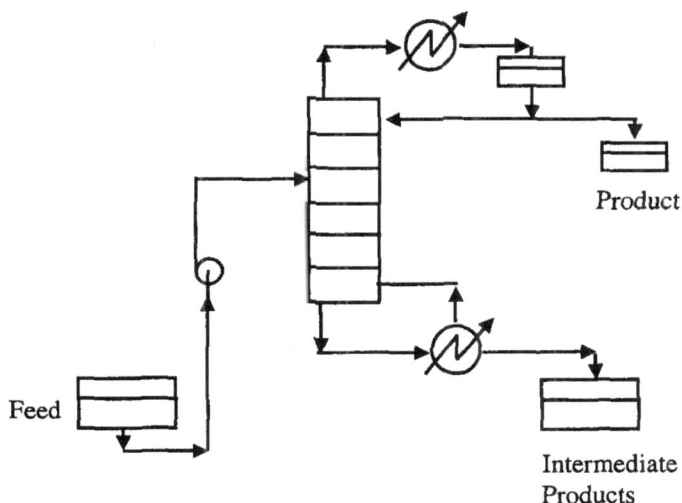

Figure 11.2. Continuous Column For Batch Distillation. [Mujtaba, 1997][a]

During each pass of a continuous column operation, the feed flow rate, feed location, vapour boilup rate, etc. will influence the recovery of component, time of operation, energy consumption, etc. Also the number of equilibrium stages and the relative volatility of the components in the feed mixture, feed location will dictate the number of *passes* which will be required to achieve a high recovery and high purity of a particular component. However, very little information regarding these was available in the literature until the work of Mujtaba (1997). Whether the CBD operation can be replaced by the continuous column operation without compromising the efficiency in terms of recovery of component, total time of operation and energy consumption is an open area. Mujtaba (1997) has addressed some of these issues using optimisation techniques.

Separation of multiple mixtures using a single column is a very common feature in CBD columns (Mujtaba and Macchietto, 1996). If a continuous column were to be used for batch distillation then it would be interesting to evaluate the performance of such column undergoing *multiple separation duties*.

Mujtaba (1997) considered *single separation duty* where a single column is used for separating only one mixture during its total operation period in a year and compared the performances of CBD with continuous column operation. The author

---

[a] Reprinted with permission from IChemE, UK. Full reference is at the end of the chapter.

also considered multiple mixtures and ***multiple separation duties*** in continuous columns and evaluated the optimum performance of such columns. Some of these works will be presented in the next few sections.

11.2.2. SPSS, SPSSS, MPSSS Operations

Mujtaba (1997) provided the following definitions of the terms SPSS, SPSSS, MPSSS frequently used for continuous column operation.

11.2.2.1. Single Pass Steady State Operation (SPSS)

When only one product (distillate) is produced in a continuous column using only one pass by processing a binary or a multicomponent mixture, the operation is defined as SPSS operation. In this type of operation the bottom product after the first pass is not processed further. Figure 11.3 shows the operation for a ternary mixture. Only one product (rich in component A) is obtained using one pass.

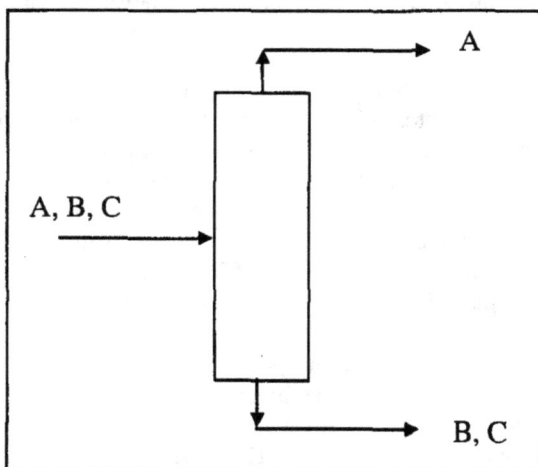

Figure 11.3. Single Pass Steady State (SPSS) operation. [Mujtaba, 1997][b]

---

[b] Reprinted with permission from IChemE, UK. Full reference is at the end of the chapter.

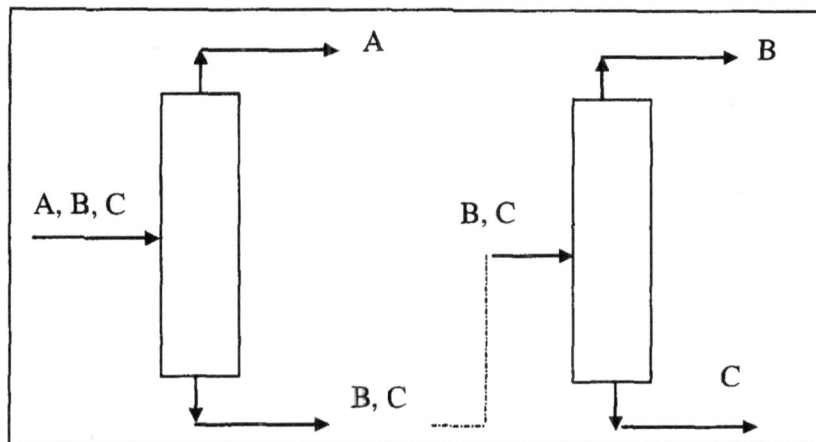

Figure 11.4. Single Pass Sequential Steady State (SPSSS) operation. [Mujtaba, 1997][c]

11.2.2.2. Single Pass Sequential Steady State Operation (SPSSS)

When a number of products are produced sequentially using one pass to produce one distillate product from a binary or a multicomponent mixture in a continuous column, the operation is defined as SPSSS operation. For binary mixtures there will be a maximum of two distillate cuts, one being a specified product cut and the other being an off-specification product cut. Under such condition, of course, the final bottom product will be of specified purity. Figure 11.4 illustrates the operation for a ternary mixture. Here two passes are allowed in the same column sequentially. The first pass separates component A and the second pass separates component B.

11.2.2.3. Multi-Pass Sequential Steady State Operation (MPSSS)

When a distillate product of specified purity is produced using a number of passes in a continuous column, the operation is defined as MPSSS operation. The number of distillate products depends on the number of components in the original mixture. This type of operation is used mainly to improve the recovery of a particular species in the feed mixture (see example section for further explanation). This type of operation can be applied to both binary and multicomponent mixtures.

---

[c] Reprinted with permission from IChemE, UK. Full reference is at the end of the chapter.

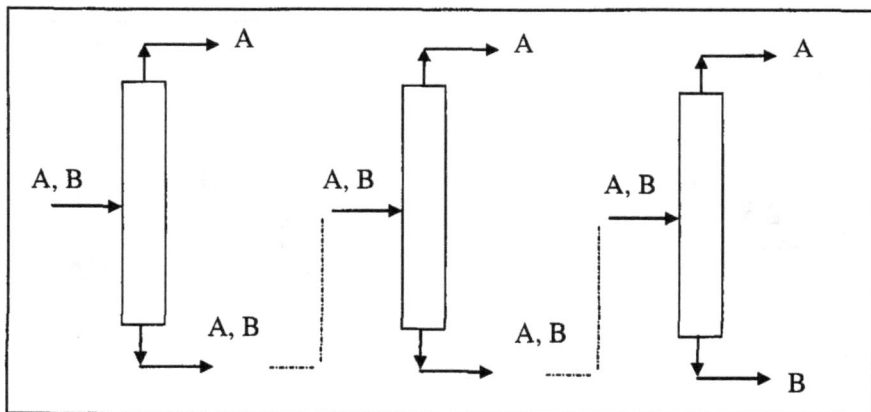

Figure 11.5. Multi-Pass Sequential Steady State (MPSSS) operation. [Mujtaba, 1997][d]

Figure 11.5 illustrates the operating sequence for a binary mixture. Successive passes (multi-pass) are used sequentially using the same column to separate only component. The purity of the distillate product remains the same for each pass but the distillate rate varies (this in turn varies the reflux ratio). This strategy is similar to a time sequenced reflux ratio operation for individual cuts in CBD operation. The total recovery of a component (say A) is calculated from the accumulated amount of distillate from all the passes.

11.2.3. Single Separation Duty in Continuous Columns

11.2.3.1. Performance Measure

Mujtaba (1997) used the *maximum distillate* problem to compare the performances of the two types of distillation columns (CBD and continuous). With the amount of initial charge and the feed flow rate fixed in a continuous column, the operation time (*pass* time) also becomes fixed. The *performance measure* using *maximum distillate problem* allows fixing of the operation time. Other types of optimisation problems such as *minimum time* or *maximum profit* problems (presented in the previous chapters) are not suitable for the purpose of comparing the performances of

---

[d] Reprinted with permission from IChemE, UK. Full reference is at the end of the chapter.

these two types of distillation columns as the operation time in these optimisation problems is not fixed but is relaxed and optimised.

Also Mujtaba (1997) considered the separation of binary mixtures into one distillate product of specified purity. The objectives were to find out whether it was possible to replace conventional dynamic operation of batch columns by steady state operation using continuous columns for a comparable recovery, energy consumption, operation time, productivity, etc. and to obtain optimal operating policy in terms of reflux ratio. The following strategy was considered to compare the performances of the two types of operations:

Given an initial feed mixture, say B_1 (kmol)-

(a) choose a feed flow rate (say F_1, kmol/hr) for the steady state operation using a continuous column.

(b) determine the operation time $(t_1 = B_1/F_1)$ for the continuous column

(c) for the operation time t_1, obtain maximum achievable amount of distillate D_1 of specified purity using (i) a CBD column (Figure 11.1) and (ii) a continuous column (Figure 11.2).

(d) compare the performances in terms of total recovery, energy consumption, etc.

11.2.3.2. Optimisation Problem Formulation and Solution

The optimisation problem for a *single cut* (in CBD) or for a *single pass* (in a continuous column) can be stated as follows:

given: the column configuration (including feed location for continuous column), the feed mixture, vapour boilup rate, feed rate (for continuous column), cut time (or pass time) and a separation task (e.g. obtain product of specified purity)

determine: the optimal reflux ratio profile for the operation

so as to maximise: the amount of distillate product

subject to: equality and inequality constraints.

Mathematically the problem (*problem OP*) can be written as:

$$OP \qquad \text{Max} \qquad D_p$$
$$R, t_s$$

s.t.   Model Equations   (equality constraints)
$$x_D^k \geq \underline{x}_D^k \qquad\qquad \text{(inequality constraints)}$$
$$t = t_p$$
Linear bounds on reflux ratio (inequality constraints)

where, R is the reflux ratio value, D_p is the total amount of distillate produced (kmol) over the operation time t_p and x_D^k is the purity of distillate achieved (molefraction) and $\underline{x}_D^k$ is the specified product purity for component k. t_s [0, t_p] is the switching time from one reflux ratio value to another in a *single cut* multiple reflux ratio operation (explained in detail in the example section) using CBD operation. For a *single cut* single reflux ratio operation in CBD or in continuous column operation (SPSS or MPSSS) the cut time or the pass time is equal to tp and therefore there is no switching time to be optimised.

For CBD operation, the problem presented above results in a non-linear dynamic optimisation problem, which is solved using the technique in Mujtaba and Macchietto (1993, 1996) as outlined in earlier chapters. For continuous column operation the problem *OP* results in a non-linear steady state optimisation problem which is solved using the computer software SRQPDV1.1 due to Chen (1988).

Piecewise constant reflux ratio levels (which are optimised) are assumed with a finite number of intervals (or passes) over the total time period concerned for CBD and continuous column operations respectively. Figure 11.6 illustrates the computation sequence for dynamic optimisation of CBD operation for a *cut* or steady state optimisation of continuous column operation for a *pass* using one reflux ratio level to be optimised within the specified time period of t_p. For each iteration of the OPTIMISER, dynamic optimisation requires full integration of the model equations from $t = 0$ to $t = t_p$. This requires the solution of the DAEs at each integration step (the number of steps depends on the type of numerical method employed). On the other hand, for continuous column operation, each iteration of the OPTIMISER requires the solution of AEs only once. This explains why the solution of dynamic optimisation problem is computationally much more expensive than that of steady state optimisation problem (as already mentioned in the earlier section). Mujtaba (1997) used an efficient DAE integrator based on Gear (1971) and Morison (1984) which uses variable step length to minimise the number of integration steps.

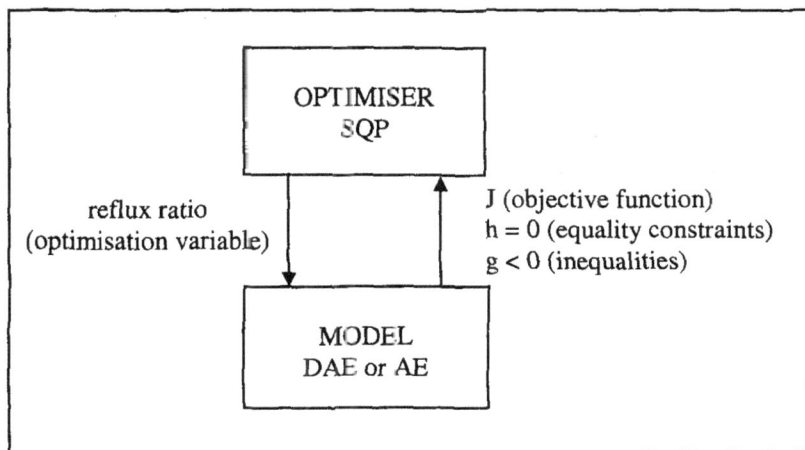

Figure 11.6. Computational Sequence of Problem *OP*. [Mujtaba, 1997][e]

Both the steady state and dynamic column models (for CBD only) used by Mujtaba (1997) are based on the assumptions of constant relative volatility and equimolal overflow and include detailed plate-to-plate calculations. This will allow a direct comparison between CBD and continuous column operation. The continuous column model is presented in section 4.3.1 and the CBD model (Type III) is presented in section 4.2.3. Some of the modelling assumptions, for example, constant molar holdup, constant pressure, equimolal overflow, etc., can be relaxed, if needed, by replacing them with more realistic assumptions and therefore by adding the relevant equations (as presented in Chapter 4).

11.2.4. Case Study – Single Separation Duty

The case study is taken from Mujtaba (1997). Here, a simple binary mixture is considered. The separation task is defined as: *Obtain distillate product with purity of 0.90 molefraction in component 1*. The column configuration and input data are given in Table 11.1.

---

[e] Reprinted with permission from IChemE, UK. Full reference is at the end of the chapter.

Table 11.1. Column Configuration and Input Data. [Adopted from Mujtaba, 1997]

Number of plates,	$N = 20$	Feed location, N_F = bottom stage
No. of components,	$n_c = 2$	Total feed, B_I, kmol = 10.0
Initial feed composition, x_{BI} = <0.6, 0.4>		Relative volatility = <2.0, 1.0>
Vapour boilup rate, V, kmol/hr = 5.0		$\underline{x}_D^1$, molefraction = 0.90
Feed rate, F_p, kmol/hr = variable		

For pass 1: feed composition, $x_{FI} = x_{BI}$

11.2.4.1. SPSS Operation

To allow a direct comparison with CBD operation, the feed is introduced at the bottom of the column so that a same number of stages (in the rectification section) are available in both cases. Therefore, the model equations given in SS.9 and SS.10 (Chapter 4, section 4.3.1) will have the following form:

$$L + F - L_B - V = 0 \qquad \text{(SS.9a)}$$

and

$$Lx_{N-1,i} + Fx_F - L_B x_{N,i} - Vy_{N,i} = 0 \qquad \text{(SS.10a)}$$

The variable L' and therefore the equation SS.6 (presented in Chapter 6) will be redundant. Note the variables are defined in Chapter 4.

The steady state optimisation problem is solved for different feed flow rates. The maximum achievable distillate rate, optimum reflux ratio (internal), total amount of distillate, pass time and recovery of key component (e.g. component 1) for the first pass are summarised in Table 11.2. For any *pass p*, the pass time (t_p, hr), total amount of distillate (D_p, kmol) and recovery of key component (Re_p) are calculated using:

$$t_p = \frac{B_p}{F_p} \qquad \text{(11.1)}$$

$$D_p = Dt_p \qquad \text{(11.2)}$$

$$Re_p = \frac{D_p \underline{x}_D^k}{B_p x_{Bp}^k} \times 100 \qquad \text{(11.3)}$$

Table 11.2. Effect of Feed Flow rate (SPSS Operation).
[Adopted from Mujtaba, 1997]

Case	F_p Kmol/hr	D kmol/hr	R	t_p hr	D_p kmol	Re_p %
1	2.0	1.25	0.750	5.00	6.25	93.75
2	3.0	1.73	0.654	3.33	5.75	86.25
3	5.0	2.23	0.554	2.00	4.46	66.90
4	10.0	2.46	0.509	1.00	2.46	36.90
5	15.0	2.48	0.503	0.67	1.66	24.90
6	25.0	2.49	0.500	0.40	1.00	15.00

respectively, where, B_p is the amount of feed (kmol) available at the beginning of the pass, F_p is the feed flow rate (kmol/hr) for the pass, x_D^k and x_{Bp}^k are the compositions (molefraction) of component k in D_p and B_p respectively, D is the distillate rate.

It is clear in Table 11.2 that the recovery of component 1 in pass 1 ($p = 1$) is strongly dependent on F_1 although for $F_1/V > 2.0$ the value of D does not change significantly. The smaller the value of F_1/V, the better the recovery is. Mujtaba (1997) has observed similar trend using other vapour boilup rates. Note that for $F_1/V < 1.0$ the operation is governed by F_1 (as D can not exceed the value of F_1 for whatever the value of V is) but for $F_1/V > 1.0$ the operation is governed by V (as D can not exceed the value of V for whatever the value of F_1 is). These findings are in agreement with Mujtaba and Macchietto (1994) which deals with the operation of an MVC column.

Comparison with CBD operation: Using different batch times (equal to the pass time of case 1 and case 4 of Table 11.2) the dynamic optimisation problem is solved and the results are presented in Table 11.3. In all cases single reflux ratio is used for the whole operation which is optimised. Column holdup is assumed negligible (steady state model does not include column hydrodynamics and therefore holdup does not appear in the model equations, Rose (1985)) and the number of stages is kept same as those used for SPSS operation for a direct comparison of the results.

Table 11.3. CBD Operation (single reflux ratio). [Adopted from Mujtaba, 1997]

Case	t_p hr	D_p kmol	R	Re_p %
1	5.0	6.29	0.748	94.27
2	1.0	2.31	0.539	34.65

It is evident that the operation in terms of product amount, reflux ratio, recovery, etc. of case 1 of Table 11.3 is very close to the operation of case 1 operation of Table 11.2. Also note that the energy consumption for CBD operation presented in Table 11.3 over a time period t_p is same as that for corresponding SPSS operation over the same time period, since the vapour boilup rate in all cases used is 5 kmol/hr. Note that the results of case 2 of Table 11.3 are not close to those of case 4 of Table 11.2. The results (Table 11.2 and Table 11.3) clearly show that if the operation is restricted to one single reflux ratio or one single pass operation, it is possible to replace the CBD operation by continuous column operation using a very low feed flow rate.

Table 11.2 shows that a low feed flow rate results in high recovery and long operation time but a high feed flow rate gives low recovery and short operation time. But from the profitability point of view it is required to achieve high recovery in shorter operation time, as this will also minimise the production of off-cuts and requirement of its further reprocessing. Mujtaba (1997), therefore, proposed to use MPSSS operation to achieve high recovery in shorter operation time.

Similar explanations and comparisons will also hold for other binary mixtures. Mujtaba (1997) considered two other binary mixtures one having a lower relative volatility ($\alpha = 1.5$, difficult separation) and other having a higher $\alpha = 2.5$ (easy separation). The results of the investigations can be found in the original reference.

11.2.4.2. MPSSS Operation

The summary of the results, using *six passes* for case 4 of Table 11.2, is presented in Table 11.4. The results are also illustrated in Figure 11.7.

The distillate purity of 0.90 molefraction in component 1 is achieved in all passes. For 6 passes, the total operation time is 3.685 hr which gives 94.59% of recovery (accumulated) for component 1. For a comparable recovery, this operation time is about 26% lower than that of case 1 of Table 11.2. For a higher recovery with larger feed flow rate the results clearly show the superiority of an MPSSS operation over a SPSS operation. The feed composition of the more volatile component in the earlier passes is higher than those in the later passes (Figure 11.7) and therefore low reflux operations in early passes are possible. However, as the composition of the more volatile component in the subsequent passes decreases, high reflux operation is required to maintain the product purity. This is quite evident in Table 11.4 (also in Figure 11.7). Having more than one reflux ratio in *multi-pass* operation shortens the overall operation time and improves the overall recovery of component. On the other hand for a similar recovery using only a *single pass* operation requires very low feed flow rate. This requires a high reflux operation throughout to maintain the product quality and therefore requires longer operation time (case 1, Table 11.2) and more energy.

Comparison with CBD operation: The dynamic optimisation problem is solved with a batch time of 3.685 hr (equal to the total operation time of the *multi-pass* case of Table 11.4). The objective is to maximise the amount of distillate (or recovery) with a purity of 0.90 molefraction in component 1.

First, a single reflux ratio is used and optimised for the whole operation. Second, six (6) reflux intervals over the entire time period of 3.685 hr is considered. The reflux ratio level in each interval together with the switching time from one interval to other are optimised. The second optimisation problem optimises a total of 11 variables (6 reflux ratio values and 5 switching times) which makes the solution of the problem computationally more expensive than the first optimisation problem and the steady state optimisation using MPSSS operation. For a direct comparison of the results, the column holdup is assumed negligible and the number of stages is kept same as those used for MPSSS operation. The results are summarised in Table 11.5.

Table 11.4. MPSSS Operation for Case 4 of Table 2.
[Adopted from Mujtaba, 1997]

Pass	B_p kmol	t_p hr	D_p kmol	R	Re_p %
1	10.00	1.0	2.46	0.509	36.90
2	7.54	0.754	1.54	0.591	60.04
3	6.00	0.600	1.00	0.668	75.00
4	5.00	0.500	0.64	0.742	84.67
5	4.36	0.436	0.41	0.812	90.80
6	3.95	0.395	0.25	0.871	94.59
Total =		3.685 hr	6.30 kmol		

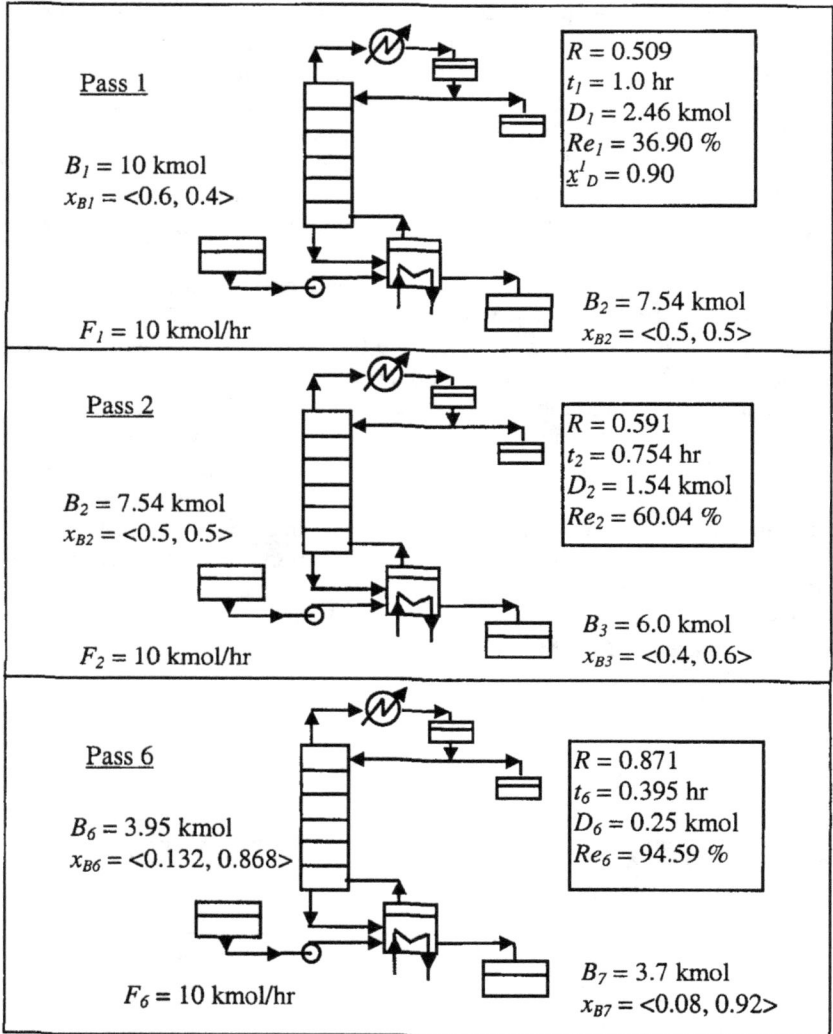

Figure 11.7. Results of MPSSS Operation (Table 11.4)[f].

---

[f] Adopted from Mujtaba (1997)

Table 11.5. CBD Operation (single and multiple reflux ratio). [Mujtaba, 1997][g]

Total Batch Time, $t_p = 3.685$ hr					

1. Single Reflux Ratio

    $D_p = 5.86$ kmol $R = 0.682$ $x_D^1 = 0.90$ $Re_p = 87.90\ \%$

2. Multiple Reflux Ratio

    $D_p = 6.13$ kmol $t_p = 3.685$ hr $x_D^1 = 0.90$ $Re_p = 91.95\ \%$

Interval	1	2	3	4	5	6
R	0.409	0.455	0.638	0.688	0.739	0.820
Interval Length, hr	0.507	0.389	0.432	0.597	0.674	1.086

For a given batch time of 3.685 hr, Tables 11.4 and 11.5 show that the CBD operation, using either a single reflux ratio or multiple reflux ratio, result in a low recovery (87.9% and 91.95% respectively, Table 11.5) compared to the accumulated recovery (94.59%) obtained using *MPSS* operation (Table 11.4). However, the use of multiple reflux ratio improves the performance of conventional column significantly as compared to single reflux ratio operation.

Note, the analogies between SPSS operation and single reflux CBD operation and between MPSSS operation and multiple reflux CBD operation when the feed is introduced at the bottom stage of the continuous column and when there is no significant column holdup in CBD column. However, at high feed flow rate, the *single pass* or *multi-pass* continuous column operation results in a better recovery compared to single or multiple reflux CBD operation (compare case 4 of Table 11.2 and case 2 of Table 11.3; also compare results of Tables 11.4 and 11.5). This is due to that fact that the separation of the light component from the top of the column becomes slightly easier because of simultaneous removal of the heavy component from the bottom of the column. Mujtaba and Macchietto (1994) noted similar behaviour.

---

[g] Reprinted with permission from IChemE, UK. Full reference is at the end of the chapter.

Table 11.6. Effect of N_F on the Performance. [Mujtaba, 1997][h]

N_F	Total time hr	Total D_p kmol	Recovery %	B_7 kmol	x_{B7} molefraction
5	4.20	4.82	72.36	5.18	<0.32, 0.68>
10	3.69	6.27	94.00	3.73	<0.09, 0.91>
20	3.685	6.30	94.59	3.70	<0.08, 0.92>

B_7 = final bottom product amount and x_{B7} = composition of B_7 (Figure 11.7)

11.2.4.3. Effect of Feed Location on the Performance of Continuous Column-MPSSS Operation

Mujtaba (1997) studied the effect of feed location (N_F) on the performance of continuous column using MPSSS operation for the same separation task with six passes as mentioned in the previous section. The results are presented in Table 11.6.

It is clear from Table 11.6 that although the feed location at the bottom half of the column does not drastically affect the performance, the feed introduced at the top half of the column significantly affects the performance. Note that the performance of CBD with six reflux ratio (Table 11.5) is better than that of continuous column with N_F =5, although both columns have the same number of plates.

Diwekar (1995) also noted a case where the performance of CBD column was better than that of a continuous column. Although there was no mention of the location of the feed plate, it is possible that the performance was not only affected by the location of the feed plate but also by the small number of plates (only six plates). In fact, Abrams et al. (1987) suggested that the comparison of a CBD operation with that of a continuous column is worth making only when the actual number of plates in both columns is at least twice the minimum. In case of Diwekar (1995) the minimum number of plates for continuous column was 6.72 (greater than the actual number of plates). See the original references for further details.

11.2.5. Multiple Separation Duties in Continuous Columns

Optimal design and operation of CBD column for single and multiple separation duties have been presented in detail by Mujtaba and Macchietto (1996) and in Chapter 7.

---

[h] Reprinted with permission from IChemE, UK. Full reference is at the end of the chapter.

11.2.5.1. The degree of difficulty of Separation for Continuous Column

Christensen and Jorgensen (1987) proposed a measure, q, for *the degree of difficulty* of separation for batch distillation (as presented in Chapter 3). Mujtaba (1997) used the ratio of minimum number of trays required to achieve a separation task to the actual number of trays available for the task to calculate, q_c, for continuous column. For a fixed distillate and bottom composition (x_D^k, x_B^k) the measure, q_c, can be defined as:

$$q_c = \frac{N_{min} + 1}{N_T + 1} \tag{11.4}$$

where,

$$N_{min} + 1 = \frac{\ln\left(\frac{x_D^k}{1 - x_D^k} \cdot \frac{1 - x_B^k}{x_B^k} \right)}{\ln \alpha} \tag{11.5}$$

and α is the relative volatility of the mixture, N_{min} is the minimum number of plates, N_T is the total number of plates $(= N - 2)$. The numerical value of q_c lies between 0 and 1. Lower the value of q_c the less difficult the separation is.

A very similar measure of *the degree of difficulty* of separation proposed by Kerkhof and Vissers (1978) could also be used to compare the performances by the two columns.

11.2.5.2. Performance Measure

Mujtaba (1997) used the *minimum time* to evaluate the performance of continuous column operation under *multiple separation duties*. However, time does not explicitly appear in continuous column model equations but the feed rate is a measure of the batch time $(t = B_l/F)$. Note, maximisation of the feed rate will therefore ensure minimisation of the batch time.

11.2.6. Case Study – Multiple Separation Duties

In the following example (taken from Mujtaba, 1997), a continuous column (instead of a CBD) is used for separating three different binary mixtures (different relative volatility) during its available operating period in a year. The column configuration and input data are given in Table 11.7. Each mixture produces a top distillate product with a purity of 90% (0.9 molefraction) in component 1 and a bottom product with a purity of 90% (0.9 molefraction) in component 2. Only SPSS type operation is considered here.

Since the top and bottom product compositions are specified, the overall mass balance using $B_1 = 10.0$ kmol and $x_{B1} = <0.6, 0.4>$ will give a total distillate product of 6.25 kmol and a total bottom product of 3.75 kmol.

Here the feed rate is maximised while the reflux ratio is optimised. The bottom product composition imposes an additional constraint to the problem. The results are summarised in Table 11.8 which gives the maximum feed rate, minimum batch time, optimum reflux ratio, and total number of batches for each mixture and total yearly profit.

It is evident in Table 11.8 that, the feed rate increases (therefore batch time decreases) with decreasing *degree of difficulty* of separation (increasing relative volatility of the mixtures). For an easy separation (high relative volatility) a high feed rate and low reflux ratio move the operating lines (in the column) upward (McCabe and Smith, 1976). This allows matching the available number of plates with a fixed feed location. The reverse is true for a difficult separation (low relative volatility). This is explained with a McCabe-Thiele diagram (not to scale) shown in Figure 11.8. The equilibrium and operating lines for two binary mixtures with low (thin line) and high (thick line) relative volatility are shown on the figure. Feed line is shown dotted. The number of stages drawn and shown on the diagram is same for both cases.

Table 11.7. Column Configuration and Input Data. [Adopted from Mujtaba, 1997]

Number of stages (including condenser and reboiler), $N = 20$

Feed location (from top, condenser counts stage 1), $N_F = 10$

No. of components, $n_c = 2$ Initial feed composition, $x_{B1} = <0.6, 0.4>$

Total feed, B_1, kmol = 10.0   Vapour rate, V, kmol/hr = 5.0

$\alpha_{Mixture-1} = <1.5, 1.0>$ $\alpha_{Mixture-2} = <2.0, 1.0>$ $\alpha_{Mixture-3} = <2.5, 1.0>$

x_D^1, molefraction = 0.90 x_B^2, molefraction = 0.90

Feed rate, F_p, kmol/hr = variable (maximised to minimise batch time)

Total operation time, H = 8000 hr/yr

Total allocated time for each mixture, H_i = 2667 hr/yr.

Table 11.8. Performance of Continuous Columns Undergoing *Multiple Separation Duties*. [Adopted from Mujtaba, 1997]

Mixture	q_c	Max F kmol/hr	Min t hr	Opt R	N_B batches/yr	Total P $/yr
1	0.57	1.86	5.38	0.768	453.5	
2	0.33	3.88	2.57	0.514	868.7	
3	0.25	5.03	1.99	0.371	1071.0	
						135365.5

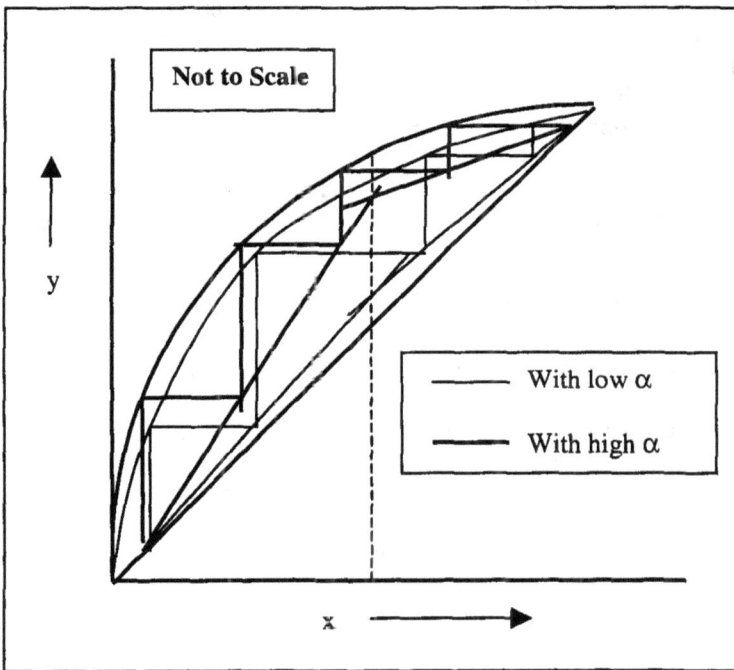

Figure 11.8. McCabe-Thiele Plot for Two Binary Mixtures

The profit shown in Table 11.8 is calculated using the profit function defined as:

$$P \text{ ($/yr)} = \sum_{i=1}^{3} (C1_i(D+B) - C2_i B_1 - OC_i)N_{B,i} - ACC \qquad (11.6)$$

$$N_{B,i} \text{ (No. of batches/yr)} = \frac{H_i}{t_i + t_s} \qquad (11.7)$$

$$OC_i \text{ (operating costs, $/batch)} = \frac{K_3 V}{A} \text{ ($/hr)} \times (t_i + t_s) \text{ (hr/batch)} \qquad (11.8)$$

$$ACC \text{ (annualised capital cost, $/hr)} = K_1 V^{0.5} (N-2)^{0.8} + K_2 V^{0.65} \qquad (11.9)$$

where, i = mixture index, $C1_i$ = product price = $10 /kmol, $C2_i$ = raw material cost = $2 /kmol, K_1 = 1500, K_2 = 9500, K_3 = 180 and A = 8000, t_i = batch time, t_s = start-up time = 0.5 hr. The annualised capital cost function and parameters are taken from Logsdon et al. (1990).

Note that since both the top and bottom product compositions are fixed (and so are the total amounts of distillate and bottoms for each mixture) the profit shown in Table 11.8 represents the maximum achievable profit using the continuous column for equal time-sharing by the mixtures. Different profit figure is expected for unequal time-sharing of the column by the mixtures and for different start-up time.

11.2.7. Notes on the Use of Continuous Columns for Batch Distillation

In the last few sections, the implications of operating batch distillation in continuous columns have been discussed. It is observed that for a given batch time and recovery of key component and energy consumption, the CBD operation can be replaced by a continuous column operation using low feed flow rate. Compared to batch columns, the modelling and optimisation tasks become easier with the use of continuous columns. For the same number of passes or reflux intervals with any feed flow rate continuous column operation results in a better recovery compared to conventional batch operation.

Higher feed flow rate in a continuous column requires shorter operation time and lower energy but achieves poor recovery while a lower feed flow rate requires longer operation time and higher energy but achieves good recovery of the key component. Therefore, if a continuous column is used for batch distillation with high feed flow rate MPSSS operation is suggested to improve productivity (amount of product/operation time) significantly.

The performance of continuous columns undergoing ***multiple separation duties*** is evaluated in terms of minimum batch time and yearly profit. By choosing an optimal feed rate existing continuous columns can be used effectively for separating multiple mixtures with different *degrees of difficulty* of separation.

Effect of transients in MPSSS operation has not been considered so far but note that the transient time for later passes can be significant compared to the operation time for those passes. The effect of transients and also the prospect of MPSSS operation using multicomponent mixtures with multiple product cuts should be further investigated.

11.3. Middle Vessel Batch Distillation Column (MVC)

The features and modelling issues of MVC columns with or without chemical reaction are presented in Chapter 2 and 4 respectively. Simulations of MVC columns with and without chemical reactions are also presented in Chapter 4. In this chapter an example with optimisation of MVC columns will be briefly presented.

Safrit and Westerberg (1997) used an MVC column for batch extractive distillation. They considered the separation of Acetone from Methanol using Water as the extractive agent. Acetone is the lightest and Water is the heaviest component in the ternary mixture. The mixture of Acetone and Methanol is charged in the middle vessel. Water is fed continuously into the column from some optimal location (usually few plates above the middle vessel). Acetone and Water are simultaneously taken out of the column as distillate and bottom product respectively, while methanol is being purified in the middle vessel. The water recovered as the bottom product is recycled back into the column as the extractive agent. The operation is shown in Figure 11.9. Safrit and Westerberg also considered production of an off-cut (an azeotropic mixture of Acetone and Methanol) when pure Methanol is to be recovered in the middle vessel. The authors defined the following (Equation 11.10) profit function (*P*) and maximised the profit by optimising the water flow rate and the cut time using repetitive simulation strategy. The cost parameters together with the optimum results are shown in Table 11.9.

$$Profit = C_{acetone}P_{acetone} + C_{methanol}P_{methanol} - (C_{steam}A_{steam} + C_{CW}A_{CW}) \quad (11.10)$$

where, $C_{acetone}$ and $C_{methanol}$ are the sales value of acetone and methanol. C_{steam} and C_{CW} are the cost of steam and cooling water. $P_{acetone}$ and $P_{methanol}$ are the amount of acetone and methanol product. A_{steam} and A_{CW} are the amount of steam and cooling water.

Note Safrit and Westerberg observed that the optimal operation policies (cut time and water flow rate) are quite sensitive to the sales value of the products and the cost parameters of steam and cooling water. Please refer to the original reference for further details. Mujtaba and Macchietto (1993) reported similar observations using CBD columns.

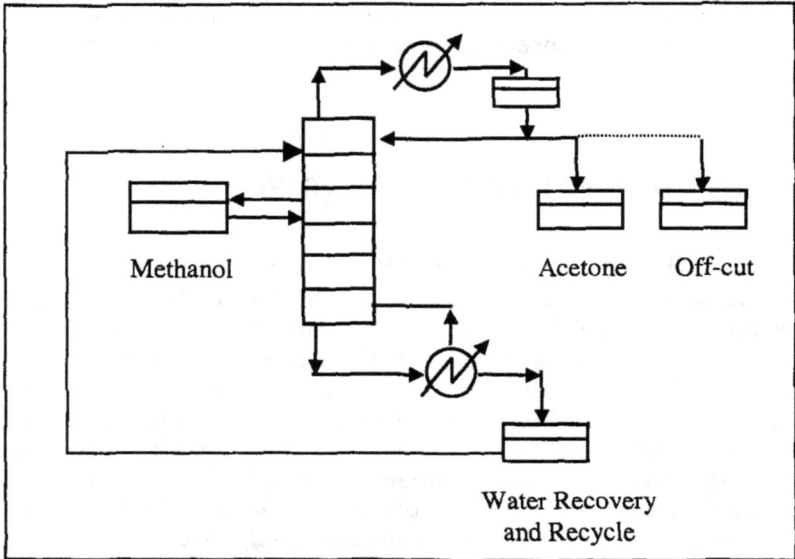

Figure 11.9. MVC Column as Batch Extractive Distillation

Table 11.9. Summary of the Results for MVC Column

$C_{Acetone}$	= \$1/mol
$C_{Methanol}$	= \$0.3/mol
C_{steam}	= \$0.0025/lbm
C_{CW}	= \$0.006/gal
Optimum Time for Acetone Cut	= 111 s
Optimum Water Flow Rate	= 6 mol/s
Maximum Profit	= \$60 per batch (approx)

11.4. Inverted Batch Distillation (IBD) Column

The features and modelling issues of IBD columns with or without chemical reaction are presented in Chapter 2 and 4 respectively. Simulation of IBD columns without chemical reactions is also presented in Chapter 4. After Robinson and Gilliland (1950) had introduced IBD columns, Abrams *et al.* (1987), Mujtaba and Macchietto (1994) and Sorensen and Skogestad (1996) used such columns for batch distillation and compared their performances with conventional columns. While Mujtaba and Macchietto (1994) studied simultaneous chemical reaction and separation using IBD columns, Sorensen and Skogestad (1996) presented the most comprehensive study on IBD columns. Some examples from these works are presented below.

11.4.1. Example 1

Mujtaba and Macchietto (1994) considered the following reversible reaction scheme in an IBD column.

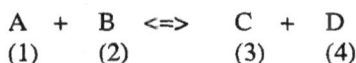

$$A + B <=> C + D$$
$$(1) \quad (2) \qquad (3) \quad (4)$$

The reaction products are C (main product) and D, the latter being the most volatile component and C being the least volatile (heavy) component in the reaction mixture. Separation of C as a bottom product permits increased conversion, while at the same time yields the product in concentrated form.

The input data defining column configurations, feed, feed composition, column holdup, etc. are given in Table 11.10. The reaction is modelled by simple rate equations (Table 11.10). The batch time is 12 hrs (t^*_f). The objective of the study was to maximise the conversion (X) of the limiting reactant and to obtain the main product with purity of 0.7 molefraction by optimising the reboil ratio defined as V/L. The following optimisation problem $(P1)$ was considered. Model type III was considered with chemical reaction.

P1   Max   X
                 $r_b(t)$

     s.t.   Model Equations   (equality constraints)

$$x_C(t_f) \geq x^*_C$$ (inequality constraints)

$$t_f = t^*_f$$
Linear bounds on reboil ratio   (inequality constraints)

Table 11.10. Input Data for Example 1

No. of plates (including reboiler, total condenser), N	= 13
Total fresh feed in the feed tank, B_0, kmol	= 5.0
Components	<A, B, C, D>
Feed composition, x_{B_0}, molefraction:	= <0.5, 0.5, 0.0, 0.0>
Column holdup, kmol:	
Condenser	= 0.07
Reboiler	= 0.07
Internal stages, total	= 0.11
Vapour load, kmol/hr	= 2.5
Flow rate from feed tank to the	
Feed plate, kmol/hr	= 5.0
Relative volatility, α	= <0.56, 0.50, 0.44, 1.0>
Rate of reaction, 1/hr: $r = k_1 x_1 x_2 - k_2 x_3 x_4$	
(with $k_1 = 2k_2$; x_i is the mole fraction of the ith component)	

where, r_b is the reboil ratio value, $x_C(t_f)$ is the purity of the bottom product achieved (molefraction) at time t_f and x_C^* is the specified product purity for the bottom product (=0.70 molefraction).

The solution of the problem **P1** results in the maximum conversion of 61.3% with 1.29 kmol of product C. The optimum reboil ratio was 0.96. Please see the original reference for further details.

11.4.2. Example 2

Sorensen and Skogestad (1996) made an extensive and most comprehensive study on IBD columns and compared the performances of IBD and CBD columns. It may seem quite obvious that the performance of IBD column will outweigh that of the CBD column when there is a small amount of heavy component present in the initial charge. Sorensen and Skogestad found the opposite true, i.e. an IBD column performs better than a CBD column when there is a small amount of light component in the feed mixture and a large amount of heavy component has to be removed by distillation. Some examples from their work are briefly presented in this section.

Minimum time is chosen as the *measure* to compare the performances of IBD and CBD columns. *Minimum time* optimisation problem similar to that presented in section 5.2.1 is considered where reflux ratio (r) is optimised for a CBD column and reboil ratio (r_b) is optimised for an IBD column. Binary separation with no off-cut strategy was considered. Instead of specifying the amount of distillate or bottom product and their compositions, distillate and bottom product compositions are specified. The optimisation variables r and r_b are assumed piecewise continuously linear over two time intervals (t_c to t_1 and t_1 to t_f). This gives a total of 5 optimisation variables [$r(t_0)$, $r(t_1)$, $r(t_1)$, t_1, t_f] for CBD and 5 optimisation variables [$r_b(t_0)$, $r_b(t_1)$, $r_b(t_1)$, t_1, t_f] for IBD columns. The optimisation problem was solved using the software DAEOPT due to Vassiliadis (1993).

The column configurations and other parameters are defined in Table 11.11. Thirteen cases were considered using different initial charge compositions and different product purity specifications (same specifications for light and heavy product). The results are summarised in Table 11.11. The results clearly show the cases (shown in bold) when an IBD column is superior to a CBD column. The optimum reflux or reboil ratio profiles for 3 cases are shown in Figure 11.10. Refer to the original reference for further details.

11.5. Multi Vessel Batch Distillation Column (MultiBD)

The features and modelling issues of MultiBD (Figure 11.11) columns are presented in Chapter 2 and 4 respectively. Simulation of MultiBD columns is also presented in Chapter 4. Furlonge et al. (1999) considered optimal operation of such column. Some of their work is briefly presented in this section.

11.5.1. Optimisation Problem Formulation

The optimisation problem considered by Furlong et al. (1999) can be described as:

given:	the column configuration, production rate, product quality
optimise:	the instantaneous energy consumption in the reboiler, (and vessel reflux flow rates), and batch time
so as to minimise:	the mean rate of energy consumption
subject to:	constraints (model equations, bounds on optimisation variables, etc.)

Table 11.11. Input Data and Summary of the Results of Example 2

Column configuration and Parameters		
Number of plates, N	= 10	
Initial charge, B_0	= 10 kmol	
Vapour boilup rate, V	= 10 kmol/hr	
Plate holdup, H_j	= 0.001 kmol	
Condenser holdup (CBD), H_c	= 0.01 kmol	
Reboiler holdup (IBD), H_r	= 0.01 kmol	
Constant relative volatility, α	= 2	

Product	Minimum batch time, t_f, hr	
specification	IBD Column	CBD Column
Initial charge composition, x_{B0} = <0.1, 0.9> molefraction		
0.925	1.04	0.41
0.950	1.37	1.02
0.975	**2.26**	2.94
0.980	**2.53**	4.45
0.990	**4.04**	> 100
Initial charge composition, x_{B0} = <0.5, 0.5> molefraction		
0.925	3.02	2.20
0.950	3.81	2.61
0.990	10.01	5.91
Initial charge composition, x_{B0} = <0.9, 0.1> molefraction		
0.925	**0.71**	1.25
0.950	1.79	1.63
0.975	4.69	2.41
0.980	7.86	2.46
0.990	>100	3.19

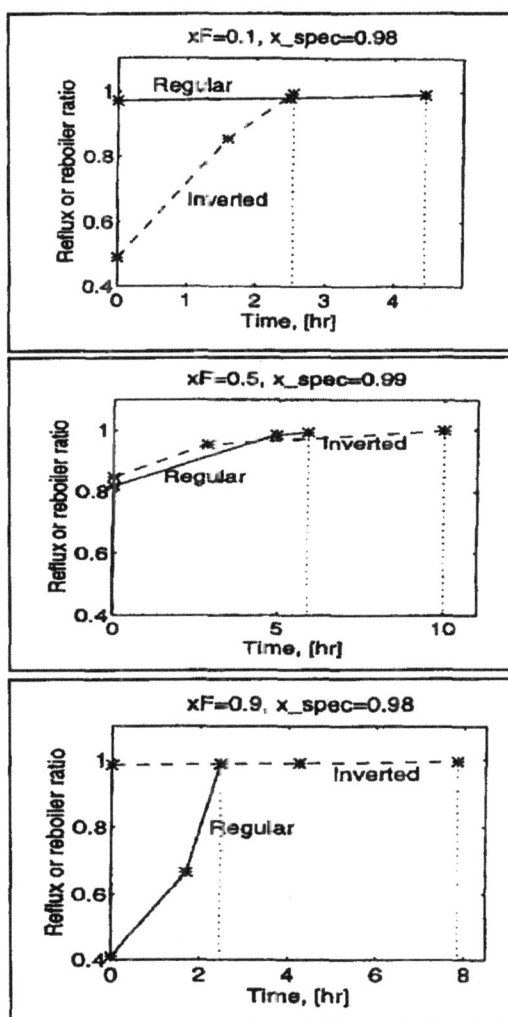

Figure 11.10. Optimum Reflux or Reboil Ratio Profiles for CBD and IBD Columns. [Sorensen and Skogestad, 1996][i]

---

[i] Reprinted from *Chemical Engineering Science*, 51, Sorensen, E. and Skogestad, S., Comparison of inverted and regular batch distillation, 4949-4962, Copyright (1996), with permission from Elsevier Science .

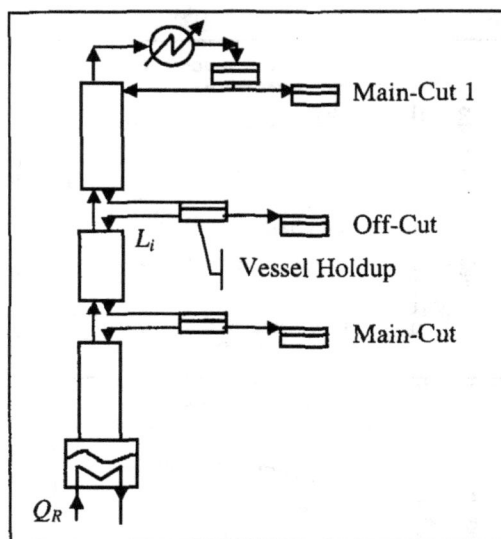

Figure 11.11. Multivessel Batch Distillation Column (MultiBD)

Mathematically the problem can be written as:

$$\underset{Q_R(t),(L_i),t_f}{Min} \quad \frac{\int_0^{t_f} Q_R(t)dt}{t_f + t_s}$$

s.t.   Model Equations   (equality constraints)

$x_{D,i}^{max} \geq x_{D,i} \geq x_{D,i}^{min}$; $i = 1, .., N_V+2$ (inequality constraints)

$P_i^{max} \geq P_i \geq P_i^{min}$; $i = 1, .., N_V+2$ (inequality constraints)

Linear bounds on vessel reflux flow rates (inequality constraints)

where, $Q_R(t)$ is the instantaneous rate of energy consumption in the reboiler, t_f is the batch time, t_s is the set-up time, L_i is vessel reflux flow rate (shown in Figure 11.11). $x_{D,i}$ is the product quality in vessel i (including the top and bottom accumulators), P_i is the mean production rate in each vessel (calculated as total amount in each vessel divided by $[t_f + t_s]$). $x_{D,i}^{max}$ and $x_{D,i}^{min}$ are the bounds on $x_{D,i}$. P_i^{max} and P_i^{min} are the bounds on P_i. The authors considered a detailed dynamic model (similar to those presented in Chapter 4) but included pressure drop model and Francis weir formula to calculate liquid flow rate from the tray. Refer to the original work for further details on modelling.

The optimisation problem was solved using CVP approach of Vassiliadis et al. (1994a,b) as implemented within gPROMS process modelling tool (Process Systems Enterprise Ltd., 1998).

11.5.2. Example

Furlonge et al. (1999) considered the separation of 100 mol of an equimolar quaternary mixture of methanol, ethanol, n-propanol and n-butanol in a column with two intermediate vessels. The results of two operating policies are presented here (refer to the original work for further examples).

In *Policy 1*, the feed is distributed equally among the reboiler, vessels and reflux drum. These holdups are kept constant and the column is run under total reflux. Only $Q_R(t)$ and t_f are optimised.

In *Policy 2*, the feed is initially charged in the reboiler and holdups in the reflux drum, vessels and reboiler are allowed to vary. Here, L_i is optimised together with $Q_R(t)$ and t_f.

The product specifications (minimum) in both policies are as follows:

- reflux drum/accumulator 1: $x_{D,methanol}^{min}$ $= 92.8$ mol%,

    P_1^{min} $= 0.209$ mol/min

- vessel 1/accumulator 2: $x_{D,ethanol}^{min}$ $= 85.4$ mol%,

    P_2^{min} $= 0.225$ mol/min

- vessel 2/accumulator 3: $x_{D,n-propanol}^{min}$ $= 91.4$ mol%,

    P_3^{min} $= 0.221$ mol/min

- reboiler/accumulator 4: $x_{D,n-butanol}^{min}$ $= 97.0$ mol%,

    P_4^{min} $= 0.213$ mol/min

The set-up time for both policies is 30 min. In both policies, all productions are achieved on specification. The instantaneous composition in each vessel and the optimum instantaneous $Q_R(t)$ profiles for Policy 1 are shown in Figure 11.12. For

Policy 2, the instantaneous holdup and composition profiles in each vessel are shown in Figure 11.13. The optimum reflux flow rates and the instantaneous $Q_R(t)$ profiles are shown in Figure 11.14.

The batch time required for each policy is 85 min. The mean rate of energy consumption for Policy 1 is 1478W and for Policy 2 is 1253W. It is noted that Policy 2 consumes 15% less energy compared to Policy 1 for producing products of same specifications.

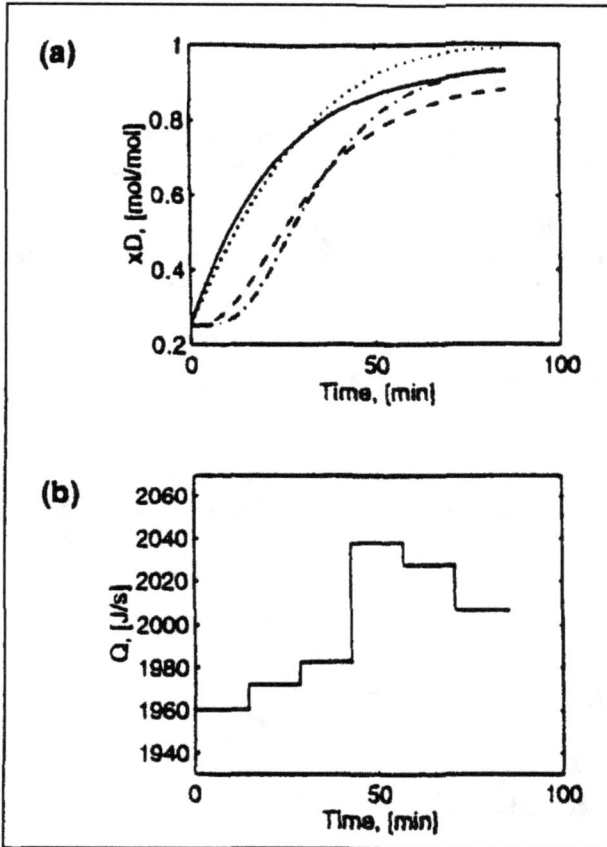

Figure 11.12. Instant Composition and Reboiler Heat Duty Profiles of Policy 1. [Furlonge et al., 1999]

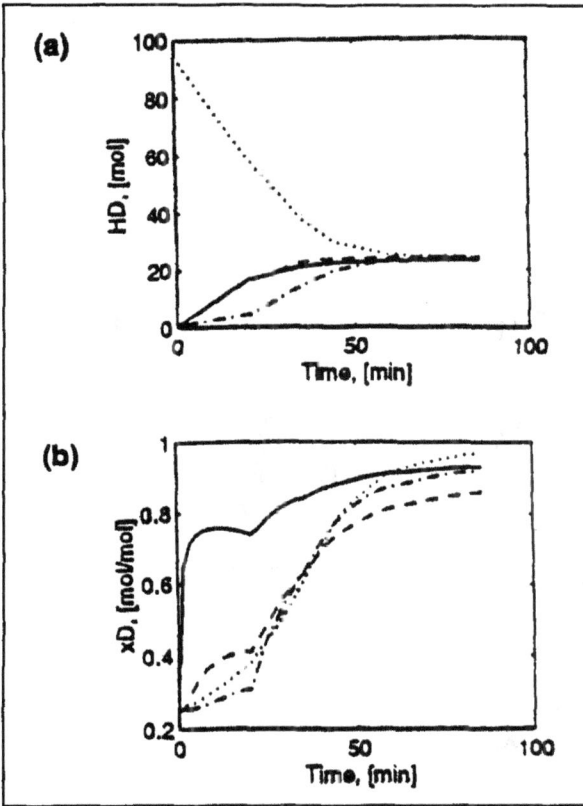

Figure 11.13. Instant Holdup and Composition Profiles of Policy 2.
[Furlonge et al., 1999][j]

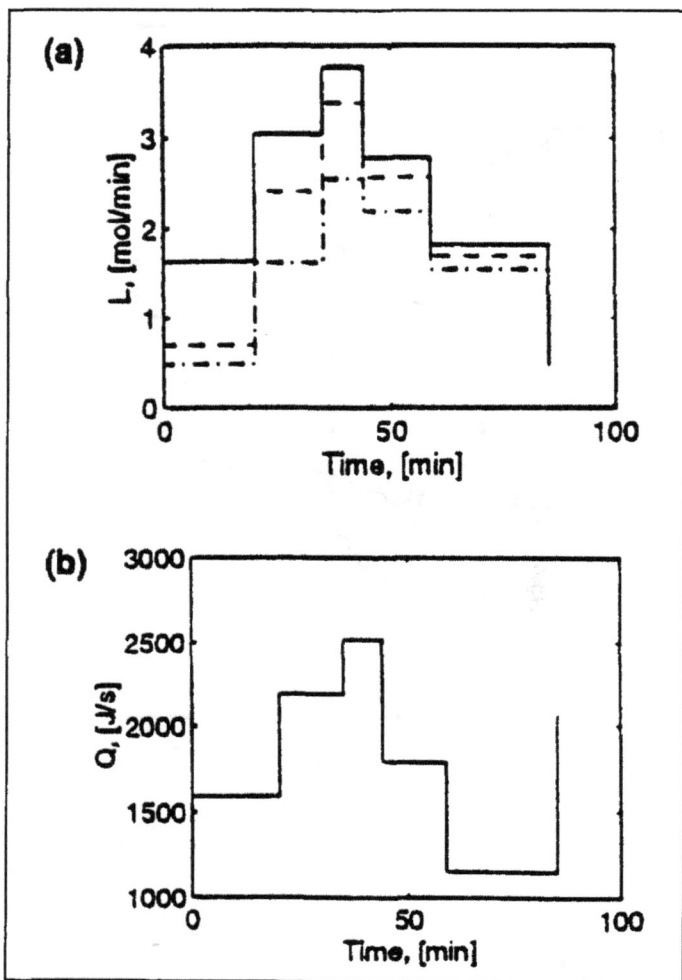

Fig. 11.14. Optimum Reflux Flow Rates and Instant Reboiler Duty Profiles of Policy 2. [Furlonge et al., 1999][j]

---

[j] Furlonge, H.I., Pantelides, C.C. and Sorensen, E., *AIChE J.* **45** (1999), 781. Reproduced with permission of the American Institute of Chemical Engineers. Copyright © 1999 AIChE. All rights reseved.

References

Abrams, H.J., Miladi, M. and Attarwala, F.T., Preferable alternatives to conventional batch distillation, Presented at the IChemE Symposium Series no. 104, Brighton, 7-9 September, 1987.

Barolo, M., Guarise, G. B., Rienzi S. A. and Trotta, A., *Comput. chem. Engng.* **22** (1998), S44.

Chen, C.L., *A Class of Successive Quadratic Programming Methods for Flowsheet Optimisation.* PhD Thesis, (Imperial College, London, 1988).

Chen, C.L., BP Research, Sunbury, UK, Private communication, 1995.

Christensen, F.M. and Jorgensen, S.B., *Chem. Eng. J.*, **34** (1987), 57.

Diwekar, U.M., *Batch Distillation: Simulation, Optimal Design and Control* (Taylor and Francis, Washington, D.C., 1995).

Furlonge, H.I., Pantelides, C.C. and Sorensen, E., *AIChE J.* **45** (1999), 781.

Gear, C.W., *IEEE Trans.* Circuit Theory, **CT-18** (1971), 89.

Greaves, M. A., *Hybrid Modelling, Simulation and Optimisation of Batch Distillation Using Neural Network Techniques.* Ph.D. Thesis, (University of Bradford, Bradford, UK, 2003).

Jenkins, M.J., Coalite Chemicals, UK. Private communication, 2000.

Kerkhof, L.H. and Vissers, H.J.M., *Chem. Eng. Sci.* **33** (1978), 961.

Logsdon, J.S., Diwekar, U.M. and Biegler, L.T., *Trans IChemE*, **68A** (1990), 434.

McCabe, W.L. and Smith, J.C., Unit Operations of Chemical Engineering, (3rd ed., McGraw Hill Kogakusha, Ltd., Japan, 1976).

Morison, K.R., *Optimal Control of Processes Described by Systems of Differential and Algebraic Equations.* PhD. Thesis, (Imperial College, University of London, 1984).

Mujtaba, I.M., *Trans. IChemE*, **75A** (1997), 609.

Mujtaba, I.M. and Macchietto, S., *Comput. chem. Engng.* **16S** (1992a), S273.

Mujtaba, I.M. and Macchietto, S., *Comput. chem. Engng.* **17**(1993), 1191.

Mujtaba, I.M. and Macchietto, S., 1994, *In Proceedings of IFAC Symposium ADCHEM'94*, Kyoto, Japan, 25 - 27 May, (1994), 415.

Mujtaba, I.M. and Macchietto, S., *J. Proc. Control.* **6** (1996), 27.

Parakrama, R., *The Chemical Engineer.* September (1985), 24.

Process Systems Enterprise Ltd., *gPROMS Advanced User's Guide*, (www.psenterprise.com, London, 1998).

Robinson, C.S. and Gilliland, E.R., *Elements of Fractional Distillation*, (4th ed., McGraw Hill, New York, 1950).

Rose, L.M., *Distillation Design in Practice* (Elsevier, New York, 1995).

Safrit, B.T. and Westerberg, A.W., *IEC Res.* **36** (1997), 436.

Sorensen, E. and Skogestad, S., *Chem. Eng. Sci.* **51** (1996), 4949.

Vassiliadis, V.S., *Computational Solution of Dynamic Optimization Problems with General Differential-Algebraic Constraints.* PhD thesis, (Imperial College, London, 1993).

Vassiliadis, V.S., Sargent, R.W. and Pantelides, C.C., *IEC. Res.* **33** (1994a), 2111.
Vassiliadis, V.S., Sargent, R.W. and Pantelides, C.C., *IEC. Res.* **33** (1994b), 2123.
Willet, J.D., 1995, Wellcome Foundation, Dartford, UK, Private communication.

CHAPTER 12

12. APPLICATION OF NEURAL NETWORKS IN BATCH DISTILLATION

12.1. Introduction

The use of Neural Networks (NNs) in all aspects of process engineering activities, such as modelling, design, optimisation and control has considerably increased in recent years (Mujtaba and Hussain, 2001; Mujtaba et al., 2003, 2004). For a given set of inputs, NNs are able to produce a corresponding set of outputs according to some mapping relationship. This relationship is encoded into the network structure during a period of training (also called learning), and is dependant upon the parameters of the network, i.e. weights and biases. Once the network is trained (with known sets of input/output data), the input/output mapping can be produced in a time that is orders of magnitude lower than the time needed for rigorous deterministic modelling (Greaves, et al., 2003). Neural networks have been known to be able to approximate non-linear continuous functions with a high degree of accuracy (Cybenko, 1989; Hussain et al., 1995). In recent years, neural network techniques have been used in batch distillation for:

- Modelling batch distillation and control (Cressy et al., 1993)
- Modelling process-model mismatches and hybrid modelling in batch distillation (Mujtaba and Hussain, 1998; Greaves et al., 2001)
- Model predictive control in batch distillation (Fileti et al., 2000)
- Composition estimator in MVC columns (Zamprogna, et al., 2001)
- Dynamic modelling and optimisation of MVC columns (Greaves et al., 2003)
- Modelling and optimisation of BREAD process (Mujtaba et al., 2004).

In this chapter, some of the recent works of Mujtaba and co-workers will be presented.

12.1.1. Neural Networks Architecture

There are various neural network architectures but all the applications considered in this chapter have utilised the feed forward multi-layered neural network. The multi-layered feed forward neural network (Figure 12.1) consists of a set of nodes, which are arranged in layers. The nodes in each layer are connected to all the nodes in the layer above/next to it and all the signals propagate in a forward direction through the network layers. There are no self-connections, lateral connections or back connections. In each node (in hidden and output layer), a constant bias is added. The outputs of nodes in one layer are transmitted to nodes in another layer through connections, which incorporate weighting factors that amplify or attenuate such outputs. The net input to each node (except for input layer nodes) is the sum of the weighted outputs of the nodes in the preceding layer. Each node is activated in accordance with the input to the node, the activation function and the bias/threshold of the node. There are various types of activation functions available (Greaves, 2003) but the log-sigmoid function has been the most used one in both the hidden and output layers.

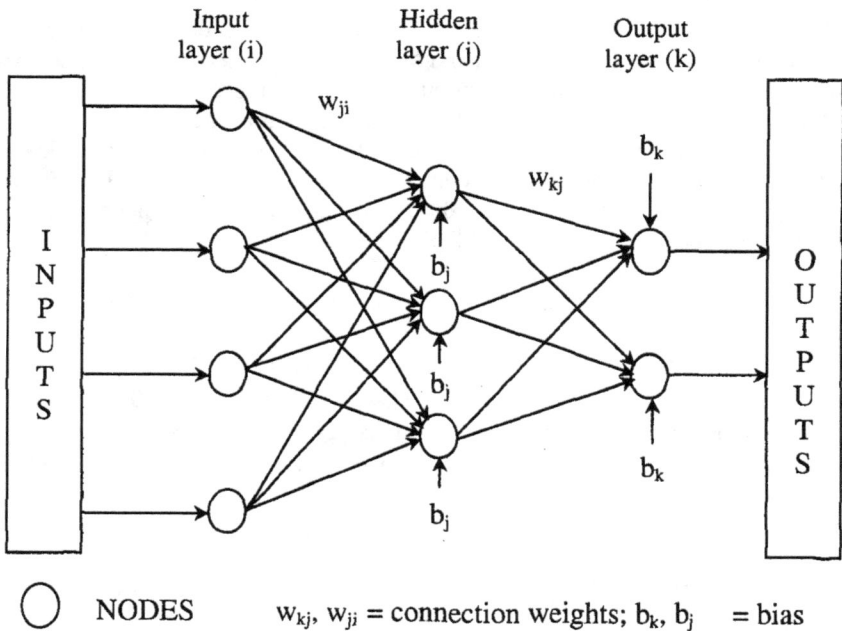

Figure 12.1. Multi-layered Feed Forward Neural Network Architecture

The numbers of hidden layers and nodes may vary in different applications and depend on the user specifications. No specific technique is available to decide the optimum number and it is usually carried out through trial-and-error procedure.

12.1.2. Neural Networks Training

The multi-layered feed forward network shown in Figure 12.1 is trained using the back-propagation method. This method is chosen because it is the most well known and widely used algorithm associated with the training of a feed forward neural network. Basically it is a gradient descent, parallel distributed optimisation technique to minimise the error between the network and target output by manipulation of the connection weights (parameters) of the network (Mujtaba and Hussain, 1998; Aziz, 2001).

12.2. Hybrid Modelling and Optimisation in CBD

It is well understood that the optimal control policies can be significantly different with and without due consideration to the *process-model mismatches*. Accurate modelling of batch distillation processes is often very difficult due to complex non-linearity of the thermophysical properties in addition to basic mass and energy balances. For example, modelling of vapour liquid equilibrium calculations is often difficult for many non-ideal and azeotropic mixtures. Availability of faster computers and sophisticated numerical methods although allow development of complex models (as presented in Chapter 4), these models are not completely free from *process-model mismatches*. Therefore, optimal control policy of dynamic processes can be significantly different with and without due consideration to the *process-model mismatches*. The nature of the mismatches of a dynamic system is also dynamic. The magnitude of the error in predicting the dynamic behaviour of the actual process using a model depends on the extent of the *process-model mismatches*. While modelling of steady state mismatches can be simple, the modelling of dynamic mismatches can be much more difficult. The use of standard regression techniques to estimate these *process-model mismatches* can be extremely difficult due to the inherent non-linearity and dynamic nature of these mismatches.

In Mujtaba and Hussain (1998), the detailed dynamic model was assumed to be the exact representation of the process while the difference in predictions of the process behaviour using a simple model and the detailed model was assumed to be the dynamic *process-model mismatches*. Theses dynamic mismatches were modelled using neural network techniques and were coupled with the simple model to give a hybrid model. They also developed a general optimisation framework based on the hybrid model for dynamic processes and applied in batch distillation.

In Greaves et al. (2001), a hybrid model for an actual pilot plant batch distillation column is developed. However, taking advantage of some of the inherent properties of batch distillation processes a simpler version (new algorithm) of the general optimisation framework is developed to find optimal reflux ratio policies which minimises the batch time for a given *separation task*.

Discrete reflux ratio used in most pilot plant batch distillation columns, including those used in industrial R&D Departments (Jenkins, 2000; Greaves, 2003), does not allow a direct implementation of the optimum reflux ratio (treated as a continuous variable) obtained using a model based technique (as presented in earlier chapters of this book). In Greaves et al. (2001), a relationship between the continuous and the discrete reflux ratio is developed. This allows easy communication between the model and the process and comparison on a common basis.

12.2.1. The Model and the Actual Process

As described in the earlier chapters, dynamic processes are often represented by a set of DAEs of the form:

$$f(t, \dot{x}(t), x(t), u(t), v) = 0 \ ; \ [t_0, t_F] \tag{12.1}$$

The above equation is same as that presented in Equation 5.4.

In many chemical processes, especially inherently dynamic processes, it is not always possible to model the actual process. Therefore, the states predicted by using the model (Equation 12.1) will be different than that of the actual process and will result in *process-model mismatches*. The implementation of the optimal operating policies obtained using the model will not result in a true optimal operation. Regardless of the nature of the mismatches, a true process can be described (Agarwal, 1996) as:

$$f(t, \underline{\dot{x}}(t), \underline{x}(t), u(t), \underline{v}, e_x(t)) = 0 \ ; \qquad [t_0, t_F] \tag{12.2}$$

where $\underline{x}(t)$ is the true set of all state variables, $\underline{\dot{x}}(t)$ denotes the derivatives $\underline{x}(t)$ with respect to time; $\underline{v}(t)$ is the true set of time independent design variables; $e_x(t)$ is the set of *process-model mismatches* for the state variables x; and the control vector u, and the function f are identical to those used in the model (Equation 12.1).

The error $e_x(t)$ is in general time dependent and describes the entire deviation due to *process-model mismatches*. Structural incompleteness in the model, reformulation of the model equations as needed by a particular solution algorithm, discrepancy between v and $\underline{v}$, inaccurate initial estimate of $x(t_0)$ of the model, inaccuracies in the measurement of u, unmeasured disturbances, simplified assumption in the estimation of thermo-physical properties of the process, etc., can result in these mismatches (Agarwal, 1996). The error $e_x(t)$ takes into account of all these sources of mismatches.

At any time t, the true estimation of the state variables requires instantaneous values of the unknown mismatches $e_x(t)$. To find the optimal control policies in terms of any decision variables (say z) of a dynamic process using the model will require accurate estimation of $e_x(t)$ for each iteration on z during repetitive solution of the optimisation problem (see Chapter 5). Although estimation of *process-model mismatches* for a fixed operating condition (i.e. for one set of z variables) can be obtained easily, the prediction of mismatches over a wide range of the operating conditions can be very difficult.

12.2.2. Hybrid Modelling of Dynamic Processes

Greaves et al. (2001) modelled the actual process (Equation 12.2) by combining a simple dynamic model (of type Equation 12.1) and the *process-model mismatches* $(e_x(t))$ model.

As the mismatches of the state variables of a dynamic system (i.e. instant distillate and reboiler compositions in batch distillation) are dynamic in behaviour, they have to be treated as such and not as static processes. To develop them from first principles would be very difficult due to their non-linear dynamic behaviour and it would also be difficult to quantify them in terms of the original state variables.

Here, neural network techniques are used to model these *process-model mismatches*. The neural network is fed with various input data to predict the *process-model mismatch* (for each state variable) at the present discrete time. The general input-output map for the neural network training can be seen in Figure 12.2. The data are fed in a moving window scheme. In this scheme, all the data are moved forward at one discrete-time interval until all of them are fed into the network. The whole batch of data is fed into the network repeatedly until the required error criterion is achieved.

The error between the actual mismatch (obtained from the simulation results) and that predicted by the network is used as the error signal to train the network (see Figure 12.3). This is a classical supervised learning problem, where the system provides target values directly to the output co-ordinate system of the learning network.

The prediction of mismatch profiles starts from discrete point 3. Time $t=0$ represents discrete point 1 where the mismatch is assumed to be zero for all state variables. At discrete point 2, mismatches are initialised with given values (obtained by judging the trend in all the data set used for the training of the neural network).

		Optimisation Variables, Z
☐ Input Data		
■ Output Data		
	State Variable $X(K-1)$	State Variable $X(K)$
Mismatch $E_x(K-2)$	Mismatch $E_x(K-1)$	Mismatch $E_x(K)$

Figure 12.2. General Input/Output Map of Neural Network

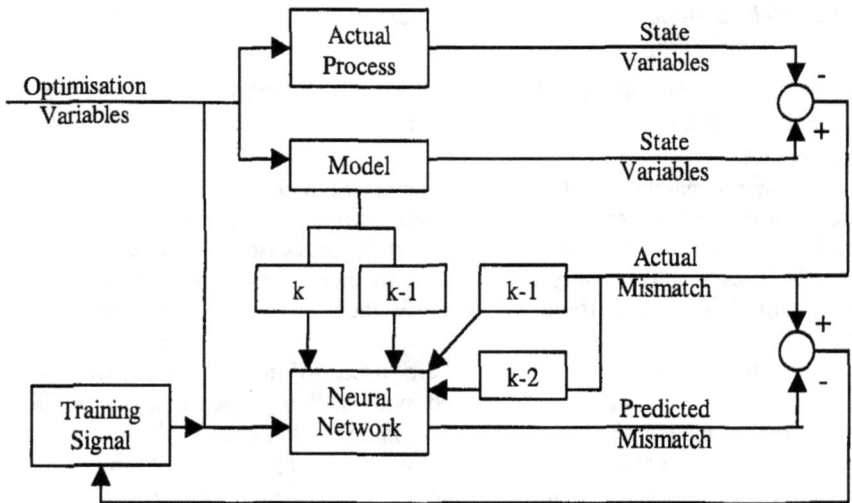

Figure 12.3. Forward Modelling of State Variable Mismatch by Neural Network

12.2.3. Dynamic Optimisation Framework Using First Principle Model

In the past many authors considered Equation 12.1 as the true representative of the actual dynamic process and developed dynamic optimisation and solution algorithms for such processes. These are presented in Chapter 5.

12.2.4. Dynamic Optimisation Framework Using Hybrid Model

12.2.4.1. General Strategy

Figure 12.4 illustrates a general optimisation framework (developed by Mujtaba and Hussain, 1998) to obtain optimal operation policies for dynamic processes with *process-model mismatches*.

Dynamic sets of *process-model mismatches* data is generated for a wide range of the optimisation variables (z). These data are then used to train the neural network. The trained network predicts the *process-model mismatches* for any set of values of z at discrete-time intervals. During the solution of the dynamic optimisation problem, the model has to be integrated many times, each time using a different set of z. The estimated *process-model mismatch* profiles at discrete-time intervals are then added to the simple dynamic model during the optimisation process. To achieve this, the *discrete process-model mismatches* are converted to continuous function of time using linear interpolation technique so that they can easily be added to the model (to make the hybrid model) within the optimisation routine. One of the important features of the framework is that it allows the use of discrete process data in a continuous model to predict discrete and/or continuous mismatch profiles.

12.2.4.2. Generation of Discrete Mismatch Profiles

The development and training of the neural network estimators for mismatches requires both the state variables (predicted by the model) and the mismatches at discrete points for a wide range of each optimisation variables. The number of sets of state variable and mismatch data for each type of state variable depends on the non-linearity and complexity of the system concerned.

The state variable profiles of the model are assumed to be continuous and are obtained by integration of the DAEs over the entire length of the time. Also efficient integration methods (as available in the literature) are based on variable step size methods and not on fixed step size method where the step sizes are dynamically adjusted depending on the accuracy of the integration required. Therefore, the discrete values of the state variables are obtained using linear interpolation technique. For example, if the values of a state variable predicted by the model are $x_{d,k}$ and $x_{d,k+1}$ at t_k and t_{k+1}, then at any discrete t_i, which lies within $[t_k, t_{k+1}]$, the state variable value $(x_{d,i})$ is calculated using the following expression:

$$x_{d,i} = \frac{x_{d,k+1} - x_{d,k}}{t_{k+1} - t_k}(t_i - t_k) + x_{d,k} \qquad (12.3)$$

Usually discrete points are of equal length $(\Delta = t_{i+1} - t_i)$, which usually represents the sampling time of the actual process. Now, if the state variable of the actual process at discrete time t_i, is given by $\underline{x}_{d,i}$, the discrete mismatch at t_i will therefore be $e_{xd,i} = \underline{x}_{d,i} - x_{d,i}$.

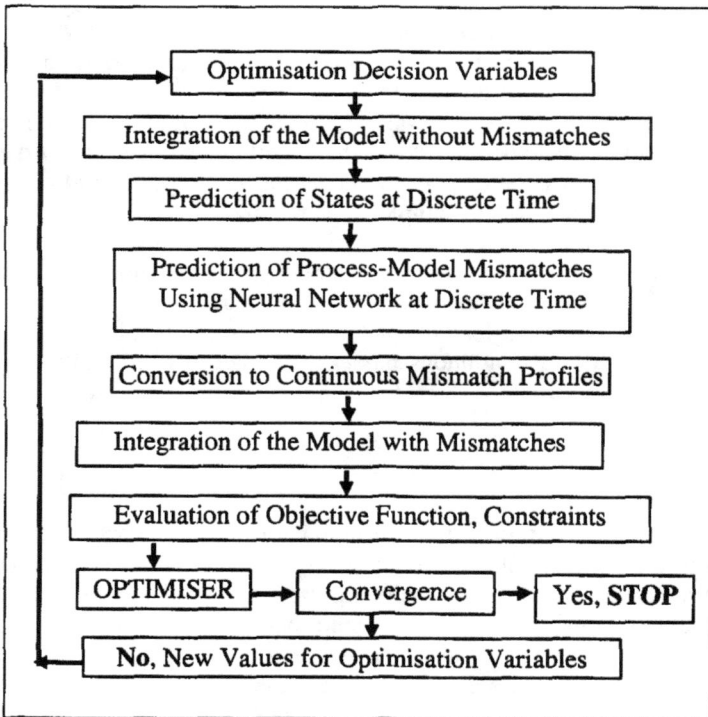

Figure 12.4. General Optimisation Framework For Dynamic Processes with Process-model Mismatch

12.2.4.3. Continuous Mismatch Profiles During Optimisation

The mismatch estimator of the neural networks estimates mismatches only at fixed discrete points. Therefore, to use the optimisation framework presented in Figure 12.4 requires estimation of mismatches at variable discrete points (these points should coincide with those by the DAE integrator). This is again achieved by interpolation techniques. For example, if the values of a mismatch predicted by the estimator are $e_{xd,i}$ and $e_{xd,i+1}$ at discrete points t_i and t_{i+1} (fixed $\Delta = t_{i+1} - t_i$) then at any variable discrete point (by the integrator) t_k, which lies within $[t_i, t_{i+1}]$, the mismatch value ($e_{xd,k}$) is calculated using the following expression:

$$e_{xd,k} = \frac{e_{xd,i+1} - e_{xd,i}}{t_{i+1} - t_i}(t_i - t_k) + e_{xd,i} \tag{12.4}$$

Note, for highly non-linear profiles of state variables, switching from continuous to discrete or from discrete to continuous using linear interpolation technique may not be efficient and non-linear interpolation technique may need to be employed.

12.2.5. Hybrid Model Development for Pilot Batch Distillation Column

Mujtaba and Hussain (1998) implemented the general optimisation framework based on the hybrid scheme for a binary batch distillation process. It was shown that the optimal control policy using a detailed process model was very close to that obtained using the hybrid model.

In Greaves et al. (2001) and Greaves (2003), instead of using a rigorous model (as in the methodology described above), an actual pilot plant batch distillation column is used. The differences in predictions between the actual plant and the simple model (Type III and also in Mujtaba, 1997) are defined as the dynamic *process-model mismatches*. The mismatches are modelled using neural network techniques as described in earlier sections and are incorporated in the simple model to develop the hybrid model that represents the predictions of the actual column.

The pilot-plant description is given in section 3.3.4. in Chapter 3. Methanol-Water system was considered with an initial charge of 900 ml of Methanol and 2100 ml of Water giving a total of 85.04 gmol of the mixture with <0.25, 0.75> molefractions for Methanol and Water respectively.

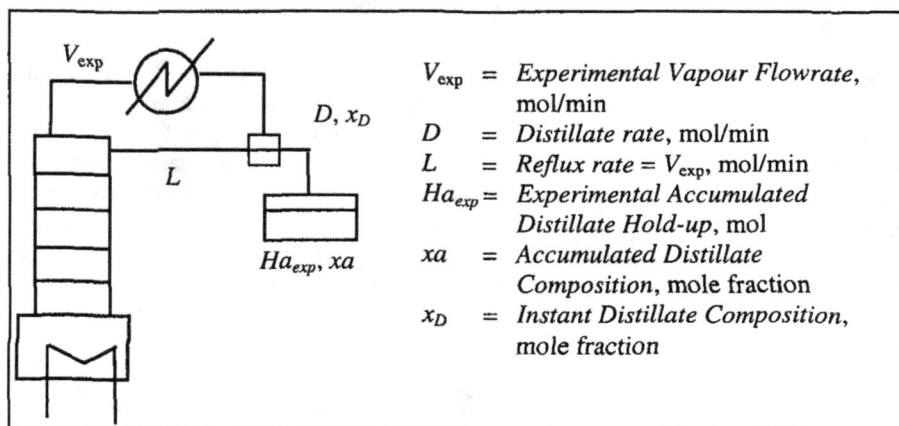

Figure 12.5. Schematic of Batch Distillation Column

12.2.5.1. Relation Between Experimental Reflux Ratio (R_{exp}) and Model Reflux Ratio (R_{model})

Many industrial users of batch distillation (Chen, 1998; Greaves, 2003) find it difficult to implement the optimum reflux ratio profiles, obtained using rigorous mathematical methods, in their pilot plants. This is due to the fact that most models for batch distillation available in the literature treat the reflux ratio as a continuous variable (either constant or variable) while most pilot plants use an on-off type (switch between total reflux and total distillate operation) reflux ratio controller. In Greaves et al. (2001) a relationship between the continuous reflux ratio used in a model and the discrete reflux ratio used in the pilot plant is developed. This allows easy comparison between the model and the plant on a common basis.

For the sake of convenience, Figure 3.11 is presented here again. The reflux in the column (Figure 12.5) is produced by a simple switching mechanism that is controlled by a solenoid in a cyclic pattern (on-off). The valve is open for a fixed period of time (to withdraw distillate) and is closed for a fixed period of time (to return the reflux to the column). In this column the valve is always open (t_{op}) for 2 seconds and then closed (t_{close}) for $2x(R_{exp})$ second, where R_{exp} is the reflux setting.

It was shown in section 3.3.4 that the discrete experimental reflux ratio R_{exp} in the cyclic operation could be viewed as the average continuous external reflux ratio

(R). As shown in section 3.3.4, the total amount of distillate collected (Ha_{exp}) over a period of t_{diff} (assuming V_{exp} is constant over that period) can be calculated using:

$$Ha_{exp} = V_{exp} \frac{1}{\left(1 + R_{exp}\right)} t_{diff} \tag{12.5}$$

However, most of the batch distillation models (e.g. Mujtaba and co-workers; Sorensen and Skogestad, 1996) relate the amount of distillate collected (Ha_{model}) with the vapour boil-up rate in the column (V_{model}), the internal reflux ratio (R_{model}) and the total operating time (t_{diff}) by,

$$Ha_{model} = V_{model} \left(1 - R_{model}\right) t_{diff} \tag{12.6}$$

In these models the reflux ratio is treated as a continuous variable. If V_{exp} is the same as V_{model} then Equation 12.5 and Equation 12.6 give the desired relationship between the discrete R_{exp} (of cyclic operation) and continuous R_{model} as,

$$R_{model} = 1 - \frac{1}{\left(1 + R_{exp}\right)} = \frac{R_{exp}}{\left(1 + R_{exp}\right)} \tag{12.7}$$

Interestingly, the relationship between the external (R) and internal reflux ratio (r) in a distillation column with continuous flow of reflux and distillate is very similar to Equation (12.7) which is:

$$r = \frac{R}{1 + R} \tag{12.8}$$

12.2.5.2. Plant-Model Simulation

Greaves et al. (2001) carried out 5 experiments in total using the pilot-plant for different R_{exp}. The accumulated distillate composition and distillate hold-up profiles are shown in Figures 3.13 and 3.14 respectively.

Greaves et al. used the simple model of Mujtaba (1997) to simulate the plant. For each R_{exp}, R_{model} and V_{model} (= V_{exp}) are calculated using Equation 12.7 and Equation 3.5. The simulated and experimental instant distillate composition profiles are shown in Figure 12.6 for $R_{exp} = 2$ (corresponding $R_{model} = 0.666$). Curves A and

B show the model and pilot plant predictions respectively. Figure 12.6 clearly shows that there are large *process-model mismatches* in the composition profiles although for a given batch time of t_{diff} = 220 min the amount of distillate achieved by the experiment was the same as that obtained by the simulation. These *process-model mismatches* can be attributed to factors such as: use of constant V_{model} instead of a dynamic one; constant relative volatility parameter used in the model and uncertainties associated with it; actual efficiency of the plates.

12.2.5.3. Hybrid Model

The four experiments done previously with R_{exp} (= 0.5, 1, 3, 4) were used to train the neural network and the experiment with R_{exp} = 2 was used to validate the system. Dynamic models of *process-model mismatches* for three state variables (i.e. X) of the system are considered here. They are the instant distillate composition (x_D), accumulated distillate composition (xa) and the amount of distillate (Ha). The inputs and outputs of the network are as in Figure 12.2. A multilayered feed forward network, which is trained with the back propagation method using a momentum term as well as an adaptive learning rate to speed up the rate of convergence, is used in this work. The error between the actual mismatch (obtained from simulation and experiments) and that predicted by the network is used as the error signal to train the network as described earlier.

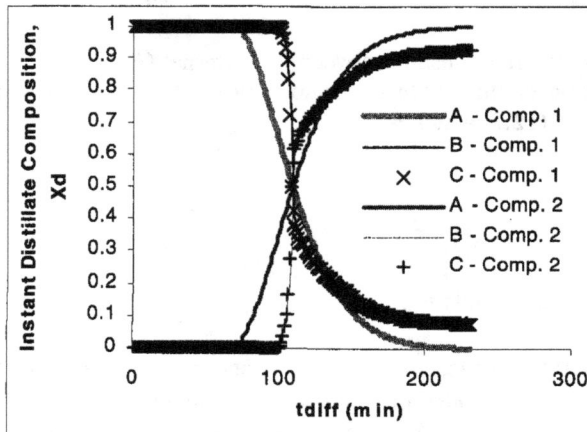

Figure 12.6. Experimental Simulation Results and Dynamic Process-model Mismatch Model (R_{exp}= 2).

Figure 12.6 also shows the instant distillate composition profile for $R_{exp} = 2$ (which is used to validate the network) using the simple model coupled with the dynamic model for the *process-model mismatches* (curve C). The predicted profile (curve C) shows very good agreement with the experimental profile (curve B). Similar agreements have been obtained for the accumulated distillate amount and composition profiles (Greaves, 2003).

12.2.6. NN Based Optimisation Algorithm

Solution of optimisation problems using rigorous mathematical methods have received considerable attention in the past (Chapter 5). It is worth mentioning here that these techniques require the repetitive solution of the model equations (to evaluate the objective function and the constraints and their gradients with respect to the optimisation variables) and therefore computationally can be very expensive.

Greaves et al. (2001) and Greaves (2003) presented two simple algorithms which is computationally less expensive to obtain the minimum batch time for a given *separation task*. These algorithms are the results of the application of some of the unique properties of batch distillation process to the general optimisation framework discussed earlier. The algorithm-1 is based on experiment and is presented in section 3.4 of Chapter 3.

12.2.6.1. Algorithm -2: Model Based

To reduce the time consuming and expensive experiments of algorithm-1 by a considerable amount, Greaves et al. (2001) and Greaves (2003) proposed a second algorithm based on simple model and neural network techniques as shown in Figure 12.7 (ε_l = Finite small positive number)

The algorithm-2 has been tested for the *separation tasks* used in algorithm-1 (Chapter 3) and they were in very good agreement (Greaves, 2003). For a given *separation task*, while the algorithm-1 requires approximately 3 to 4 set of experiments for a total period of 18-22 hours, the algorithm-2 requires only about half an hour of computation time and about 4-5 hours of experiment to achieve the given *separation task*.

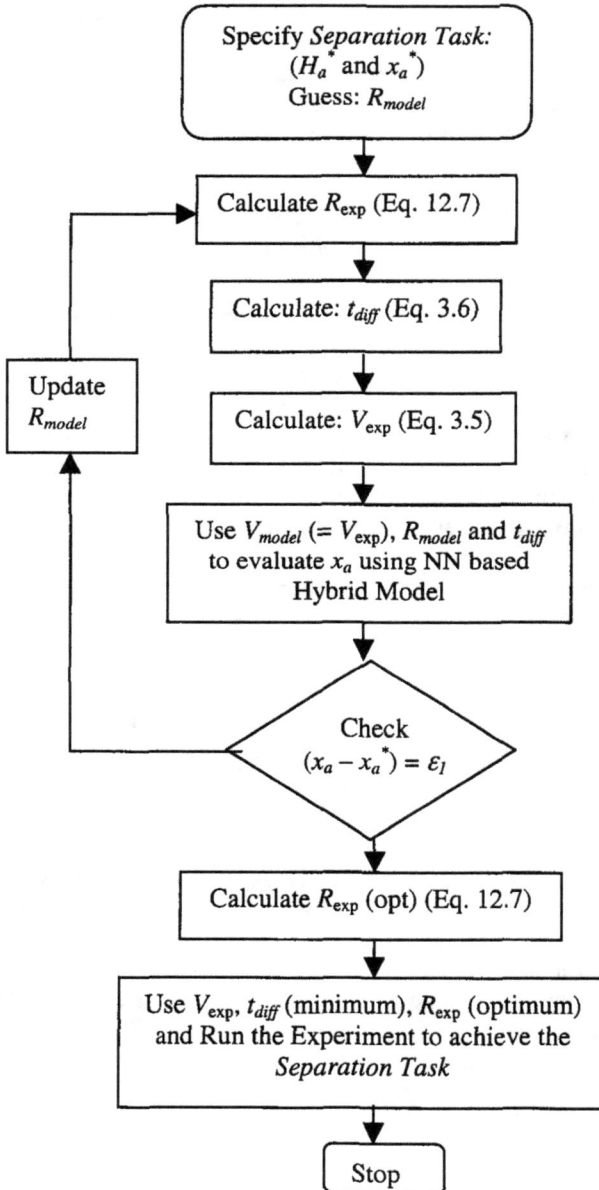

Figure 12.7. Model Based Algorithm -2 for Calculating Minimum Batch Time

12.3. NN Based Modelling and Optimisation in MVC

Greaves et al. (2003) proposed a framework to optimise the operation of MVC columns with substantial reduction of the computational power needed to carry out the optimisation calculations. The framework relies on the use of NN based process model. The optimisation of a pilot-plant middle-vessel batch column (MVC) was considered to test the viability of the proposed framework. The *maximum product* problem was considered and solved by optimising the column operating parameters, such as the reflux and reboil ratios and the batch time. The NN based model is found to be capable of reproducing the actual plant dynamics with good accuracy, and that the proposed framework allows a large number of optimisation studies to be carried out with little computational effort.

The separation of a binary mixture (ethanol-water) with strong nonideal thermodynamics was considered. This makes the column behaviour significantly nonlinear. For a given feed, the objective of the distillation is to obtain two main cuts at different ethanol purities. The lighter cut is obtained from the top of the column as distillate, while the heavier cut is segregated in the middle vessel and recovered at the end of the batch. The stripping section of the column is used to remove the heavy impurity (water), paying attention to avoid any ethanol losses from the bottom. The operating procedure is the same as those described by Barolo et al. (1998), i.e. the distillate and bottoms flows are started and stopped intermittently, depending on the top and bottom compositions; the product and impurity withdrawal is accomplished at a constant reflux ratio and a constant reboil ratio. The top and bottom levels are maintained by proportional-only controllers, and no measurement is available on-line for the middle-vessel level.

A detailed model of the pilot-plant MVC was derived and validated against experimental data in a previous study (Barolo et al., 1998 and also see Chapter 4). The model consists of material and energy balances, vapour liquid equilibrium on trays (with Murphree tray efficiency to account for tray nonideal behaviour), liquid hydraulics based on the real tray geometry, reflux subcooling, heat losses, and control-law calculations based on volumetric flows. The model provides a very accurate representation of the real process behaviour, but is computationally expensive for direct use within an optimisation routine. Greaves et al. (2003) used this model as a substitute of the process.

12.3.1. Neural Network Based Modelling

With reference to an MVC column, at any time instant one would be interested to know the values of the amount (mole) and composition (ethanol mole fraction) of the light cut (M_{DE} and x_{DE} respectively), and of the amount and composition of the heavy cut (M_F and x_F respectively). These four variables will therefore correspond to the NN output vector (indicated by Z in Figure 12.8).

A certain amount of information has to be made available to the network which will be processed to obtain the output vector at the current time instant. The inputs can come either from direct measurements from the process, or from the network itself (i.e., NN outputs from previous time instants). The choice of the inputs should be made with engineering judgment. Too many inputs would overload the network and introduce unnecessary correlations among data, and can therefore disrupt the network performance; on the other hands, too few inputs could be not enough for the network to "learn" the actual process behaviour. In Greaves et al. (2003), at time t_k the following inputs were fed to the network: t_k, output vector Z_0 at initial time (t_0), and current values for the internal reflux and reboil ratios (R_D and R_B, respectively), distillate rate (D), and bottoms rate (B).

A feedforward NN topology (Figure 12.1) was selected. This structure is robust and performs well in approximating the complex behaviour of distillation processes (Greaves *et al.*, 2001). Four NNs were used to calculate the elements of the output vector at any given time as a function of the NN inputs. For each network, the output evaluation is accomplished as illustrated in Figure 12.8. Note that for a given system, with the input parameters known at time t_0, the dynamic product holdups and compositions profiles can be calculated in a straightforward way. It should also be noted that the bottom impurity amount and composition M_B and x_B respectively) can be evaluated at any time by overall mass balances, assuming negligible tray holdup. This assumption is good at the beginning of the operation, but may not be satisfied at the end of the batch, if the amount of liquid remaining in the middle vessel is low.

12.3.1.1. Network Training and Validation

Greaves et al. (2003) used a single-hidden-layer feedforward network. The network was trained with the backpropagation method using a momentum term as well as an adaptive learning rate to speed up the rate of convergence (Demuth and Beale, 1998). The difference between the actual output value (as obtained from the process simulator) and that calculated by the network was used as the error signal to train the network. The stopping criteria for the training was set to 2000 epochs or if the overall error on the outputs was lower than 10^{-8} (where the error was evaluated as the mean-squared error function). To ensure that all the values were of the same magnitude, all the inputs and outputs were scaled between 0.1 and 0.9.

NN Inputs	**NN Outputs**
R_D, R_B, D, B, Z_0, t_1	Z_1
R_D, R_B, D, B, Z_0, t_2	Z_2
$\vdots$	$\vdots$
R_D, R_B, D, B, Z_0, t_F	Z_F

where:
$$Z_0 = \{M_{DE}; x_{DE}; M_F; x_F\}|_{t=0}$$
$$Z_1 = \{M_{DE}; x_{DE}; M_F; x_E\}|_{t=t_1}$$
$$\vdots$$
$$Z_F = \{M_{DE}; x_{DE}; M_F; x_F\}|_{t=t_F}$$

Figure 12.8. Input and Output Specifications for Each NN. [Greaves et al., 2003][a]

Using the detailed model of Barolo et al. (1998) 25 datasets (Greaves, 2003) were generated for a grid of reflux (R_D) and reboil ratios (R_B) in such a way so as to span a large set of possible operating conditions. In each data set, the time profiles of the variables that are used as network inputs or outputs were recorded for a given total batch time (with no assigned specifications for the light cut and heavy cut compositions). A typical dataset is shown in Table 12.1. Five of these datasets were singled out and grouped into a Validation Set, while the remaining twenty datasets were grouped into two Training Sets (which differ only in terms of the order according to which the datasets are grouped).

Each NN (Figure 12.8) has 6 inputs and 1 output; a number of 3 to 12 nodes were considered in the hidden layer. The optimal number of nodes, together with the best activation function, was selected according to an algorithm given in Greaves et al. (2003). Note that an array of all the combinations of 11 activation functions was used to help in finding the most appropriate neural network for each output. The

---

[a] Reprinted with permission from IChemE, UK. Full reference is at the end of the chapter.

final configuration of each net is shown in Table 12.2, as well as the final errors associated with training and validation. x_F is the only output that seems to be somewhat less accurately reproduced than the others. The impact of the error on x_F will be discussed later.

Figure 12.9 and Figure 12.10 show the results related to two datasets belonging to the Validation Set. These results are representative of the best (Figure 12.9) and worst (Figure 12.10) results obtained during the NN validation. In both cases, it is apparent that the NN does a satisfactory job in reproducing the actual process behaviour. The saving in calculation time is significant. For example, the simulation of dataset #20 with the detailed tray-to-tray model took about 29 s on a Pentium III PC, while it was almost instantaneous with the NN model. This reduction in calculation time leads to a parallel reduction in the time needed to carry out the optimisation studies.

Table 12.1 Typical Dataset to be Reproduced by the NN (dataset #20). [Greaves et al., 2003][b]

Time (min)	M_{DE} (kmol)	x_{DE}	M_F (kmol)	x_F
0	0	0	4.921	0.322
10	0.093	0.865	4.532	0.331
20	0.185	0.862	4.143	0.343
:	:	:	:	:
120	1.111	0.856	0.251	0.672

Table 12.2. Neural Network Configuration and Overall Final Errors. [Greaves et al., 2003][b]

Output	Layer Activation Function		No. of hidden nodes	Error	
	Hidden	Output		Training	Validation
M_{DE}	Hyperbolic Tangent	Linear	11	9.94E-09	5.21E-05
x_{DE}	Hyperbolic Tangent	Symmetric Saturating Linear	6	1.00E-08	4.64E-08
M_F	Logarithmic	Logarithmic	10	4.16E-07	6.62E-08
x_F	Hyperbolic Tangent	Hyperbolic Tangent	10	1.48E-05	6.05E-04

---

[b] Reprinted with permission from IChemE, UK. Full reference is at the end of the chapter.

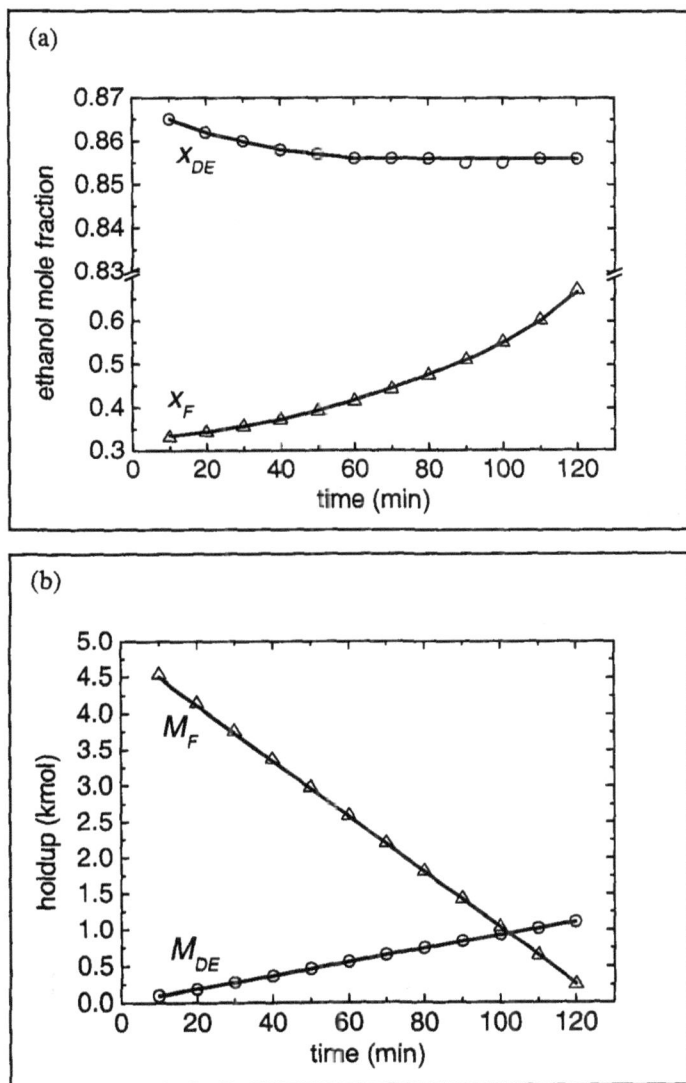

Figure 12.9. Profiles of (a) Product Composition and (b) Product Holdup for Dataset #20 (symbols: detailed model; lines: NN model).
[Greaves et al., 2003][c]

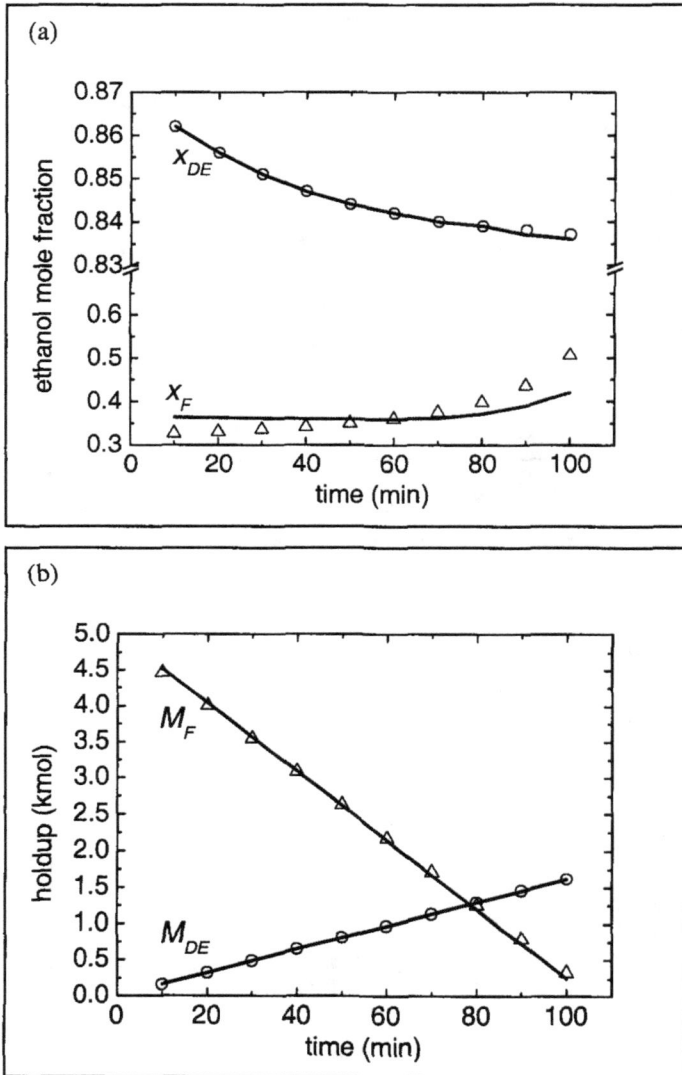

Figure 12.10. Profiles of (a) Product Composition and (b) Product Holdup for Dataset #09 (symbols: detailed model; lines: NN model). [Greaves et al., 2003][c]

---

[c] Reprinted with permission from IChemE, UK. Full reference is at the end of the chapter.

12.3.2. Optimisation Problem Formulation

Greaves et al. (2003) considered the following dynamic optimisation problem, with the objective of maximizing the amount of distillate product and middle-vessel product which can be described as follows:

given: the column configuration, the feed mixture, a separation task
determine: the optimal reflux ratio, reboil ratio and batch time
so as to maximise: the amount of distillate product and middle vessel product
subject to: equality and inequality constraints (e.g. process model, linear bounds on optimisation variables, etc.)

Mathematically the problem can be written as:

$$\underset{R_D(t);R_B(t);t}{\text{Max}} \quad (M_{DE} + M_F)$$

subject to:

process model	(equality constraint)
$x_F \geq Xf^* - \varepsilon_2$	(composition constraint)
$x_{D\bar{z}} \geq Xa^* - \varepsilon_1$	(composition constraint)
$x_{DE} \leq Xa^* + \varepsilon_1$	(composition constraint)
$x_F \leq Xf^* + \varepsilon_2$	(composition constraint)
$M_F \geq 0$	(holdup constraint)
$M_{F0}\, x_{F0} - (M_{DE}\, x_{DE} + M_F\, x_F) \geq 0$	(component balance constraint)
$R_D^L \leq R_D \leq R_D^U$	(reflux ratio bounds)
$R_B^L \leq R_B \leq R_B^U$	(reboil ratio bounds)
$t^L \leq t \leq t^U$	(time bounds)

M_{DE}, x_{DE}, M_F and x_F are evaluated at the end of the operation time; Xa^* and Xf^* are the purity targets (separation tasks) for the light cut and for heavy cut (respectively); ε_1 and ε_2 are purity tolerances (both set to 0.001); R_D^L and R_D^U are the lower and upper bounds of R_D within which it is optimised; R_B^L and R_B^U are the lower and upper bounds of R_B within which it is optimised; t^L and t^U are the lower and upper bounds of t within which it is optimised. Holdup and component balance constraints were included to avoid numerical difficulties in terms of negative values of M_F, or ethanol mole fractions greater than 1 or smaller than 0 in the column.

Finite but small values ε_1 and ε_2 are used to bracket the desired composition region to ensure that the products are not under or over specified. Although it would be justified to leave the upper bound open for the optimiser to adjust freely, it was found that the optimiser was biased towards the top product, and therefore it was decided that the upper bound should be enforced. Greaves et al. (2003) used the NN based process model as described earlier.

The calculation sequence is started with an initial estimate of vector z (reflux and reboil ratios). Each iteration of the OPTIMISER (see section 5.5 in Chapter 5) requires the evaluation of the process model at $t = t_F$ to evaluate the objective function and the constraints. The NN model can predict these instantaneously reducing the computational time considerably in comparison to a DAE process model, which requires full integration from t_0 to t_F. The values of the objective function and of the constraints are passed to the OPTIMISER, which then takes a step in z, and the process is repeated until convergence is achieved within an acceptable accuracy. The optimisation problem was solved using the efficient Successive Quadratic Programming (SQP) based method due to Chen (1988). Greaves et al. used the method of forward finite difference to generate the required first derivatives of the objective function and of the constraints (required by the OPTIMISER) with respect to the optimisation variables. The reason for choosing a finite difference method to evaluate the gradients as opposed to the analytical based method is purely due to computational time and difficulty in evaluating an analytical based method for a neural network model. To solve an analytical based method, the integration and solution of the full DAE model is required before the gradients can be determined (Morrison 1984; Vassiliadis *et al.*, 1994). However, given the initial states and other operating parameters, the proposed NN model can predict the outputs at any time $t \in [t_0, t_F]$ without requiring evaluation of the states at any other previous time step. This is certainly the strength of the proposed NN based dynamic process model (Greaves, 2003).

The solution algorithm developed by Greaves et al. (2003) is shown in Figure 12.11 (where the desired accuracy for the optimiser is $Acc^* = 1.0E\text{-}06$ and the actual accuracy, Acc, is the value of the objective function). If optimisation is not required, the model will follow the Simulation route, and output the results for a given set of input parameters at given time intervals. If optimisation is required, the model will follow the Optimisation route, optimising the variables for a given set of input parameters, and will output the optimised values of the operating variables as well as the outputs.

12.3.3. Results and Discussions

Greaves et al. (2003) chose 13 datasets (out of the 25 datasets available) to test the algorithm sequence. To assess the possible improvements achievable through optimisation, the purity targets of each dataset (belonging either to the Training Set or to the Validation one) were set equal to the final x_{DE} and x_F of that dataset; in this way, improvements over a given dataset can be appreciated easily by considering the values of the total amount of product collected and of the total batch time. A collective set of results is summarised in Table 12.3 and Table 12.4 (adopted from Greaves et al., 2003). The capacity factor (CAP; Luyben, 1988) is included in these Tables as performance index. This is defined as follows:

$$CAP = \frac{M_{DE} + M_F}{t_F} \qquad (12.9)$$

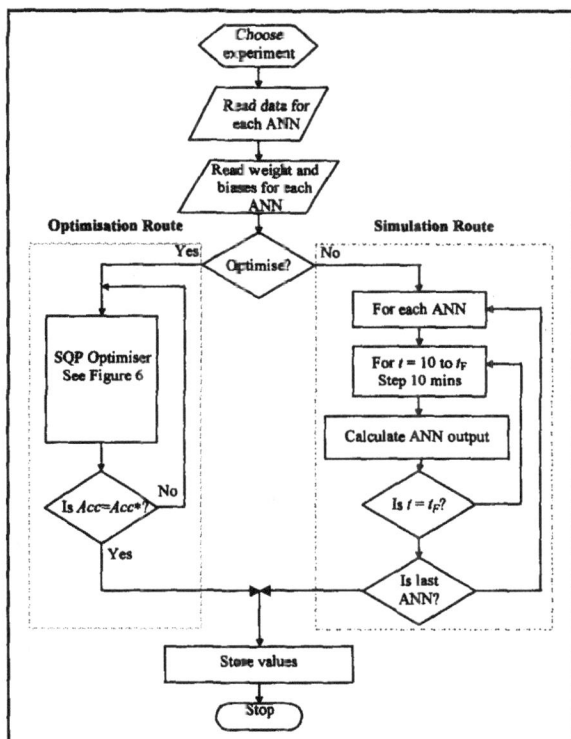

Figure 12.11. The Calculation Algorithm Sequence. [Greaves et al., 2003][d]

---

[d] Reprinted with permission from IChemE, UK. Full reference is at the end of the chapter.

Table 12.3. Summary of operation data for a subset of the available datasets.
[Adopted from Greaves et al., 2003]

Run	Dataset #	M_{DE} (kmol)	x_{DE}	M_F (kmol)	x_F	R_D	R_B	t_F (min)	CAP (kmol/min)
1	6	1.512	0.827	0.294	0.496	0.639740	0.641600	70	0.02580
2	15	1.426	0.846	0.233	0.611	0.787601	0.726048	110	0.01508
3	19	1.389	0.854	0.195	0.682	0.849927	0.778421	150	0.01056
4	21	0.833	0.858	0.083	0.676	0.852133	0.642685	90	0.01018
5	22	0.653	0.862	0.263	0.647	0.884149	0.643194	90	0.01018
6	23	0.944	0.861	0.122	0.732	0.885776	0.727466	130	0.00820
7	24	1.162	0.859	0.199	0.728	0.885799	0.778706	160	0.00851
8	25	1.597	0.856	0.061	0.728	0.869817	0.844330	220	0.00754
9	11	2.808	0.736	0.185	0.000	0.661707	0.888221	130	0.02302
10	18	1.852	0.849	0.103	0.303	0.845639	0.857818	200	0.00978
11	1	1.872	0.853	0.157	0.140	0.880173	0.891129	260	0.00780
12	9	1.620	0.837	0.335	0.506	0.701377	0.735610	100	0.01955
13	20	1.111	0.856	0.251	0.672	0.854951	0.726855	120	0.01135

Table 12.4. Summary of the operation data for optimised operations.
[Adopted from Greaves et al., 2003]

Run	Dataset #	M_{DE} (kmol)	x_{DE}	M_F (kmol)	x_F	R_D	R_B	t_F (min)	CAP (kmol/min)
1	6	1.4510	0.828	0.4966	0.495	0.639598	0.642363	67.18	0.02899
2	15	0.9430	0.847	1.2882	0.610	0.802010	0.703980	72.90	0.03061
3	19	1.2958	0.853	0.4071	0.681	0.843149	0.771487	139.74	0.01219
4	21	0.8137	0.857	0.4552	0.675	0.847207	0.641600	87.88	0.01444
5	22	0.6620	0.861	1.5706	0.646	0.865164	0.647146	91.26	0.02446
6	23	0.8473	0.860	0.8278	0.731	0.876247	0.721198	116.57	0.01437
7	24	1.0505	0.858	0.6740	0.727	0.872620	0.771157	144.77	0.01191
8	25	1.2858	0.855	0.6674	0.727	0.879714	0.821547	176.93	0.01104
9	11	1.9835	0.735	2.5674	0.000	0.758788	0.915785	90.94	0.05004
10	18	0.9469	0.850	2.5653	0.304	0.834002	0.880999	102.04	0.03442
11	1	1.7757	0.852	0.5154	0.139	0.883142	0.896059	246.68	0.00929
12	9	1.3702	0.838	0.8641	0.505	0.729563	0.724203	84.45	0.02646
13	20	0.9141	0.855	1.1967	0.671	0.842071	0.715294	98.62	0.02140

The results mostly show that the optimiser has been able to adjust the reflux and reboil ratios and the distillation time in such a way as to increase the total amount collected (i.e., the objective function), the capacity factor (productivity). Greaves et al. (2003) included another performance index (%Recovery of key component) in addition to CAP and the conclusions remained the same.

The dynamic product composition and product amount profiles for dataset #09 (i.e., the worst-case validation dataset considered in Figure 12.10) are shown in Figure 12.12. It is clear that, according to the NN calculations (full lines), the optimised operating parameters were able to achieve the separation task a lot faster than the simulated experiments (symbols), with considerable savings in batch time. It should be noted that, using the approach of Greaves et al. (2003), the optimisation can be carried out in a very short computer time.

Finally, to check the reliability of the simulation results provided by the NN model, the optimal reflux and reboil ratios were implemented into the detailed (i.e., deterministic) MVC simulator (Barolo et al., 1998). The broken lines in Figure 12.12 represent the relevant time profiles obtained with the detailed MVC simulator for an operation carried out using the optimal values of the reflux and reboil ratios as calculated by the NN. Greaves et al. (2003) noted that the actual process is indeed represented quite accurately by the NN model, which confirmed that the optimisation results were reliable.

The results for the other datasets were found to be similar to those shown in Figure 12.12. In general, the actual profiles of M_{DE}, M_F, and x_{DE} were reproduced quite accurately by the NN, whereas it was harder to get the same accuracy in reproducing x_F. This is also clear from Table 12.2, where the error on x_F in the training set is shown to be markedly larger than the error in the other variables. This would probably suggest strengthening the tolerance on x_F when the network is being trained.

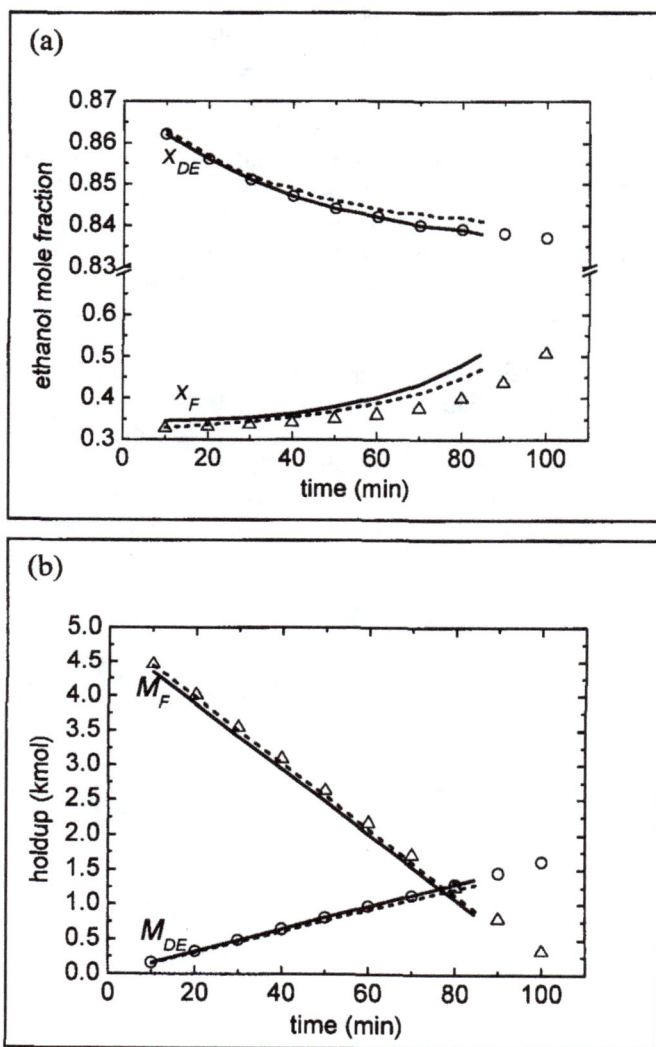

Figure 12.12. Profiles of (a) product composition and (b) product holdup for dataset #09 and: nominal operation with the detailed model (symbols), optimal operation with the NN model (full lines), optimal operation with the detailed model using values obtained from NN optimisation (broken lines). [Greaves et al., 2003][e]

[e] Reprinted with permission from IChemE, UK. Full reference is at the end of the chapter.

References

Agarwal, M, In *Batch Processing Systems Engineering: Fundamentals and Applications for Chemical Engineering*, eds. G.V. Reklaitis et al., Series F: Computer and Systems Sciences, (Springer Verlag, Berlin). **143** (1996), 295.

Aziz, N., *Dynamic Optimization and Control of Batch Reactors*. PhD Thesis, (University of Bradford, 2001).

Barolo, M., Guarise, G. B., Rienzi S. A. and Trotta, A., *Comput. chem. Engng.* **22** (1998), S37.

Chen, C.L., *A Class of Successive Quadratic Programming Methods for Flowsheet Optimisation*. PhD Thesis, (Imperial College, London, 1988).

Cressy, D.C., Nabney, I.T., Simper, A.M,. *Neural Coput & Applic.* **1** (1993), 115.

Cybenko, G., *Math. Cont. Sig. Syst.* **2** (1989), 303.

Demuth, H. and Beale, M., *Neural Network Toolbox User Guide for Use with Matlab – Version 3* (The Mathworks, Inc., Natick MA, USA, 1998).

Diwekar, U.M., *Batch Distillation: Simulation, Optimal Design and Control* (Taylor and Francis, Washington, D.C., 1995).

Fileti, A.M.F., Cruz, S.L. and Pereira, J.A.F.R., *Chem. Eng. Proc.* **39** (2000), 121.

Greaves, M. A., *Hybrid Modelling, Simulation and Optimisation of Batch Distillation Using Neural Network Techniques*. Ph.D. Thesis, (University of Bradford, Bradford, UK, 2003).

Greaves, M. A., I. M. Mujtaba and M. A. Hussain (2001). in *Application of Neural Network and Other Learning Technologies in Process Engineering*, eds. Mujtaba, I.M. and Hussain, M.A. (Imperial College Press, London, 2001), 149.

Greaves, M.A., Mujtaba, I. M., Barolo, M., Trotta, A. and Hussain, M. A., *Trans. IChemE*, **81A** (2003), 393.

Hussain, M.A., Allwright, J.C. and Kershenbaum, L.S., in *Proceedings of IChemE - Advances in Process Control 4*, York, 27-28 September, (1995), 195.

Jenkins, M.J., Coalite Chemicals, UK. Private communication, 2000.

Luyben, W. L., *IEC Res.* **27** (1988), 642.

Morison, K.R., *Optimal Control of Processes Described by Systems of Differential and Algebraic Equations*. PhD. Thesis, (Imperial College, University of London, 1984).

Mujtaba, I.M., *Trans. IChemE*, **75A** (1997), 609.

Mujtaba, I.M., Aziz, N. and Hussain, M.A., *The Chemical Engineer*, May, (2003), 36.

Mujtaba, I.M., Greaves, M.A. and Hussain, M.A., *In Proceedings of the 8th International Symposium on Process Systems Engineering*, Kunming, China, 5 – 10 January, (2004).

Mujtaba, I.M., Hussain, M.A., *Comput. chem. Engng.* **22** (1988), S621.

Mujtaba, I. M. and Hussain, M.A., *Application of Neural Networks and Other Learning Technologies in Process Engineering* (Imperial College Press, London, 2001).

Vassiliadis, V. S., Sargent, R. W. H. and Pantelides, C. C., *IEC Res.* **33** (1994), 2123.

Zamprogna, E., Barolo, M., Seborg, D.E., *Trans. IChemE*, **79A** (2001), 689.

Index

www.ingramcontent.com/pod-product-compliance
Lightning Source LLC
Chambersburg PA
CBHW050633190326
41458CB00008B/2255